Smart Delivery Systems

Solving Complex Vehicle Routing Problems

Smart Delivery Systems

Solving Complex Vehicle Routing Problems

Edited by
Jakub Nalepa

Series editor
Fatos Xhafa

ELSEVIER

Elsevier
Radarweg 29, PO Box 211, 1000 AE Amsterdam, Netherlands
The Boulevard, Langford Lane, Kidlington, Oxford OX5 1GB, United Kingdom
50 Hampshire Street, 5th Floor, Cambridge, MA 02139, United States

Library of Congress Cataloging-in-Publication Data
A catalog record for this book is available from the Library of Congress

British Library Cataloguing-in-Publication Data
A catalogue record for this book is available from the British Library

ISBN: 978-0-12-815715-2

Publisher: Joe Hayton
Acquisition Editor: Brian Romer
Editorial Project Manager: Redding Morse
Production Project Manager: Bharatwaj Varatharajan
Designer: Alan Studholme

Typeset by VTeX

This book is in memory of Dr. Grzegorz Nalepa, an extraordinary scientist and pediatric hematologist/oncologist at Riley Hospital for Children, Indianapolis, USA, who helped countless patients and their families through some of the most challenging moments of their lives.

Contents

Contributors

Dominika Bankiewicz Faculty of Mathematics, Informatics and Mechanics, University of Warsaw, Warsaw, Poland

Miroslaw Blocho ABB, Krakow, Poland

Agata Buchcik Department of Mining Mechanization and Robotisation, Faculty of Mining, Safety Engineering and Industrial Automation, Silesian University of Technology, Gliwice, Poland

Rajeev Kr. Goel C.S Deptt., Govt. College, Naraingarh, India

Paweł Gora Faculty of Mathematics, Informatics and Mechanics, University of Warsaw, Warsaw, Poland

Tomasz Jastrzab Institute of Informatics, Faculty of Automatic Control, Electronics and Computer Science, Silesian University of Technology, Gliwice, Poland

Katarzyna Karnas Faculty of Mathematics, Informatics and Mechanics, University of Warsaw, Warsaw, Poland

Wojciech Kaźmierczak Faculty of Mathematics, Informatics and Mechanics, University of Warsaw, Warsaw, Poland

Michał Kutwin Faculty of Mathematics, Informatics and Mechanics, University of Warsaw, Warsaw, Poland

Paulo S. Martins School of Technology (FT), University of Campinas (UNICAMP), Limeira, SP, Brazil

Luis A.A. Meira School of Technology (FT), University of Campinas (UNICAMP), Limeira, SP, Brazil

Mauro Menzori School of Technology (FT), University of Campinas (UNICAMP), Limeira, SP, Brazil

Jakub Nalepa Silesian University of Technology, Gliwice, Poland

Przemysław Perkowski Faculty of Mathematics, Informatics and Mechanics, University of Warsaw, Warsaw, Poland

Szymon Płotka Faculty of Mathematics, Informatics and Mechanics, University of Warsaw, Warsaw, Poland

Sandhya Rani Bansal C.S.E. Deptt., M.M. Deemed University, Mullana, India

Anna Szczurek Faculty of Mathematics, Informatics and Mechanics, University of Warsaw, Warsaw, Poland

Jacek Widuch Institute of Informatics, Silesian University of Technology, Gliwice, Poland

Guilherme A. Zeni School of Technology (FT), University of Campinas (UNICAMP), Limeira, SP, Brazil

Damian Zięba Faculty of Mathematics, Informatics and Mechanics, University of Warsaw, Warsaw, Poland

Chapter 1

Current and emerging formulations and models of real-life rich vehicle routing problems

Jacek Widuch

Institute of Informatics, Silesian University of Technology, Gliwice, Poland

1.1 Introduction

The class of vehicle routing problems encompasses discrete optimization problems concerned with the determination of routes for a given fleet of vehicles according to defined objectives and constraints. Vehicle routing problems are the subject of intensive research for more than 50 years. It is the class of problems of real-life importance, and its applications include logistics, travel, communications, manufacturing, transportation, distribution, civil, and military systems, among others. All mentioned domains have a direct impact on the modern economy and the cost of goods. The models surveyed in this chapter are based on the transportation networks where the real objectives are considered. In many cases a few objectives are considered simultaneously, and thus the problem is resolved as the NP-hard Multicriteria Optimization problem.

Recently, more and more constraints are being introduced into vehicle routing problems, and new types of vehicles are being considered, resulting in several new variants of the problem. The most vehicles run on diesel engines, which are major sources of Greenhouse Gas emissions and pollution. Therefore, ecological aspects and the reduction of pollution are taken into account in vehicle routing problems, and vehicles with alternative energy source are considered as a mean of transport. The routing problem for the electric vehicles and hybrid electric vehicles is studied, and problems related to using this type of vehicle are analyzed. An unmanned vehicle is the next new type of vehicle. Over the past few years it has become more and more popular, and the problem of determining the route for this type of vehicle has also become the subject of research.

In this chapter, we present the most popular and important vehicle routing problems. A number of variants of researched problems based on real-life com-

Smart Delivery Systems. https://doi.org/10.1016/B978-0-12-815715-2.00006-3

munication and transportation networks are considered with particular emphasis put on new types of vehicles and ecofriendly means of transport. The problems on determining vehicle routes concerned with a transport of products and people are presented. The chapter surveys their characteristics, alongside the methods exploited to tackle such optimization problems and highlight differences between those formulations.

The structure of this chapter is as follows. Section 1.2 presents Vehicle Routing Problem and its variants. In Subsection 1.2.1 the classical Vehicle Routing Problem is presented, and its variants are presented in Subsection 1.2.2. The ecofriendly Vehicle Routing Problem, that is, Green Vehicle Routing Problem and Electric Vehicle Routing Problem are given in Subsections 1.2.3 and 1.2.4, respectively. Subsection 1.2.5 describes the methods used for solving mentioned problems. In Section 1.3 the following variants of Bus Routing Problem are presented: Bicriterion Bus Routing Problem (Subsection 1.3.1), Multicriteria Bus Routing Problem (Subsection 1.3.2), and School Bus Routing Problem (Subsection 1.3.3). Additionally, other selected variants of Bus Routing Problem are presented in Subsection 1.3.4. In Section 1.4, Unmanned Vehicle Routing Problem is described. Section 1.5 presents the other routing problems of electric vehicles that do not belong to the group of problems presented in Subsections 1.2.3 and 1.2.4. Finally, Section 1.6 contains concluding remarks.

1.2 Vehicle Routing Problem and its variants

1.2.1 The classical Vehicle Routing Problem

The classical Vehicle Routing Problem (VRP), also known as the Capacitated VRP (CVRP), first appeared in 1959 [49] and can be defined as follows [29]. Let $G = (V, E)$ be a weighted graph with weight function $d : E \to \mathbb{R}_{\geq 0}$. The graph contains the set of arcs E and the set of vertices $V = 1, \ldots, n$, where vertex 1 represents the depot, and the other vertices represent cities or customers to be served. With the graph, the matrix $D = (d_{ij})$ is associated, where d_{ij} is equal to the weight of arc (i, j) and can be interpreted as a travel cost. A fleet of vehicles, based at the depot, is available for serving the customers and the cities, and each vehicle has the same characteristics, that is, we consider a homogeneous fleet. With each vertex $i > 1$, a demand $q_i \geq 0$ is associated, and the sum of demands on any vehicle routed should not exceed the vehicle capacity. The goal of the VRP is to determine a set of least-cost vehicle routes satisfying the following conditions:

- each vertex $v \in V \backslash \{1\}$ is served exactly once by exactly one vehicle,
- each route starts and ends at the depot, that is, in the vertex $v = 1$,
- the capacity of the vehicles is not exceeded.

The goal of classical problem is to find a single solution. The problem is modified, and many its variants are studied. Talarico et al. [195] considered the VRP the aim of which is to find a set of k-dissimilar solutions. Zhang et al. [222]

assumed three-dimensional loading constraints, which accounts for the actual needs of businesses in the logistics industry such as the delivery of consumer goods and agricultural products. Each item is described be a three-dimensional cuboid of length, width, and height. There is a fleet of vehicles available for carrying goods, and each vehicle has a fixed loading space (a container) defined by the length, width, and height of the loading space. In addition, each vehicle is specified with a weight capacity.

The VRP is an NP-hard problem [46,113]. It is extended in many ways by introducing additional real-life aspects or characteristics, resulting in numerous variants of the VRP (see Subsection 1.2.2).

Braekers et al. [29] presented a taxonomic review of the VRP literature published between 2009 and June 2015 and the variants of VRP. It contains a classification of 277 papers. In the next subsections, there are references to works published since 2015 except publication from 2015 referenced in [29] and publication containing a survey of the variants of VRP. Another review of the VRP literature is presented in [19] and [74].

1.2.2 Variants of the VRP

The VRP is extended by varying the capacities of vehicles, which results in the Heterogeneous Fleet VRP (HFVRP), also known as the Mixed Fleet VRP (MFVRP) [156,216]. The HFVRP was introduced around 30 years ago, and Koç et al. [105] present a survey of the metaheuristic algorithms for solving it.

The popular extension of the VRP is the VRP with Time Windows (VRPTW), where is assumed that the time of delivery to a given customer must occur in a certain time interval named as “time window”, which varies from customer to customer [196,197,131]. The variant of the VRPTW with two objectives was considered by Nalepa and Blocho [138,139], who minimized the number of vehicles and the total distance in the routing plan. Hernandez et al. [80] considered the Multi-Trip VRP with Time Windows (MTVRPTW) as a variant of the VRPTW. In MTVRPTW, multiple trips are allowed for vehicles with in the planning time horizon. Multiple trips are beneficial to the carrier by limiting the number of vehicles to the deliveries. In classical MTVRPTW a single time window is assumed for each customer. The VRP with Multiple Time Windows (VRPMTW) is also considered [18]. The VRPTW with Driver-Specific Times (VRPTWDST) uses driver-specific travel and service times to model the familiarity of the different drivers with the customers to visit [179]. In the robust VRPTW, uncertain travel times are considered [85].

In Cumulative Capacitated Vehicle Routing Problem (CCVRP) the objective is the minimization of the sum of arrival times at the customers instead of the total routing cost [99,170,189,194].

In the Dynamic Multi-Period VRP (DMPVRP), customers place orders dynamically over a planning horizon consisting of several periods or days [200]. Each request specifies a demand quantity, a delivery location, and a set of consecutive periods during which delivery can take place. The distributor must plan its delivery routes over several days so as to minimize the routing cost and the customer waiting time, and to balance the daily workload over the planning horizon. A literature review of the problem is presented by Ouaddi et al. [145].

Another extension of the VRP is the VRP with Pickup and Delivery (VRPPD) [7]. In the VRPPD the goods need to be picked up from a certain location point and must be dropped off at their destination point. The restriction that the goods are picked up and dropped off by the same vehicle is assumed, and therefore the pick-up and drop-off points belong to the same route. The VRP with Backhauls (VRPB) assumes that in the each route all deliveries must be made before any pickups of the goods [106].

Another variant of the problem, the Multi-Depot VRP (MDVRP), assumes that multiple depots are geographically spread among the customers. Montoya-Torres et al. [136] presented a review of papers published between 1988 and 2014, with several variants of the MDVRP.

In the Periodic VRP (PVRP), customers can be visited more than once, though often with limited frequency [8]. There is assumed that the deliveries to the customer can be made in different days and a planning is made over a certain horizon.

In the Open VRP (OVRP) variant, vehicles are not required to return to the depot after visiting customers [180]. If they do return to the depot, then the vehicles must visit the same customers in the reverse order. In the most cases of the OVRP, two objectives are minimized: the total traveled distance and the number of used vehicles.

The next variant of the problem assumes that the input data are revealed or updated continuously. The vehicle routes are adapted dynamically based on the actual input data. This variant of the VRP is named as the Dynamic VRP (DVRP) [1,62,109]. A survey of proposed methods for solving the problem is presented in [157,161].

Most variants of VRP assume that the travel times between depots and customers are deterministic and constant. In the Time-Dependent VRP (TDVRP), it is assumed that the travel times are not constant and they are functions of current time [86]. The work of Gendreau et al. [68] presents a review of the TDVRP and the algorithms for solving it. The VRP with Stochastic Demands (VRPSD) assumes that customer demands are stochastic variables [120,127].

In most cases the type of transported products is not considered. The Multi-Compartment VRP (MCVRP) assumes that different products are transported together in one vehicle with multiple compartments [2]. The products are stored in different compartments because they cannot be mixed together due to differences in their individual characteristics. Another variant of the problem in which the type of goods is taken into account is the VRP for Hazardous Materials transportation (VRPHazMat). The objective of the VRPHazMat is to determine a set of routes that minimizes the total expected routing risk [34]. The state-of-the-art related to the VRPHazMat is presented by Hamdi et al. [76].

In the classical VRP, each customer is required to be visited by exactly one vehicle. In the Split Delivery VRP (SDVRP), this restriction is removed, and split deliveries are allowed, that is, the customer can be visited by many vehicles [158,185].

The Clustered VRP (CluVRP) is a variant of the VRP in which customers are partitioned into clusters, and it is assumed that each cluster must have been served completely before the next cluster is served [82]. It decomposes the problem into three subproblems: the assignment of clusters to routes, the routing inside each cluster, and the sequencing of the clusters in the routes.

In real transportation networks the theft of transported goods sometimes occurs. The Cargo Theft Weighted VRP (CTWVRP) considers a model regarding physical distribution of goods in areas where the probability of thefts cannot be neglected. The goal of the problem is to minimize total transportation and theft costs [166].

For several years, ecological aspects and environmental impact have been taken into account in planning of vehicle routes. This problem is important and will certainly be the direction of further research. An important problem in the modern world is the reduction of pollution, the prevention of smog, and the use of alternative energy sources. An example of using alternative energy source are electric vehicles and hybrid electric vehicles. Two variants of the "ecological" VRP are considered: the Green VRP and the Electric VRP (see Subsections 1.2.3 and 1.2.4).

Many variants of the VRP are researched, but combinations of these variants are often considered. A combination of several variants of the VRP more precisely describes the real problems. In the literature the following combinations of variants of the VRP are considered:

- Dynamic VRP with Time Windows (DVRPTW) [15]: a combination of the DVRP and VRPTW,
- Open VRP with Time Windows (OVRPTW) [30]: a combination of the OVRP and VRPTW,
- Multi Depot VRP with a Heterogeneous Fleet (MDHFVRP) [24]: a combination of the MDVRP and HFVRP,
- Multi-Depot VRP with Time Windows (MDVRPTW) [12]: a combination of the MDVRP and VRPTW,
- Multi-Depot Multi-Period VRP with a Heterogeneous Fleet (MDMPHFVRP) [124]: a combination of the MDVRP, PVRP, and HFVRP,
- Multi-Period and Multi-Depot Dynamic VRP with Time Windows (MPMDVRPTW) [118]: a combination of the DMPVRP, MDVRP, and VRPTW,
- Time-Dependent Multi-Depot VRP with Time Windows and Heterogeneous Fleet (TDMDHFVRPTW) [5]: a combination of the TDVRP, MDVRP, HFVRP and VRPTW,
- Split Delivery VRP with Time Windows (SDVRPTW) [129]: a combination of the SDVRP and VRPTW,

- VRP with Pickup and Delivery with Time Windows (VRPPDTW) [20,21, 123,140]: a combination of the VRPPD and VRPTW,
- VRP with Pickup and Delivery with Time Windows and Handling Operation (VRPPDTWH) [205]: a combination of the VRPPD and VRPTW, where in addition a handling operations are considered. A handling operation can refer to a rehandling operation (the unloading and reloading operations of an item at a pickup or delivery location), loading an item at its pickup location, or unloading an item at its delivery location.

In this subsection, several variants of the VRP have been described. The problem that incorporates multiconstraints for tackling real-life scenarios is named as the Rich VRP (RVRP) [110].

1.2.3 Green Vehicle Routing Problem (GVRP)

In the last decade the Green VRP (GVRP) is studied. This variant of the problem aims at including different environmental issues in the optimization process, such as Greenhouse Gas emissions, pollution, waste, noise, fuel consumption, the effects of using "greener" fleet configurations, and so on.

The most popular objective in this context is the Greenhouse Gas emission. In [84,90,142], as the objective, the total fuel emissions cost and CO_2 emissions cost per liter are minimized. The problems of minimizing CO_2 [47,199], the other Greenhouse Gas emissions [121,162], and the fuel consumption [97,159, 172,227] are also considered. In [89] the total emission cost reduction model of the GVRP is assumed.

Sawik et al. [175] present the multiobjective GVRP, where minimization of amount of money paid as externality cost for noise, pollution and costs of fuel versus minimization of noise, and pollution and fuel consumption themselves are assumed as optimality criteria.

In [4,225] the refueling stations and the limit of fuel tank capacity are considered for the construction of a route. It assumes that each all refueling stations have unlimited capacity, the tank is filled to capacity when refueling of the vehicle is performed, and the fuel level is limited to the defined minimum volume when vehicles arrive at depot or at refueling station.

The problem of refueling or recharging of vehicles is also considered by Yavuz [219]. The fueling or charging station is divided into two groups, internal and external stations. An internal station is located in the company or in the customer, and it can be used while the operator is performing the job at that site. An external station is an outside location, it is public or private, where the operator can take the vehicle anytime during the workday but has to wait idle while the vehicle is being brought up to full energy or fuel.

In [69,134] the transportation fleet works with alternative ecofriendly fuels, where each vehicle has limited fuel tank capacity, and the minimum amount of fuel remaining in the tank of the vehicle is defined. In addition, there is limited access to alternative fuel stations.

Yu et al. [221] proposed the Hybrid Vehicle Routing Problem (HVRP), which is an extension of the Green Vehicle Routing Problem (GVRP). The vehicles that use a hybrid power source are assumed. The proposed model considers the utilization of electric and fuel power depending on the availability of either electric charging or fuel stations.

The problem of recycling is also considered as the GVRP. Soleimani et al. [190] presented a case of redistribution of products that are repairable or reusable. The distributor collects the secondhand products from the market to repair them. Next, it delivers the firsthand products and repaired secondhand ones (with lower price) to the market. While minimizing the distribution costs the emission produced by vehicle involved in the distribution system is minimized too. The first objective is to minimize air pollution, and it minimizes the Greenhouse Gas emitted by reducing fuel consumption. The second objective is to minimize the cost of fuel, cost of setting up distribution centers, and cost of preparing the vehicles.

Excellent and updated surveys on the GVRP are presented in [51,115,198].

1.2.4 Electric Vehicle Routing Problem (EVRP)

This variant of the VRP designs routes to serve a set of customers using a fleet of electric vehicles (EV). It assumes using ecofriendly vehicles, and therefore it can be viewed as a variant of the GVRP. There are important differences between using the EVs and traditional combustion vehicles (CV). The CVs have a long driving range; furthermore, the petrol stations are available almost everywhere, and the refueling takes a negligible time. Therefore the routing algorithms for CVs do not need to care for scheduling visits to refueling stations. In contrast, the EVs have a short driving range and a long battery recharging time, and availability of charging stations is limited. Therefore the time of recharging cannot be omitted. The routing algorithms for the EVs need to consider the visiting of the EVs in the charging stations and should minimize the number of visits. In addition, the nonlinear charging function, the battery degradation cost, and the compatibility constraint between the EV and charging stations must be considered. In the literature, there are many variants of the EVRP and different objectives are considered.

Leggieri and Haouari [111] presented the problem of designing a set of routes for a homogeneous fleet of electric vehicles. Each vehicle has defined the following parameters related to power supply: an energy consumption rate, the maximum storage of energy and a minimum reserve of energy $E_{\min}$. Some of vertices of the graph G represent the charging stations with unlimited capacities. The goal of the problem is to determine a set of vehicle routes satisfying the following green restrictions:

- each charging station can be visited more than once by the vehicles,
- between two consecutive customers, at most one visit to a charging station is allowed,

- upon the arrival at each travel point, the energy level of each vehicle should be at least $E_{\min}$.

Lin et al. [116] considered the EVRP in which a fleet of electric commercial vehicles with a limited range may recharge at a charging station during their daily operations. Each charging station may be visited more than once by the same and different vehicles, and the locations of all charging stations are within the service area. There is a cost associated with electricity, and the routes are determined minimizing the total cost, that is, the sum of travel time cost and energy cost.

Yang et al. [217] proposed a model where an electric vehicle visits a set of customers and returns to the depot. The charging station can be visited many times or never be visited, but each customer must be visited only once. The proposed objective function is a sum of three costs: the fast-charging cost, the cost of battery loss of life during the fast-charging, and the regular-charging cost. The aim of the problem is to determine the route to minimize the objective function while satisfying the constraints of battery capacity, charging time and delivery (or pickup) demands, and the impact of vehicle loading on the unit electricity consumption per mile.

Barco et al. [14] proposed the model where the aim is to determine the routes with minimum energy consumption, the route assignment for each vehicle, and the recharge scheduling for electric vehicles and to minimize the recharge cost and the cost associated with the battery degradation caused by route assignation and recharge cycles.

In [93,228] a heterogeneous fleet of electric vehicles is assumed. The vehicles differ with respect to battery capacity, battery charge rate, battery consumption rate, load capacity, fixed cost, and variable cost.

In the most cases, it is assumed that the battery-charge level is a linear function of the charging time, but in reality the function is nonlinear. In [65,135] the nonlinear charging functions are assumed, and the routes that minimize the total time, which is the sum of travel times and charging times, are determined.

In [167] the possibility of recharging the electric modules at customer locations is assumed. The available fleet of vehicles differ from the battery ones since the modules are autonomous in terms of consumption and electric charging. The objective is to minimize the acquisition cost, the total distance traveled, and the recharging costs.

The charging problem was studied by Paz et al. [154], who considered two energy supply technologies, the "Plug-in" conventional charge technology and Battery Swapping Stations (BSS). The location of the recharging stations is only possible at special charge stations due to the complex structure required for the battery swapping, which makes the application of this technology at customer vertices impossible. It assumes that the recharging time is a function of the amount of energy to charge and a fixed time.

In [183] the aim of the problem is to provide an optimal operation scheme consisted of the routes, a charging plan, and driving paths to totally fulfill customer demands, ensure the vehicles operation safety, and reduce the cost to the greatest extent. The driving paths are regarded as the most energy efficient paths between any two adjacent visited nodes in the route. The charging plan is to solve the problem of when and where the vehicle with charging demands should be recharged.

In [42] the EVRP with Time Windows and Battery Swapping Stations (EVRPTW-BSS) is researched. It is assumed that the only way to supplement the energy is a battery swapping in the BSS, in which the used battery could be exchanged with a full one. The battery swapping has some advantages over the conventional battery recharging. Comparing to the long recharging time, the battery swapping takes less. The swapped batteries could be recharged collectively.

Many researches assume that charging stations can simultaneously charge an unlimited number of electric vehicles, which is not met in practice. Froger et al. [64] take into consideration the limited capacity of charging station and present the EVRP with nonlinear charging function. The objective of this problem is to minimize the total time.

Shao et al. [184] present the EVRP with Charging Time and Variable Travel Time (EVRP-CTVTT). The distribution area may be relatively large. If the battery-level cannot satisfy the distance demand of completing the trip, then the vehicle must be recharged at charging stations in transit. The time is lost during recharging, and therefore the goal is to resolve the recharging planning problem, that is, how the charging stations are assigned to the vehicles and when the vehicles recharge. The previous EVRP research focuses on the static traffic environment, in which the travel time is regarded as a constant factor. The traffic environment is dynamic in real road networks. If the static model were used for the real road network, then the obtained results would be significantly flawed. Therefore Shao et al. [184] concentrates on the dynamic traffic environment, where the travel time is a variable factor. It is the first work that concentrates on the variable travel time. The routes are determined minimizing the total costs, which consist of the vehicle fixed cost, the travel cost, the penalty cost due to early or late arrivals by customers, and the charging cost.

Other objectives of determining the routes are considered in [33,32,52,107], where the EVRP with Time Windows (EVRPTW) is presented. The goal of the problem is to minimize the number of used vehicles and the total time spent by the vehicle. The total time is the sum of travel times, charging times, and waiting times due to the customer time windows. In [226] the goal of the EVRPTW is to minimize the total travel costs and the penalty costs for violating the time window of each customer. Another variant of the EVRPTW was studied by Keskin and Çatay [100], who assumed the partial charge (EVRPTW-PR). In the partial charge case the battery is charged to a specified level such as 80% of battery capacity. In [204] the objective is to minimize the sum of fixed and travel costs. In [81] the available vehicle types differ in their transport capacity, battery size, and acquisition cost. The aim of the problem is to minimize acquisition costs and the total distance traveled.

The VRPPDTW where the fleet of electric vehicles is used is studied by Grandinetti et al. [72]. The routes are determined to minimize three objectives simultaneously: the total distance, the cost of using the vehicles, and the penalties due to delays in delivering the service to the customers. Shao and Bi [182] describe the VRPPD for the fleet of electric vehicles.

Hof et al. [83] and Yang and Sun [218] propose a solution of the BSS-EV-LRP (Battery Swap Station Location-Routing Problem with Capacitated Electric Vehicles). The aim of the problem is to determine simultaneously:

- the routes to serve a set of customers minimizing the sum of construction and routing cost,
- the battery swap stations selected from a set of candidate locations.

The problem degenerates to the CVRP. A similar problem is considered by Arias et al. [9], where the aim is to determine the route minimizing the traveling cost and determine the location of charging stations, which indirectly increases the range of electric vehicles.

1.2.5 Algorithms for solving the VRP and its variants

The VRP and its variants are widely studied and reviewed in the literature [29, 105,136,145,161]. The works on the solution of the variants of VRP have continued by the adaptation of various methods such as simulated annealing [69, 144,166,221], tabu search [142,179,199], ant colony system [225] and fuzzy ant colony system [109], genetic algorithm [1,12,15,24,47,88,133,226], forward dynamic programming [123], iterative penalty method [195], the PSO (Particle Swarm Optimization) [88,159,185], a column generation-based heuristic (CGB-heuristic) [227], memetic algorithms [138,139,167], guided ejection search [20,140], the ALNS (Adaptive Large Neighborhood Search) [82,84,100, 120,124,216], the ILS (Iterated Local Search) [30,131,170], the VNS (Variable Neighborhood Search) [4,33,34,180] metaheuristics, the Glowworm Swarm Optimization (GSO) [127]. The problems are tackled using the following branching methods: the branch-cut-and-price [156], the branch-price-and-cut [52,205], the branch-and-price [197,80,93], and the branch-and-cut [7] algorithms.

Blocho and Nalepa [21] proposed the longest common subsequence-based selective route exchange crossover (LCS-SREX) operator and applied this operator in the memetic algorithm. Tan et al. [196] proposed the ACLBFO (Adaptive Comprehensive Learning Bacterial Foraging Optimization) algorithm, which is a variant of the BFO (Bacterial Foraging Optimization) algorithm with time-varying chemotaxis step length and comprehensive learning strategy.

The following hybrid algorithms for solving the variants of VRP are proposed: a hybrid of tabu search and the artificial bee colony algorithm [222], a hybrid of tabu search and the VNS [18], a hybrid of ant colony optimization and the VNS [90], a hybrid of ant colony optimization and the LS (Local Search)

[129], a hybrid of ant colony optimization and 2-opt LS [2,62], a hybrid of simulated annealing and the branch-and-bound [107], a hybrid of genetic algorithm and dynamic Dijkstra algorithm [184], a hybrid of genetic algorithm and the LS [183], a hybrid of the ILS and the HC (Heuristic Concentration) [135], a hybrid of a multistart ILS and a set partitioning [204], a hybrid of the ALNS and the ELS (Embedded Local Search) [81], a hybrid of the IBS (Iterated Beam Search) and the branch-and-bound [219].

The problems have been also tackled using two-stage algorithms. Sze et al. [194] proposed a two-stage AVNS (Adaptive Variable Neighborhood Search) algorithm that incorporates the LNS as a diversification strategy. Hu et al. [85] proposed a two-stage algorithm, where the first stage minimizes the total number of routes (the number of vehicles used), and the second stage minimizes the total travel distance of all determined routes. Another two-stage algorithm was proposed by Froger et al. [64]. In the first stage the ILS is used to generate a pool of high-quality routes while not taking the capacity constraints into account. In the second stage a Benders decomposition is used to determine the solution of the problem by assembling routes from the pool.

A three-phase matheuristic, combining an exact method with the VNSB (Variable Neighborhood Search local Branching) is proposed by Bruglieri et al. [32]. The first two phases based on Mixed Integer Linear Programs are exploited to generate feasible solutions, used in the last phase by a VNSB algorithm.

1.3 Bus Routing Problem and its variants

1.3.1 Bicriterion Bus Routing Problem (BBRP)

1.3.1.1 Formulation of the BBRP

The Bicriterion Bus Routing Problem (BBRP) is connected with the choice of means of transport and finding the route of travel between two given points. For the first time, it was defined and described by Boryczka [25], and its model wasbased on real bus and tram network in Upper Silesia in Poland. Widuch [211] reformulated and extended the definition of the problem so that it more accurately corresponds to the real bus network. The problem is formulated as follows [211]: we have given the bus network consisting of n stops $s_1, \ldots, s_n$, where buses of M bus lines numbered from 1 to M are run. The bus network is divided into zones that determine the cost of travel.

The route for each bus line is defined by the sequence of stops through which the bus runs from the start stop to the final stop of the line. The travel between a pair of bus stops is directed, and the bus of a given bus line can run in both directions, but these routes can be different. Buses of each line run between stops with a given frequency. A simplified model of the bus network is assumed, where the times of getting on and off the bus by passengers are omitted.

To define the bus network structure, the bus line routes, and the timetable, we are given: the start stop s_s and the final stop s_e ($s_s \neq s_e$) between which we want to travel and the time of starting travel T_s at the start stop. The goal of the problem is to find a route from the start stop s_s to the final stop s_e minimizing the time and the cost of travel simultaneously. The stops belonging to the route, the stops of changes the bus, the times of departure from all stops belonging to the route, and the bus lines along which the buses run between stops should be determined.

The time of the travel depends on the chosen route and the possible stops of changes. It is the sum of the travel times between stops belonging to the route, time of waiting at the start stop s_s, and times of waiting for changes. The cost of travel depends on the location of the stops in the area of zones and is calculated as follows. A ticket for a travel without a bus change within the area of a single zone equals c_1 $(0 < c_1)$ units, within two zones it equals c_2 $(c_1 < c_2)$ units, and within the confines of more than two zones it equals c_3 $(c_2 < c_3)$ units. Therefore the cost of travel from the start stop s_s to the final stop s_e is equal to the sum of costs of travel between the stops of bus changes. Among bus lines there are fast lines. The cost of the travel by a fast line is twice as large as the cost of the travel by a regular line.

1.3.1.2 Analysis of the BBRP

The BBRP can be modeled as a problem of graph theory [95]. The bus network is represented by the directed weighted multigraph $G = (V, E)$. The vertices represent the bus stops, and the arc $e_k = (v_i, v_j)$ represents the route of travel of a given bus line whose buses run directly from a stop represented by v_i to a stop represented by v_j. The zone to which the bus stop represented by v_i belongs is denoted by $z(v_i)$. Each arc e_k of the multigraph G has three weights: $l(e_k)$, $c(e_k)$, and $t(e_k)$. The weight $l(e_k)$ equals the line number the bus runs from v_i to v_j and takes values from the range $1, \dots, M$. The weight $t(e_k)$ takes positive values and is not constant. It is equal to the sum of the travel time from v_i to v_j and the possible time of waiting at a bus stop represented by v_i. The weight $c(e_k)$ is variable, it takes nonnegative values, and it is equal to the cost of travel from v_i to v_j. Its value depends on a possible change at the stop represented by v_i and location bus stops represented by vertices v_i and v_j in the same zone or in different zones. The determination of the values of $c(e_k)$ and $T(e_k)$ is illustrated with an example in [211–213].

The BBRP belongs to the Multicriteria Optimization (MO) problems, where $k > 1$ criterion functions are given. In the MO problems the criterion functions can be minimized or maximized, and in most cases the criterion functions are in conflict, because to decrease the value of any of the functions, we need to increase the values of other functions. Therefore the solution of the problem is a set of solutions called the set of nondominated (Pareto optimal) solutions [58, 57,148,207].

The BBRP solving is reduced to solving the Bicriterion Shortest Path Problem (BSPP) between the start vertex v_s (it represents the start stop) and the final vertex v_e (it represents the final stop) in a multigraph G with variable weights.

The solution of the BBRP consists of a set of paths in the multigraph G forming the set of nondominated solutions, where the time and the cost of travel are the criteria to be minimized.

The BSPP is a particular case of the Multicriteria Shortest Path Problem (MSPP). The BSPP and MSPP are known to be NP-complete by transformation from a 0–1 knapsack problem [67,78,188]. Both problems are widely studied, and a survey on the MSPP is presented in [45,58]. A review of methods used to solve BSPP problems is proposed by Skriver [188]. In the most cases a graph with constant weights is assumed, that is, the value of the weight function does not change for the given arc. A method for solving the BBRP and MSPP where a graph with variable weights is assumed is proposed in [75,122,164,206].

There are important differences between the properties of the paths belonging to the set of nondominated solutions and the methods used to solve the problem with constant and variable weights. If the weights of arcs are variable, then the set of nondominated solutions can contain paths that are not loopless [211,213]. The properties of nonloopless paths belonging to the set of nondominated solutions are defined by Widuch [211,213]. The path from the start vertex v_s to the vertex v_i ($v_i \neq v_e$) representing the *partial solution* can be extended to the *final solution* by determining the path from v_i to the final vertex v_e. If the weights of arcs take nonnegative and variable values, then the monotonicity assumption does not hold, and it is possible to obtain a nondominated *final solution* by extending a dominated *partial solution*. It is important to know whether it is possible to extend a dominated *partial solution* and obtain from it a nondominated *final solution*. Widuch [213] analyze all possible cases and present all conditions required to obtain a nondominated *final solution* from a dominated.

1.3.1.3 Algorithms for solving the BBRP

The first algorithm for solving the BBRP was proposed by Boryczka [25], who assumed a simplified bus network model that does not take into account the division of the network into zones and the time of starting travel T_s at the start stop. In addition, the total travel time and the number of bus changes are considered as criteria functions. In [25] an ant algorithm is presented. In [26,27] the modified ant algorithm is presented. The mentioned algorithms are heuristic methods.

Widuch [211,213] proposed two exact label correcting algorithms. The algorithms make possible to compute all paths from v_s to v_e belonging to the set of nondominated solutions. In [211] a label correcting algorithm with deleting *partial solutions* is presented. During the process of finding the solutions by the algorithm, only a single *partial solution* is stored, and it represents a path from the start vertex v_s to the given vertex v_i. The second exact algorithm belongs to a group of label-correcting algorithms with storing the *partial solutions* [213]. During the process of finding the solutions for each vertex of the multigraph, a list of *partial solutions* is stored where the lists may contain dominated solutions.

1.3.2 Multicriteria Bus Routing Problem (MBRP)

The BBRP presented in Subsection 1.3.1 was modified by adding the next criterion, and in [212] the Multicriteria Bus Routing Problem (MBRP) is described. In the MBRP the length of the route is additionally taken into consideration, where it equals the number of bus stops belonging to the route. Thus, in the MBRP, we determine the path minimizing three criteria, that is, simultaneously, the time and cost of travel and the length of the route.

The differences between the properties of the paths and the methods used to solve the MBRP and BBRP in [212] are shown. In [212] a label correcting algorithm with storing *partial solutions* for solving the MBRP is presented.

1.3.3 School Bus Routing Problem (SBRP)

The School Bus Routing Problem (SBRP) was formulated in 1969 [141]. It is a problem in the management of school bus fleet and seeks to plan an efficient schedule for a fleet of school buses that pick up students from various bus stops and deliver them to the school. At the end of the school day the student is transported again for same location where she was picked up. The SBRP aims to optimize the school bus transport by satisfying various constraints, such as the bus capacity, where all students are picked, and each student must be assigned to a particular bus. The objective of bus route planing is to visit all bus stops while minimizing the number of used school buses and the total bus travel distance and satisfying service qualities such as student maximum riding time on a bus.

The SBRP is an NP-hard problem, and it is widely studied. In the literature, we can find several variants of the problem, methods, and constraints to solve the SBRP. In [150], many variants of the problem and different methods for solving it are presented, and it contains references to works issued up to 2010. In the following subsection, there are references to works published after 2010.

The SBRP consists of five subproblems:

1. Data preparation.
2. Bus stop selection.
3. Bus route generation.
4. School bell time adjustment.
5. Route scheduling.

Díaz-Parra et al. [54] present a set of test instances of the SBRP. The test instances were created using an algorithm called SBRPGen, and it can be downloaded for other researchers for experimentation. The SBRP is also resolved based on the model of real school bus networks (see Subsection 1.3.3.9).

1.3.3.1 Data preparation

In this subproblem the following types of data are specified: the road network, students, schools, buses, and time/distance matrix. The data for students describe the location of their homes, the location of school, and the type of student. In the literature, we can find two types of students, general students and special-education students. The general students are picked up and dropped off from the bus stops, but the special-education students are picked up and dropped off directly at their homes and not at their bus stops. There is a stricter restriction on the maximum riding time in a bus for a special-education student. Each special-education student should be served differently depending on the severity of the student particular disability. Some of special-education students should be assigned to a special bus, which is able to serve them a special equipment. The special-education students in the most cases are omitted, and only the general students are considered. The SBRP for the special-education students is considered by Kamali et al. [96].

The school data contains information about the number of schools and their location. In the most cases the starting and ending times for school bus arrivals are given, but in the literature a problem of determination of the starting and ending times of schools is considered (see Subsubsection 1.3.3.4). In the most cases the SBRP with single school is considered [10,11,22,31,38,39,43,53,59, 61,66,79,87,102–104,114,119,130,137,146,152,168,169,177,176,181,202,201, 203,208,214,220,224]. In [23,35,36,147,191,60,63,96,98,108,192,125,126,128, 132,143,151,160,171,6,173,187,186,210] the SBRP with multiple schools is taken into consideration.

The data for buses define their types and capacity for general students and special-education students if special-education students are considered. The time or distance matrix contains the shortest travel times or distances between all pairs of nodes, where the nodes represent school, student location, and the bus origin location.

1.3.3.2 Bus stop selection

The subproblem of bus stop selection is often omitted and is not considered. In the most solutions, it is assumed that the locations of the bus stops are given and the students take a bus at the bus stop and all bus stops in the network must be visited by the buses.

The SBRP with bus stop selection (SBRPBSS) is considered in [23,147,63, 87,98,104,114,152,155,168,169,177,176,208]. In SBRPBSS a set of potential bus stops is given in such a way that each student lives within in maximum defined distance of at least one stop. Some constraints may also be used such as the minimum distance walked by each student to the bus stop, and it also assigns each student to the nearest stop [23,63,98,104,169], the minimum total traveled distance, and allocating students to stops in such a way that the capacity of the buses is not exceeded [168,176] and the maximum walking distance to the bus stop [177]. The assignment of students to the bus stops takes into consideration not only the distances the students need to walk, but also how often they walk across streets, particularly high-traffic streets [208]. Thus, determining the set of bus stops to actually visit by the school buses is a part of the SBRPBSS. The problem of bus stop selection is also resolved as a separate problem [174].

1.3.3.3 Bus route generation

In the subproblem of bus route generation the school buses routes are generated. The subproblem is very similar to the VRP. In the first proposed algorithm, the "route-first, cluster-second" method was implemented [141]. The method creates a large route using a Traveling Salesman Problem algorithm that considers all the stops and partitions it into smaller routes considering the constraints. Another proposed method is "cluster-first, route-second", which groups the students into clusters so that each cluster can be served as a route satisfying the constraints [10,191,53,87,114,119,171,186]. In other methods a set of paths is created, which are then optimized considering the constraints.

In the problem with multiple schools, two kinds of constraints are considered named adequately as the "mixed load" and the "single load". The "mixed load" feature allows students from different schools to ride the same bus at the same time [23,36,191,96,98,192,132,151,171,6,186]. In the "single load" the bus transports students from a single school [35,60,63,125,126,128,160,171, 173,187,210].

1.3.3.4 School bell time adjustment

The subproblem of school bell time adjustment is associated with the starting and ending times of schools, and they are constraints. The subproblem is important in the SBRP with multiple schools, because starting and ending times for each school may be different. Park and Kim [150] consider the integrated coordination of the school starting times and the public bus services.

1.3.3.5 Route scheduling

The subproblem of route scheduling specifies the exact starting and ending times of each route and defines a sequence of routes that can be executed successively by the same bus. In [11,10,22,31,36,147,59,61,103,114,119,192,125,126,130, 137,143,146,6,176,181,186,203,220,214], it is assumed that a bus performs exactly one route. In the second case, it is assumed that a bus may perform many routes [35,53,60,79,108,128,132,151,171,173,187,210].

The problem of school bus route scheduling is also considered as a separate problem [101], where the set of routes for each school is given. A school bus can serve multiple trips for multiple schools, and the goal of the problem is to optimize bus schedules to serve all the given routes considering the school time windows.

1.3.3.6 Means of transport

In the SBRP, two modes of transport are considered, the urban or the rural transport. It is assumed that students in the urban areas walk from their homes to the bus stops to take a bus [10,35,36,147,53,61,96,98,102,103,126,128,137,151, 152,168,171,214]. In the rural areas the number of students is small, and it is common to pick them up at their homes [191,60,63,192,132,146,160,187,186]. In [38,39,119], both means of transport are considered.

Another classification is based on the type of vehicles, and two kinds are considered, a homogeneous or heterogeneous fleet of buses. The problem with a heterogeneous fleet of buses assumes that each bus has different characteristics, such as various capacities, maximum allowable riding time, fixed cost, and per unit distance variable cost [11,36,38,39,147,191,63,66,96, 98,192,126,130,132,151,176,187,186,203,224]. In the problem with a homogeneous fleet of buses, each bus has the same characteristics [10,22,23,31,35,53, 59–61,79,87,103,104,108,114,119,128,137,143,146,152,160,168,169,171,173, 181,210,214]. Both the homogeneous and heterogeneous fleets of buses are considered in [43,102].

1.3.3.7 Objectives of the problem

There are many variants of the problem that differ from each other in objectives, and the following objectives are considered:

- minimizing the total cost of school transport (MC),
- minimizing the total traveled time (MT),
- minimizing the number of buses (MNB),
- minimizing the number of routes (MNR),
- minimizing the total direct distance that students travel to their school (MD),
- minimizing the total length of all routes (ML),
- minimizing duration of the longest route (MDLR),
- minimizing the maximum regret (MR),
- minimizing the assignment cost of students to bus stops (MAC),
- safety mark (SM).

In the literature, single-objective and multiobjective problems are described. Table 1.1 shows number of variants of this problem with their objectives.

1.3.3.8 Algorithms for solving the SBRP and its variants

The problem is widely studied, and a review of papers on SBRP solutions is presented by Park and Kim [150]. The work on solving the problem has continued by the adaptation of various methods such as the branch-and-cut algorithm [168], the branch-and-bound algorithm [108], ant colony optimization [11,10,59,61,87,202,220], simulated annealing [126,173], the genetic algorithm [38,53,98,130,143,181,202,201], tabu search [137,146,6], the GRASP (Greedy Randomized Adaptative Search Procedure) metaheuristic [187], the time-saving heuristic [214], the harmony search heuristic [103], the bundling

TABLE 1.1 Literature review on the school bus routing problem according to the objectives.

References	Objectives
[10], [61], [147], [53], [59], [66], [87], [103], [128], [192], [132], [152], [160], [171], [181], [203], [214], [224]	MC
[63], [96], [177], [176]	MD
[146]	MDLR
[11], [104], [186], [201]	ML
[36], [60], [114], [151], [6], [208]	MNB
[23], [31], [102], [119], [125], [126], [137], [173]	MT
[22]	MR
[168]	MAC, MC
[92], [169]	MC, ML
[108]	MC, MT
[38]	MC, SM
[35], [43], [130], [143]	ML, MNB
[79], [210]	MNB, MT
[191]	MNR, MT
[39]	MT, SM
[202]	MC, MNB, MT
[143]	ML, MNB, MNR

technique [203], the column-generation-based algorithm [35,36,104,119,169], the mixed load algorithms [60,151], the MILP (Mixed Integer Linear Programming) method [23,128,210], the MIP (Mixed Integer Programming) method [171], a hub-and-spoke network strategy [102], the K-means clustering method [202], the greedy strategy [38], the Quantum-behaved Particle Swarm algorithm [66,224], the Intelligent Water Drops (IWD) metaheuristic that simulates natural water drops [92], the ILS (Iterated Local Search) metaheuristic, [114,132], the multiobjective ILS metaheuristic [191], polynomial time 4-approximation algorithm [22], and memetic algorithm [147]. A modification of the Clarke and Wright savings heuristic algorithm [44] is proposed by Hashi et al. [79].

The problem is resolved by hybrid algorithms: a hybrid of tabu search and the VNS [176], a hybrid of tabu search and the VND [177], a hybrid of the MILP and the GRASP [63], a hybrid of a genetic algorithm and a local search method [130]. In the work of Souza Lima et al. [192], five metaheuristic-based algorithms were devised to solve the SBRP. The first proposed heuristic is the Record-to-Record Travel (RRT) method, and the other four algorithms stem from combining the Iterated Local Search (ILS) and the Variable Neighborhood Search (VNS) strategies with the Variable Neighborhood Descent (VND) local search and Random Variable Neighborhood Descend (RVND) algorithm.

In [152] an exact method named EX-EX-EX and two hybrid metaheuristic algorithms named GA-EX-TS and SA-EX-TS have been proposed. The EX-EX-EX method has the capability to determine the exact solution of the problem by explicit complete enumeration of routes and bus stops. Therefore, this method is only applicable for bus networks with small sizes. The metaheuristic GA-EX-TS is a hybrid algorithm that combines genetic algorithm and tabu search, and the SA-EX-TS is a hybrid algorithm of simulated annealing and tabu search. Chen et al. [43] proposed two algorithms for solving the SBRP, an exact method of mixed integer programming (MIP) and a hybrid simulated annealing with local search metaheuristic.

1.3.3.9 Real school bus networks

In many cases the problem is tackled based on the model of real school bus networks. In the literature the following real school bus networks are considered: Ankara, Turkey [202,201,220], Bogotá, Colombia [10,119,171], Dar es Salaam, Tanzania [125,126,137], Espírito Santo, Brazil [132], Great Tunis, Tunisia [61], Guthrie, Oklahoma, USA [208], Lisbon, Portugal [128], Minas Gerais, Brazil [63,192,160], New York, USA [35,36], Sekondi-Takoradi, Ghana [11], St. Louis, Missouri, USA [60], Thessaloniki, Greece [38], Zagazig, Egypt [59], the Flemish Region, Belgium [176], a rural area in the North of Spain [146], the Wonkwang University located in Iksan, South Korea [102], Gansu Normal University for Nationalities, China [66,224], Scholastica School in Dhaka, Bangladesh [79], Blooming Dales School situated in Ganga Nagar, India [130], the secondary school in Beijing, China [87].

1.3.4 Other selected variants of the Bus Routing Problem

Optimal planning for public transportation is one of the keys to better quality of life in urban areas. Leksakul et al. [112] considered the LRP (Location Routing Problem) connected with transporting employees to their workplaces. The objective is to minimize the number of bus stop locations and the total traveling distance of employees while satisfying employees at maximum radius. This problem is a variant of the SBRP with Bus Stop Selecton. The proposed algorithm was tested using real data from the bus transport for employees of a large-scale industrial factory in Thailand.

A night-bus route planning by leveraging taxi GPS traces is approached by Chen et al. [40]. There is need to identify the candidate bus stops that are located in the area having a big number of taxi passenger pick-up and drop-off. In the next step a bus route connecting the bus origin and a sequence of bus stops to the destination, carrying the maximum number of passengers within a defined time duration, are determined. The taxi GPS traces contain information about all taxi trips, and it defines the "hot" areas for taxi passengers and potentially the number of traveled passengers along a given route. The problem was researched on the base of real data. A taxi GPS traces data was generated from 7600 taxis in the city of Hangzhou (China) for one month, and more than 1.57 million of night passenger-delivering trips were used. A similar problem is described by Ling et al. [117].

Chen et al. [41] proposed a problem of limited-stop bus service, which allows a bus to skip some bus stops in the bus route. It assumes that in the bus network, different buses serving the same bus line may have different stopping plans. The goal of bus stop skipping is to reduce the bus dwell time and increasing in this way the operating speed. To avoid a long waiting time at the bus stop, the bus stop is not allowed to be skipped by two consecutive buses of the same line. The proposed methods of resolving the problem are validated using real data of bus lines in the city of Suzhou (China).

A specific type of public transportation is researched by Schmid [178]. The paper presents the design of routes and their frequencies for the BRT (Bus Rapid Transit) systems. There is given a network consisting of a set of stations, where buses may be operated on designated lanes. The total number of buses in use is limited. The aim of the problem is to design a set of routes and determine their frequencies minimizing the total travel time by passengers taking into account capacity restrictions. The frequency of the route determines how often the route is served by the bus, and it can be in regular intervals or in an aperiodic way.

Wu et al. [215] considered a problem of bus bunching. It occurs when at least two buses along the same route arrive at the same bus stop simultaneously. The bus bunching is undesirable for passengers and the bus operator. The problem is researched on the base of the real network of the city of Guangzhou in China.

Another problem is considered by Ceder et al. [37]. There is given a set of bus routes. For the given bus route, a predetermined set of potential bus stop locations is given. The aim of the problem is to determine the optimal set of bus stops to serve demand accessing the route in both service directions taking into account the effects of topography on access time and operational cost. It assumes that passenger arrivals at stops are uniformly distributed. The proposed solution method was tested using real bus network in the central business district of Auckland City (New Zealand).

As a variant of Bus Routing Problem, an evacuation plan before a disaster occurs is studied by Goerigk et al. [70]. General various kinds of disasters are considered. The evacuation is assumed to be done by buses to relocate people outside the endangered area. There are given a set of buses with the same capacity and the set of points within the endangered area. For each point, the number of evacuees who need to be picked up at this point is given. The set of destination points outside the endangered area, where the evacuees need to be brought is given. For each destination point, a maximum capacity is defined. The goal of the problem is to determine the set of routes minimizing the evacuation time defined as the maximum of all bus travel times. It assumes that each bus can only make one trip per round. In this work an evacuation of people in Kaiserslautern (Germany) is considered. Dikas and Minis [55] consider two variants of the evacuation problems. The first variant assumes an evacuation in anticipation of major natural threats. The second variant assumes transporting casualties after an emergency or terrorist incident, or from a battlefield.

An adoption of electric vehicles as means of transport in the EVRP is studied. A problem of adoption electric buses for people transportation is also studied. Paul and Yamada [153] analyze a problem where the bus diagram that shows the departure and arrival times of various bus trips at the bus terminals is given. The aim of the problem is to maximize the travel distance of the electric buses. The proposed method was tested using the real bus diagram data of four bus routes of a one of cities in Japan. An electric bus has shorter driving range than a traditional combustion bus, and therefore an important problem is the location of the charging stations. Wang et al. [209] research a problem where the electric bus network and the bus routes are defined. The goal of the problem is to determine the location of charging stations at the bus stops minimizing the total installation cost. The problem is studied on the base of the real bus network of a big city Qingdao in China.

1.4 Unmanned Vehicle Routing Problem

In the last few years, Unmanned Aerial Vehicles (UAV) colloquially known as drones have become more and more popular and found application in many fields. A problem of path planing for the UAV is one of the researched problems.

The Multi-Trip VRP for the UAVs is presented by Dorling et al. [56], where the goal is to determine a set of the routes flown by a fleet of drones minimizing the time and cost required to deliver packages to a set of destination points.

Sundar and Rathinam [193] proposed a single UAV routing problem. There is given a set of target points, and the UAV can be refueled at any of them. The goal of the problem is to determine the route for the UAV such that each target point is visited at least once by some UAV where the total fuel consumption by the UAV is minimal. In addition, the fuel constraint for the UAV is defined. The model where the fuel required to travel is directly proportional to the length of the path is assumed. A similar problem is considered by David Levy and Rathinam [50], but the objective is to minimize the total travel cost of the vehicles.

Coutinho et al. [48] describe the UAV routing and trajectory optimization problem. A fleet of the UAVs and the set of points that need to be visited by the UAVs are given. For each UAV, the cost of using it, its mass, wing area, and aerodynamics coefficients are given. The goal of the problem is to minimize the total cost of using the UAVs, the routing cost of flying between a pair of points, and a measure of the quality of the trajectories at the end points for each pair of points.

The work of Guerriero et al. [73] focuses on a distributed system of the UAVs, where each device is required to move toward a certain destination point in a given time. It is considered as a multicriteria optimization problem that takes into account three objectives simultaneously: the total distances traveled by the UAVs (to be minimized), the number of used UAVs (to be minimized), and the customer satisfaction (to be maximized). The customer satisfaction is defined by soft time windows constraints. In the paper the sport event filming problem using the UAVs is studied. In the first case the refueling of the cars during a car race is considered. Each race team has a pit area for refueling of its car, and the UAV flying over the pit are filming all the refueling events of the race cars. In the second case a filming of events during a football match is considered.

Zhang et al. [223] presented a problem of monitoring recurring and nonrecurring traffic conditions and special events on transportation networks by the UAVs. The network contains fixed and emerging mobile sensors to detect traffic incidents. An UAV routing model integrated with fixed traffic monitoring sites for road segments with various frequencies of incidents is considered. The aim of researched problem is to minimize the detection delay cost and operational cost, subject to feasible flying route constraints.

Rabta et al. [163] investigated a problem of using the UAVs in humanitarian logistics. An optimization model for the delivery of multiple packages of lightweight relief items via the UAVs to a set of remote locations is presented. The aim of the problem is determining a set of routes minimizing the total traveling distance of the UAV.

Another problem of the UAV path planning is presented by Razzaq et al. [165]. The goal of the problem is to determine route that satisfies some set of constraints minimizing the total cost of traversing the route. In addition, collision avoidance with other moving objects is considered. As the result, a three-dimensional route is defined by a set of points, again represented by latitude, longitude, and altitude.

Yet another kind of unmanned vehicle is the UCV (Unmanned Combat Vehicle). The UCVs are used in military applications such as border patrol and reconnaissance and surveillance expeditions. Han et al. [77] considered the multiobjective shortest path problem between two points in a terrain for the UCVs. The aim of the problem is to find Pareto-optimal paths minimizing the traverse time, risk level, and jamming level. In addition, a single-objective shortest path problem is considered, where the time of travel is minimized.

A problem of finding a route of the UCV in the military battlefield is researched by Bae et al. [13], who **modeled** a terrain as a grid. For each cell of the grid, the traverse time to pass through it and risk level associated with this cell are given. A set of cells that must be visited by the UCV are given in advance as in real situation of military operation. The goal of the problem is to determine a path minimizing the total travel time. The sum of risk levels associated with cells on the path should not exceed a given upper limit.

Park et al. [149] considered a problem of determining a path of the UCV that patrols a defined area. The UCV patrols the area by visiting a given set of points where the objective is to minimize possibility of infiltration by the enemy.

1.5 The other routing problems of electric vehicles

The electric vehicles are an ecofriendly solution for reducing Greenhouse Gas emissions. The problem of routing problem for electric vehicles is researched, and the problems that do not belong to the any variant of VRP are presented in the literature. The important problem concerned with the electric vehicles is a localization of charging stations. A review of researched problems related to the electric vehicles is presented by Jing et al. [94].

Boysen et al. [28] considered a route of single electric vehicle that executes transport request between given start and target points. The vehicle capacity and the vehicle's battery capacity are given. The battery capacity limits the travel distance until the battery requires its recharging. A set of charging stations are located along the route, one of which must be visited before the battery reaches the minimum energy level. The aim of the problem is to determine a set of visited charging stations by the vehicle along the route minimizing the total driving distance.

Baum et al. [16] assumed a concave charging function. According to the function, the charging speed decreases as the battery state of charge increases and monotonically increases with charging time, that is, charging the battery for a longer time does not decrease its state of charge. For the given network and the initial state of charge of electric vehicle, the goal is to determine a feasible route between the origin and the destination points. It is resolved as the Bicriterion Shortest Path Problem (BSPP), where the travel time and the current battery state of charge are used as criterion functions. The travel time is equal to the sum of driving time and the total charging time at charging stations. A set of routes forming a set of Pareto optimal solutions is determined. Another variant of the BSPP **is proposed** in [17,71]. The aim of the problem is to determine a set of Pareto optimal routes minimizing the travel time and the energy consumption simultaneously.

Jafari and Boyles [91] described a routing problem of electric vehicles in stochastic network. The charging stations with different electricity prices and charging rates are located in the network. A travel time between two given points of the network is represented by a stochastic variable. The energy consumption by the vehicle between two points is modeled as a linear function of distance and travel time between these points. The goal of the problem is to determine the route between the origin and destination points minimizing the cost of travel. The cost is formulated as a linear function of travel time and charging cost, subject to a minimum reliability threshold, representing the level of risk that a traveler can take when choosing a route with a lower cost. The route contains a sequence of points between the origin and destination points and information about charging stations to be visited by the vehicle and how much charge is to be taken at each of them.

The battery exchange as the method of refueling is assumed by Adler et al. [3]. The aim of the problem is to determine the shortest route from the origin to the destination point, where the maximal number of battery exchanges is defined as a constraint. The route must have no the subroute without refueling with a length greater than the distance the vehicle can travel on a full battery. Thus, determination of the route is considered as the problem of determination several subroutes between different battery exchange stations.

1.6 Conclusions

This chapter surveys the current formulations of vehicle routing problems in the context of real-life smart-delivery systems. In many cases the network models are defined to match real networks as closely as possible. A number of variants of the vehicle routing problems are presented. In the literature, new trends of transportation means are becoming increasingly popular and studied. The most important problem concerned with using of vehicles is reduction of environment pollution. The solution of the problem is using vehicles with alternative energy sources, and thus the green and electric vehicle routing problems were formulated.

Researchers consider many variants of routing problems that include the real-life properties and make their models more realistic and their solution approaches more applicable in practice. The proposed models are increasingly reflecting real-life networks, which makes methods for solving these problems more and more difficult. A future research could focus on problems by simultaneously considering real-life objectives and different types of vehicles and developing efficient methods to solve these problems. However, the presented literature demonstrates that the problems known and studied for many years are still being researched, and researchers try to develop more efficient methods for solving them.

Transport of goods and people is widely studied and optimized for many years. During the past decades, different optimization criteria and different means of transport were taken into account. Modern cities are ready to move forward with self-driving vehicles and electrification of fleet vehicles. Transportation systems are evolving toward smart delivery systems, which play a pivotal role in a smart cities. Concept of smart city integrates existing city infrastructure with transportation systems and data collected from citizens, monitoring infrastructure, and other devices to improve the efficiency of city services, safety, and relationship with its citizens. Thus, new solutions are researched by scientific and governmental organizations.

Acknowledgment

This research was supported by statutory funds of the Institute of Informatics, Silesian University of Technology (Gliwice, Poland).

References

[1] A.M.F.M. AbdAllah, D.L. Essam, R.A. Sarker, On solving periodic re-optimization dynamic vehicle routing problems, Applied Soft Computing 55 (2017) 1–12.

[2] M.M.S. Abdulkader, Y. Gajpal, T.Y. ElMekkawy, Hybridized ant colony algorithm for the multi compartment vehicle routing problem, Applied Soft Computing 37 (2015) 196–203.

[3] J.D. Adler, P.B. Mirchandani, G. Xue, M. Xia, The electric vehicle shortest-walk problem with battery exchanges, Networks and Spatial Economics 16 (2016) 155–173.

[4] M. Affi, H. Derbel, B. Jarboui, Variable neighborhood search algorithm for the green vehicle routing problem, International Journal of Industrial Engineering Computations 9 (2018) 195–204.

[5] B. Afshar-Nadjafi, A. Afshar-Nadjafi, A constructive heuristic for time-dependent multi-depot vehicle routing problem with time-windows and heterogeneous fleet, Journal of King Saud University - Engineering Sciences 29 (2017) 29–34.

[6] P. Aparicio Ruiz, J. Muñuzuri Sanz, J. Guadix Martín, Mixed trips in the school bus routing problem with public subsidy, in: Enhancing Synergies in a Collaborative Environment, Springer International Publishing, 2015, pp. 105–113, chapter 12.

[7] C. Archetti, M. Christiansen, M.G. Speranza, Inventory routing with pickups and deliveries, European Journal of Operational Research 268 (2018) 314–324.

[8] C. Archetti, E. Fernández, D.L.H.M. noz, The flexible periodic vehicle routing problem, Computers & Operations Research 85 (2017) 58–70.

[9] A. Arias, J.D. Sanchez, M. Granada-Echeverri, Integrated planning of electric vehicles routing and charging stations location considering transportation networks and power distribution systems, International Journal of Industrial Engineering Computations 9 (2018) 535–550.

[10] J.S. Arias-Rojas, J.F. Jiménez, J.R. Montoya-Torres, Solving of school bus routing problem by ant colony optimization, Revista EIA (2012) 193–208.

[11] J. Awuahaddor, S.K. Amponsah, J. Annan, C. Sebil, School bus routing: a case study of wood bridge school complex, Sekondi-Takoradi, Ghana, International Journal of Business and Social Research 3 (2013) 26–36.

[12] H. Bae, I. Moon, Multi-depot vehicle routing problem with time windows considering delivery and installation vehicles, Applied Mathematical Modelling 40 (2016) 6536–6549.

[13] K.Y. Bae, Y.D. Kim, J.H. Han, Finding a risk-constrained shortest path for an unmanned combat vehicle, Computers & Industrial Engineering 80 (2015) 245–253.

[14] J. Barco, A. Guerra, L. Munoz, N. Quijano, Optimal routing and scheduling of charge for electric vehicles: a case study, Mathematical Problems in Engineering 2017 (2017).

[15] M. Barkaoui, J. Berger, A. Boukhtouta, Customer satisfaction in dynamic vehicle routing problem with time windows, Applied Soft Computing 35 (2015) 423–432.

[16] M. Baum, J. Dibbelt, A. Gemsa, D. Wagner, T. Zündorf, Shortest feasible paths with charging stops for battery electric vehicles, in: The 23rd SIGSPATIAL International Conference on Advances in Geographic Information Systems, SIGSPATIAL'15, Seattle, Washington, USA, 2015, pp. 44:1–44:10.

[17] M. Baum, J. Dibbelt, L. Hübschle-Schneider, T. Pajor, D. Wagner, Speed-consumption tradeoff for electric vehicle route planning, in: 14th Workshop on Algorithmic Approaches for Transportation Modelling, Optimization, and Systems, ATMOS'14, Wroclaw, Poland, 2014, pp. 138–151.

[18] S. Belhaiza, R. M'Hallah, A Pareto non-dominated solution approach for the vehicle routing problem with multiple time windows, in: 2016 IEEE Congress on Evolutionary Computation, CEC, Vancouver, Canada, 2016, pp. 3515–3524.

[19] M. Bhuvaneswari, S. Eswaran, S.P. Rajagopalan, A survey of vehicle routing problem and its solutions using bio-inspired algorithms, International Journal of Pure and Applied Mathematics 118 (2018) 259–264.

[20] M. Blocho, J. Nalepa, A parallel algorithm for minimizing the fleet size in the pickup and delivery problem with time windows, in: The 22nd European MPI Users' Group Meeting, Bordeaux, France, 2015, pp. 15:1–15:2.

[21] M. Blocho, J. Nalepa, LCS-based selective route exchange crossover for the pickup and delivery problem with time windows, in: 17th European Conference, EvoCOP 2017, Springer International Publishing, Amsterdam, The Netherlands, 2017, pp. 124–140.
[22] A. Bock, E. Grant, J. Könemann, L. Sanitá, The school bus problem on trees, Algorithmica 67 (2013) 49–64.
[23] M. Bögl, K.F. Doerner, S.N. Parragh, The school bus routing and scheduling problem with transfers, Networks 65 (2015) 180–203.
[24] R. Bolaños, J. Escobar, M.G. Echeverri, A metaheuristic algorithm for the multi-depot vehicle routing problem with heterogeneous fleet, International Journal of Industrial Engineering Computations 9 (2018) 461–478.
[25] U. Boryczka, Ant colony system and bus routing problem, in: International Conference on Computational Intelligence for Modelling, Control and Automation, CIMCA'99, Vienna, Austria, 1999.
[26] U. Boryczka, M. Boryczka, Multi-cast ant colony system for bus routing problem, in: MIC'2001 – 4th Metaheuristics International Conference, Porto, Portugal, 2001, pp. 521–525.
[27] U. Boryczka, M. Boryczka, Multi-cast ant colony system for the bus routing problem, in: Metaheuristics: Computer Decision-Making, Springer US, Boston, MA, 2004, pp. 91–125, chapter 5.
[28] N. Boysen, D. Briskorn, S. Emde, Scheduling electric vehicles and locating charging stations on a path, Journal of Scheduling 21 (2018) 111–126.
[29] K. Braekers, K. Ramaekers, I.V. Nieuwenhuyse, The vehicle routing problem: state of the art classification and review, Computers & Industrial Engineering 99 (2016) 300–313.
[30] J. Brandão, Iterated local search algorithm with ejection chains for the open vehicle routing problem with time windows, Computers & Industrial Engineering 120 (2018) 146–159.
[31] E.M. Bronshtein, D.M. Vagapova, A.V. Nazmutdinova, On constructing a family of student delivery routes in minimal time, Automation and Remote Control 75 (2014) 1195–1202.
[32] M. Bruglieri, S. Mancini, F. Pezzella, O. Pisacane, S. Suraci, A three-phase matheuristic for the time-effective electric vehicle routing problem with partial recharges, Electronic Notes in Discrete Mathematics 58 (2017) 95–102.
[33] M. Bruglieri, F. Pezzella, O. Pisacane, S. Suraci, A variable neighborhood search branching for the electric vehicle routing problem with time windows, Electronic Notes in Discrete Mathematics 47 (2015) 221–228.
[34] G.A. Bula, C. Prodhon, F. Augusto, Variable neighborhood search to solve the vehicle routing problem for hazardous materials transportation, Journal of Hazardous Materials 324 (2017) 472–480.
[35] H. Caceres, R. Batta, Q. He, School bus routing with stochastic demand and duration constraints, Transportation Science 51 (2017) 1349–1364.
[36] H. Caceres, R. Batta, Q. He, Special need students school bus routing: consideration for mixed load and heterogeneous fleet, Socio-Economic Planning Sciences (2018).
[37] A. Ceder, M. Butcher, L. Wang, Optimization of bus stop placement for routes on uneven topography, Transportation Research Part B: Methodological 74 (2015) 40–61.
[38] E. Chalkia, J.M.S. Grau, E. Bekiaris, G. Ayfandopoulou, C. Ferarini, E. Mitsakis, Routing algorithms for the safe transportation of pupils to school using school buses, in: Transport Research Arena 2014, Paris, France, 2014, pp. 7–16.
[39] E. Chalkia, J.M.S. Grau, E. Bekiaris, G. Ayfandopoulou, C. Ferarini, E. Mitsakis, Bus routing safety for the transportation of children to school, in: CONAT 2016 International Congress of Automotive and Transport Engineering, Brasov, Romania, 2016, pp. 676–683.
[40] C. Chen, D. Zhang, N. Li, Z.H. Zhou, B-planner: planning bidirectional night bus routes using large-scale taxi GPS traces, IEEE Transactions on Intelligent Transportation Systems 15 (2014) 1451–1465.
[41] J. Chen, Z. Liu, S. Zhu, W. Wang, Design of limited-stop bus service with capacity constraint and stochastic travel time, Transportation Research Part E: Logistics and Transportation Review 83 (2015) 1–15.

[42] J. Chen, M. Qi, L. Miao, The electric vehicle routing problem with time windows and battery swapping stations, in: 2016 IEEE International Conference on Industrial Engineering and Engineering Management, IEEM, Bali, Indonesia, 2016, pp. 712–716.
[43] X. Chen, Y. Kong, L. Dang, Y. Hou, X. Ye, Exact and metaheuristic approaches for a bi-objective school bus scheduling problem, PLoS ONE 10 (2015) 1–20.
[44] G. Clarke, J.W. Wright, Scheduling of vehicles from a central depot to a number of delivery points, Operations Research 12 (1964) 568–581.
[45] J.C. Climaco, M.M.B. Pascoal, Multicriteria path and tree problems: discussion on exact algorithms and applications, International Transactions in Operational Research 19 (2012) 63–98.
[46] J.F. Cordeau, G. Laporte, M.W.P. Savelsbergh, D. Vigo, Vehicle routing, in: Handbooks in Operations Research and Management Science: Transportation, vol. 14, Elsevier, 2007, pp. 367–428, chapter 6.
[47] P. de Oliveira da Costa, S. Mauceri, P. Carroll, F. Pallonetto, A genetic algorithm for a green vehicle routing problem, Electronic Notes in Discrete Mathematics 64 (2018) 65–74.
[48] W.P. Coutinho, M. Battarra, J. Fliege, The unmanned aerial vehicle routing and trajectory optimisation problem, a taxonomic review, Computers & Industrial Engineering 120 (2018) 116–128.
[49] G.B. Dantzig, J.H. Ramser, The truck dispatching problem, Management Science 6 (1959) 80–91.
[50] K.S. David Levy, S. Rathinam, Heuristics for routing heterogeneous unmanned vehicles with fuel constraints, Mathematical Problems in Engineering 2014 (2014) 12.
[51] E. Demir, T. Bektaş, G. Laporte, A review of recent research on green road freight transportation, European Journal of Operational Research 237 (2014) 775–793.
[52] G. Desaulniers, F. Errico, S. Irnich, M. Schneider, Exact algorithms for electric vehicle-routing problems with time windows, Operations Research 64 (2016) 1388–1405.
[53] O. Díaz-Parra, J.A. Ruiz-Vanoye, A. Buenabad-Arias, F. Cocón, A vertical transfer algorithm for the school bus routing problem, in: 2012 Fourth World Congress on Nature and Biologically Inspired Computing, NaBIC, Mexico City, Mexico, 2012, pp. 66–71.
[54] O. Díaz-Parra, J.A. Ruiz-Vanoye, J.C. Zavala-Díaz, School bus routing problem library-SBRPLIB, International Journal of Combinatorial Optimization Problems and Informatics 2 (2011) 23–26.
[55] G. Dikas, I. Minis, Solving the bus evacuation problem and its variants, Computers & Operations Research 70 (2016) 75–86.
[56] K. Dorling, J. Heinrichs, G.G. Messier, S. Magierowski, Vehicle routing problems for drone delivery, IEEE Transactions on Systems, Man, and Cybernetics: Systems 47 (2017) 70–85.
[57] M. Ehrgott, Multicriteria Optimization, 2 ed., Springer-Verlag, Berlin Heidelberg, Germany, 2005.
[58] M. Ehrgott, X. Gandibleux, A survey and annotated bibliography of multiobjective combinatorial optimization, OR Spektrum 22 (2000) 425–460.
[59] K.A. Eldrandaly, A.F. Abdallah, A novel GIS-based decision-making framework for the school bus routing problem, Geo-spatial Information Science 15 (2012) 51–59.
[60] W.A. Ellegood, J.F. Campbell, J. North, Continuous approximation models for mixed load school bus routing, Transportation Research Part B: Methodological 77 (2015) 182–198.
[61] J. Euchi, R. Mraihi, The urban bus routing problem in the Tunisian case by the hybrid artificial ant colony algorithm, Swarm and Evolutionary Computation 2 (2012) 15–24.
[62] J. Euchi, A. Yassine, H. Chabchoub, The dynamic vehicle routing problem: solution with hybrid metaheuristic approach, Swarm and Evolutionary Computation 21 (2015) 41–53.
[63] M.F. Faraj, J.F.M. Sarubbi, C.M. Silva, M.F. Porto, N.T.R. Nunes, A real geographical application for the school bus routing problem, in: 17th International IEEE Conference on Intelligent Transportation Systems, ITSC, Qingdao, China, 2014, pp. 2762–2767.
[64] A. Froger, J.E. Mendoza, O. Jabali, G. Laporte, A Matheuristic for the Electric Vehicle Routing Problem With Capacitated Charging Stations, Research Report, Centre interuniversitaire de recherche sur les reseaux d'entreprise, la logistique et le transport (CIRRELT), 2017.

[65] A. Froger, J.E. Mendoza, G. Laporte, O. Jabali, New Formulations for the Electric Vehicle Routing Problem With Nonlinear Charging Functions, Research Report, Centre interuniversitaire de recherche sur les reseaux d'entreprise, la logistique et le transport (CIRRELT), 2017.
[66] S. Fulin, L. Yueguang, An improved quantum-behaved particle swarm algorithm and its application in school bus problem, in: 2012 Third International Conference on Digital Manufacturing Automation, ICDMA 2012, Guilin, China, 2012, pp. 198–201.
[67] M.R. Garey, D.S. Johnson, Computers and Intractability: A Guide to the Theory of NP-Completeness, W. H. Freeman & Co., New York, NY, USA, 1990.
[68] M. Gendreau, G. Ghiani, E. Guerriero, Time-dependent routing problems: a review, Computers & Operations Research 64 (2015) 189–197.
[69] V. Ghezavati, A. Sahihi, A. Barzegar, Using an the intelligent self-modifier of probability of section approach to study the revenue influence of the pricing scheme of recyclable items in a green vehicle routing problem, Simulation 94 (2018) 359–372.
[70] M. Goerigk, K. Deghdak, V. T'Kindt, A two-stage robustness approach to evacuation planning with buses, Transportation Research Part B: Methodological 78 (2015) 66–82.
[71] M.T. Goodrich, P. Pszona, Two-phase bicriterion search for finding fast and efficient electric vehicle routes, in: The 22nd ACM SIGSPATIAL International Conference on Advances in Geographic Information Systems, SIGSPATIAL'14, Dallas, Texas, USA, 2014, pp. 193–202.
[72] L. Grandinetti, F. Guerriero, F. Pezzella, O. Pisacane, A pick-up and delivery problem with time windows by electric vehicles, International Journal of Productivity and Quality Management 18 (2016) 403–423.
[73] F. Guerriero, R. Surace, V. Loscrí, E. Natalizio, A multi-objective approach for unmanned aerial vehicle routing problem with soft time windows constraints, Applied Mathematical Modelling 38 (2014) 839–852.
[74] A. Gupta, S. Saini, On solutions to vehicle routing problems using swarm optimization techniques: a review, in: Advances in Computer and Computational Sciences, Singapore, 2017, pp. 345–354.
[75] H.W. Hamacher, S. Ruzika, S.A. Tjandra, Algorithms for time-dependent bicriteria shortest path problems, Discrete Optimization 3 (2006) 238–254.
[76] K. Hamdi, N. Labadie, A. Yalaoui, Vehicle routing problem for hazardous materials transportation: an overview, in: 2014 IEEE International Conference on Industrial Engineering and Engineering Management, Malaysia, 2014, pp. 632–636.
[77] D.H. Han, Y.D. Kim, J.Y. Lee, Multiple-criterion shortest path algorithms for global path planning of unmanned combat vehicles, Computers & Industrial Engineering 71 (2014) 57–69.
[78] P. Hansen, Bicriterion path problems, in: G. Fandel, T. Gal (Eds.), Multiple Criteria Decision Making: Theory and Application, Springer-Verlag, Berlin, Germany, 1980, pp. 109–127.
[79] E.K. Hashi, M.R. Hasan, M.S.U. Zaman, GIS based heuristic solution of the vehicle routing problem to optimize the school bus routing and scheduling, in: 19th International Conference on Computer and Information Technology, ICCIT, Dhaka, Bangladesh, 2016, pp. 56–60.
[80] F. Hernandez, D. Feillet, R. Giroudeau, O. Naud, Branch-and-price algorithms for the solution of the multi-trip vehicle routing problem with time windows, European Journal of Operational Research 249 (2016) 551–559.
[81] G. Hiermann, J. Puchinger, S. Ropke, R.F. Hartl, The electric fleet size and mix vehicle routing problem with time windows and recharging stations, European Journal of Operational Research 252 (2016) 995–1018.
[82] T. Hintsch, S. Irnich, Large multiple neighborhood search for the clustered vehicle-routing problem, European Journal of Operational Research 270 (2018) 118–131.
[83] J. Hof, M. Schneider, D. Goeke, Solving the battery swap station location-routing problem with capacitated electric vehicles using an AVNS algorithm for vehicle-routing problems with intermediate stops, Transportation Research Part B: Methodological 97 (2017) 102–112.
[84] S.M. Hosseini-Motlagh, S. Majidi, S. Yaghoubi, A. Jokar, Fuzzy green vehicle routing problem with simultaneous pickup-delivery and time windows, RAIRO - Operations Research 51 (2017) 1151–1176.

[85] C. Hu, J. Lu, X. Liu, G. Zhang, Robust vehicle routing problem with hard time windows under demand and travel time uncertainty, Computers & Operations Research 94 (2018) 139–153.
[86] Y. Huang, L. Zhao, T.V. Woensel, Jean-PhilippeGross, Time-dependent vehicle routing problem with path flexibility, Transportation Research Part B: Methodological 95 (2017) 169–195.
[87] L. Huo, G. Yan, B. Fan, H. Wang, W. Gao, School bus routing problem based on ant colony optimization algorithm, in: IEEE Conference and Expo Transportation Electrification Asia-Pacific, ITEC Asia-Pacific, Beijing, China, 2014, pp. 1–5.
[88] T. Iswari, A.M.S. Asih, Comparing genetic algorithm and particle swarm optimization for solving capacitated vehicle routing problem, IOP Conference Series: Materials Science and Engineering 337 (2018).
[89] E. Jabir, V. Panicker, R. Sridharan, Design and development of a hybrid ant colony-variable neighbourhood search algorithm for a multi-depot green vehicle routing problem, Transportation Research Part D: Transport and Environment 57 (2017) 422–457.
[90] E. Jabir, V.V. panicker, R. Sridharan, Modelling and analysis of a green vehicle routing problem, in: The 12th AIMS International Conference on Management, Kozhikode, India, 2015, pp. 1310–1318.
[91] E. Jafari, S.D. Boyles, Multicriteria stochastic shortest path problem for electric vehicles, Networks and Spatial Economics 17 (2017) 1043–1070.
[92] A.S. Jaradat, M.Q. Shatnawi, Solving school bus routing problem by intelligent water drops algorithm, Journal of Computer Sciences (2017), https://doi.org/10.3844/jcssp.2017.
[93] W. Jie, J. Yang, C. Yang, Branch-and-price algorithm for heterogeneous electric vehicle routing problem, Xitong Gongcheng Lilun yu Shijian/System Engineering Theory and Practice 36 (2016) 1795–1805.
[94] W. Jing, Y. Yan, I. Kim, M. Sarvi, Electric vehicles: a review of network modelling and future research needs, Advances in Mechanical Engineering 8 (2016) 1687814015627981.
[95] D. Jungnickel, Graphs, Networks and Algorithms, 4 ed., Springer-Verlag, Berlin Heidelberg, Germany, 2013.
[96] B. Kamali, S. Mason, E.A. Pohl, An analysis of special needs student busing, Journal of Public Transportation 16 (2013) 21–45.
[97] S. Kancharla, G. Ramadurai, Incorporating driving cycle based fuel consumption estimation in green vehicle routing problems, Sustainable Cities and Society 40 (2018) 214–221.
[98] M. Kang, S.K. Kim, J.T. Felan, H.R. Choi, M. Cho, Development of a genetic algorithm for the school bus routing problem, International Journal of Software Engineering and Its Applications 9 (2015) 107–126.
[99] L. Ke, A brain storm optimization approach for the cumulative capacitated vehicle routing problem, Memetic Computing (2018) 1–11.
[100] M. Keskin, B. Çatay, Partial recharge strategies for the electric vehicle routing problem with time windows, Transportation Research Part C: Emerging Technologies 65 (2016) 111–127.
[101] B.I. Kim, S. Kim, J. Park, A school bus scheduling problem, European Journal of Operational Research 218 (2012) 577–585.
[102] J.H. Kim, S. Soh, Designing hub-and-spoke school bus transportation network: a case study of Wonkwang University, Promet-Traffic & Transportation 24 (2012) 389–394.
[103] T. Kim, B.J. Park, Model and algorithm for solving school bus problem, Journal of Emerging Trends in Computing and Information Sciences 4 (2013) 596–600.
[104] J. Kinable, F. Spieksma, G. Vanden Berghe, School bus routing – a column generation approach, International Transactions in Operational Research 21 (2014) 453–478.
[105] Ç. Koç, T. Bektaş, O. Jabali, G. Laporte, Thirty years of heterogeneous vehicle routing, European Journal of Operational Research 249 (2016) 1–21.
[106] Ç. Koç, G. Laporte, Vehicle routing with backhauls: review and research perspectives, Computers & Operations Research 91 (2018) 79–91.
[107] I. Küçükoglu, D. Cattrysse, A branch-and-bound integrated simulated annealing algorithm for the electric vehicle routing problem with time windows, in: International Conference on Computers and Industrial Engineering, CIE 47, Lisbon, Portugal, 2017.

[108] Y. Kumar, S. Jain, School bus routing based on branch and bound approach, in: 2015 International Conference on Computer, Communication and Control, IC4, Indore, India, 2015, pp. 1–4.

[109] R. Kuo, B. Wibowo, F. Zulvia, Application of a fuzzy ant colony system to solve the dynamic vehicle routing problem with uncertain service time, Applied Mathematical Modelling 40 (2016) 9990–10001.

[110] R. Lahyani, M. Khemakhem, F. Semet, Rich vehicle routing problems: from a taxonomy to a definition, European Journal of Operational Research 241 (2015) 1–14.

[111] V. Leggieri, M. Haouari, A practical solution approach for the green vehicle routing problem, Transportation Research Part E: Logistics and Transportation Review 104 (2017) 97–112.

[112] K. Leksakul, U. Smutkupt, R. Jintawiwat, S. Phongmoo, Heuristic approach for solving employee bus routes in a large-scale industrial factory, Advanced Engineering Informatics 32 (2017) 176–187.

[113] J.K. Lenstra, A.H.G. Rinnooy Kan, Complexity of vehicle and scheduling problems, Networks 11 (1981) 221–227.

[114] R. Lewis, K. Smith-Miles, K. Phillips, The school bus routing problem: an analysis and algorithm, in: Combinatorial Algorithms, in: Lecture Notes in Computer Science (including subseries Lecture Notes in Artificial Intelligence and Lecture Notes in Bioinformatics) LNCS, vol. 10765, 2018, pp. 287–298.

[115] C. Lin, K.L. Choy, G.T.S. Ho, S.H. Chung, H.Y. Lam, Survey of green vehicle routing problem: past and future trends, Expert Systems with Applications 41 (2014) 1118–1138.

[116] J. Lin, W. Zhou, O. Wolfson, Electric vehicle routing problem, Transportation Research Procedia 12 (2016) 508–521.

[117] Y. Ling, J. Zong-Fu, J. Shou-Xu, R. Xiang-Min, Z. Fu-Sheng, Urban night bus routes planning with taxi traces, in: ICCSE 2017 - 12th International Conference on Computer Science and Education, Houston, TX, USA, 2017, pp. 375–379.

[118] Y. Liu, I.H. Khalifa, A.E. Kamel, The multi-period and multi-depot dynamic vehicle routing problem with time windows, in: 3rd International Conference on Logistics Operations Management, GOL'16, Fez, Morocco, 2016.

[119] E.R. Lopéz, J. Romero, A hybrid column generation and clustering approach to the school bus routing problem with time windows, Ingeniería 20 (2015) 111–127.

[120] Z. Luo, H. Qin, D. Zhang, A. Lim, Adaptive large neighborhood search heuristics for the vehicle routing problem with stochastic demands and weight-related cost, Transportation Research Part E: Logistics and Transportation Review 85 (2016) 69–89.

[121] S. Madankumar, C. Rajendran, Mathematical models for green vehicle routing problems with pickup and delivery: a case of semiconductor supply chain, Computers & Operations Research 89 (2018) 183–192.

[122] I. Mahdavi, N. Mahdavi-Amiri, S. Nejati, Algorithms for biobjective shortest path problems in fuzzy networks, Iranian Journal of Fuzzy Systems 8 (2011) 9–37.

[123] M. Mahmoudi, X. Zhou, Finding optimal solutions for vehicle routing problem with pickup and delivery services with time windows: a dynamic programming approach based on state-space-time network representations, Transportation Research Part B: Methodological 89 (2016) 19–42.

[124] S. Mancini, A real-life multi depot multi period vehicle routing problem with a heterogeneous fleet: formulation and adaptive large neighborhood search based matheuristic, Transportation Research Part C: Emerging Technologies 70 (2016) 100–112.

[125] D.M. Manumbu, E. Mujuni, D. Kuznetsov, Mathematical formulation model for a school bus routing problem with small instance data, Mathematical Theory and Modeling 4 (2014) 121–132.

[126] D.M. Manumbu, E. Mujuni, D. Kuznetsov, A simulated annealing algorithm for solving the school bus routing problem: a case study of Dar es Salaam, Computer Engineering and Intelligent Systems 5 (2014) 44–58.

[127] M. Marinaki, Y. Marinakis, A glowworm swarm optimization algorithm for the vehicle routing problem with stochastic demands, Expert Systems with Applications 46 (2016) 145–163.
[128] L.M. Martínez, J.M. Viegas, Design and deployment of an innovative school bus service in Lisbon, Procedia - Social and Behavioral Sciences 20 (2011) 120–130.
[129] M.E. McNabb, J.D. Weir, R.R. Hill, S.N. Hall, Testing local search move operators on the vehicle routing problem with split deliveries and time windows, Computers & Operations Research 56 (2015) 93–109.
[130] B. Minocha, S. Tripathi, Solving school bus routing problem using hybrid genetic algorithm: a case study, Advances in Intelligent Systems and Computing 236 (2014) 93–103.
[131] D.M. Miranda, S.V. Conceição, The vehicle routing problem with hard time windows and stochastic travel and service time, Expert Systems with Applications 64 (2016) 104–116.
[132] D.M. Miranda, R.S. de Camargo, S.V. Conceição, M.F. Porto, N.T. Nunes, A multi-loading school bus routing problem, Expert Systems with Applications 101 (2018) 228–242.
[133] M.A. Mohammed, M.K.A. Ghani, R.I. Hamed, S.A. Mostafa, M.S. Ahmad, D.A. Ibrahim, Solving vehicle routing problem by using improved genetic algorithm for optimal solution, Journal of Computational Science 21 (2017) 255–262.
[134] A. Montoya, C. Guéret, J.E. Mendoza, J.G. Villegas, A multi-space sampling heuristic for the green vehicle routing problem, Transportation Research Part C: Emerging Technologies 70 (2016) 113–128.
[135] A. Montoya, C. Guéret, J.E. Mendoza, J.G. Villegas, The electric vehicle routing problem with nonlinear charging function, Transportation Research Part B: Methodological 103 (2017) 87–110.
[136] J.R. Montoya-Torres, J.L. Franco, S.N. Isaza, H.F. Jiménez, N. Herazo-Padilla, A literature review on the vehicle routing problem with multiple depots, Computers & Industrial Engineering 79 (2015) 115–129.
[137] A. Mushi, B. Ngonyani, E. Mujuni, Optimizing schedules for school bus routing problem: the case of Dar es Salaam schools, International Journal of Advanced Research in Computer Science 6 (2015) 132–136.
[138] J. Nalepa, M. Blocho, Co-operation in the parallel memetic algorithm, International Journal of Parallel Programming 43 (2015) 812–839.
[139] J. Nalepa, M. Blocho, Adaptive memetic algorithm for minimizing distance in the vehicle routing problem with time windows, Soft Computing 20 (2016) 2309–2327.
[140] J. Nalepa, M. Blocho, Enhanced guided ejection search for the pickup and delivery problem with time windows, in: Intelligent Information and Database Systems: 8th Asian Conference, ACIIDS 2016, Da Nang, Vietnam, 2016, pp. 388–398.
[141] R.M. Newton, W.H. Thomas, Design of school bus routes by computer, Socio-Economic Planning Sciences 3 (1969) 75–85.
[142] Y. Niu, Z. Yang, P. Chen, J. Xiao, Optimizing the green open vehicle routing problem with time windows by minimizing comprehensive routing cost, Journal of Cleaner Production 171 (2018) 962–971.
[143] S.A. Oluwadare, I.P. Oguntuyi, J.C. Nwaiwu, Solving school bus routing problem using genetic algorithm-based model, International Journal of Intelligent Systems and Applications 10 (2018) 50–58.
[144] A. Omidvar, E. Erman Ozguven, O. Arda Vanli, R. Tavakkoli-Moghaddam, A two-phase safe vehicle routing and scheduling problem: formulations and solution algorithms, 2017, ArXiv e-prints.
[145] K. Ouaddi, Y. Benadada, F.Z. Mhada, Multi period dynamic vehicles routing problem: literature review, modelization and resolution, in: 3rd International Conference on Logistics Operations Management, GOL'16, Fez, Morocco, 2016.
[146] J. Pacheco, R. Caballero, M. Laguna, J. Molina, Bi-objective bus routing: an application to school buses in rural areas, Transportation Science 47 (2013) 397–411.
[147] L. de Pádua Agripa Sales, C.S. Melo, T. de Oliveira e Bonates, B. de Athayde Prata, Memetic algorithm for the heterogeneous fleet school bus routing problem, Journal of Urban Planning and Development 144 (2018).

[148] V. Pareto, Manual of Political Economy (Manuale d'economia politica), Kelley, New York City, NY, USA, 1906, translated into English by A.S. Schwier and A.N. Page, 1971.

[149] C.H. Park, Y.D. Kim, B. Jeong, Heuristics for determining a patrol path of an unmanned combat vehicle, Computers & Industrial Engineering 63 (2012) 150–160.

[150] J. Park, B.I. Kim, The school bus routing problem: a review, European Journal of Operational Research 202 (2010) 311–319.

[151] J. Park, H. Tae, B.I. Kim, A post-improvement procedure for the mixed load school bus routing problem, European Journal of Operational Research 217 (2012) 204–213.

[152] S.P. Parvasi, M. Mahmoodjanloo, M. Setak, A bi-level school bus routing problem with bus stops selection and possibility of demand outsourcing, Applied Soft Computing 61 (2017) 222–238.

[153] T. Paul, H. Yamada, Operation and charging scheduling of electric buses in a city bus route network, in: 17th International IEEE Conference on Intelligent Transportation Systems, ITSC, Qingdao, China, 2014, pp. 2780–2786.

[154] J. Paz, M. Granada-Echeverri, J. Escobar, The multi-depot electric vehicle location routing problem with time windows, International Journal of Industrial Engineering Computations 9 (2018) 123–136.

[155] R. Pérez-Rodríguez, A. Hernandez-Aguirre, S. Jöns, I. Cruz, J. Velasco, J. Perez-Gallardo, A probability model for the school bus routing problem with bus stop selection, Dyna (Spain) 92 (2017) 138.

[156] A. Pessoa, R. Sadykov, E. Uchoa, Enhanced branch-cut-and-price algorithm for heterogeneous fleet vehicle routing problems, European Journal of Operational Research 270 (2018) 530–543.

[157] V. Pillac, M. Gendreau, C. Guéret, A.L. Medaglia, A review of dynamic vehicle routing problems, European Journal of Operational Research 225 (2013) 1–11.

[158] C. Ping, G. Bruce, W. Xingyin, W. Edward, A novel approach to solve the split delivery vehicle routing problem, International Transactions in Operational Research 24 (2017) 27–41.

[159] G. Poonthalir, R. Nadarajan, A fuel efficient green vehicle routing problem with varying speed constraint (F-GVRP), Expert Systems with Applications 100 (2018) 131–144.

[160] M.F. Porto, J.F.M. Sarubbi, S. Thiéry, N.T.R. Nunes, I.R. Vianna, Developing a GIS for rural school transportation in Minas Gerais, Brazil, Systemics, Cybernetics and Informatics 13 (2015) 89–94.

[161] H.N. Psaraftis, M. Wen, C.A. Kontovas, Dynamic vehicle routing problems: three decades and counting, Networks 67 (2016) 3–31.

[162] M. Rabbani, S. Bosjin, R. Yazdanparast, N. Saravi, A stochastic time-dependent green capacitated vehicle routing and scheduling problem with time window, resiliency and reliability: a case study, Decision Science Letters 7 (2018) 381–394.

[163] B. Rabta, C. Wankmüller, G. Reiner, A drone fleet model for last-mile distribution in disaster relief operations, International Journal of Disaster Risk Reduction 28 (2018) 107–112.

[164] A. Raith, M. Schmidt, A. Schöbel, L. Thom, Extensions of labeling algorithms for multi-objective uncertain shortest path problems, Networks 36 (2018) 1–43.

[165] S. Razzaq, C. Xydeas, M. Everett, A. Mahmood, T. Alquthami, Three-dimensional UAV routing with deconfliction, IEEE Access 6 (2018) 21536–21551.

[166] H. Repolho, J. Marchesi, O.S.S. Júnior, R.R.R. Bezerra, Cargo theft weighted vehicle routing problem: modeling and application to the pharmaceutical distribution sector, Soft Computing (2018) 1–18.

[167] D. Rezgui, J. Chaouachi-Siala, W. Aggoune-Mtalaa, H. Bouziri, Application of a memetic algorithm to the fleet size and mix vehicle routing problem with electric modular vehicles, in: The Genetic and Evolutionary Computation Conference Companion, GECCO 2017, Berlin, Germany, 2017.

[168] J. Riera-Ledesma, J.J. Salazar-González, Solving school bus routing using the multiple vehicle traveling purchaser problem: a branch-and-cut approach, Computers & Operations Research 39 (2012) 391–404.

[169] J. Riera-Ledesma, J.J. Salazar-González, A column generation approach for a school bus routing problem with resource constraints, Computers & Operations Research 40 (2013) 566–583.
[170] J. Rivera, H. Afsar, C. Prins, A multistart iterated local search for the multitrip cumulative capacitated vehicle routing problem, Computational Optimization and Applications 61 (2015) 159–187.
[171] G.R. Rodríguez-Parra, W.J. Guerrero, A. Sarmiento-Lepesqueur, Cooperation strategies featuring optimization in the school transportation system in Bogota, DYNA (Colombia) 84 (2017) 164–174.
[172] G.K.D. Saharidis, Environmental externalities score: a new emission factor to model green vehicle routing problem, Energy Systems 8 (2017) 673–691.
[173] S. Samadi-Dana, M.M. Paydar, J. Jouzdani, A simulated annealing solution method for robust school bus routing, International Journal of Operational Research 28 (2017) 307–326.
[174] J.F.M. Sarubbi, C.M.R. Mesquita, E.F. Wanner, V.F. Santos, C.M. Silva, A strategy for clustering students minimizing the number of bus stops for solving the school bus routing problem, in: NOMS 2016 - 2016 IEEE/IFIP Network Operations and Management Symposium, Istanbul, Turkey, 2016, pp. 1175–1180.
[175] B. Sawik, J. Faulin, E. Pérez-Bernabeu, A multicriteria analysis for the green VRP: a case discussion for the distribution problem of a Spanish retailer, Transportation Research Procedia 22 (2017) 305–313, 19th EURO Working Group on Transportation Meeting, EWGT2016, 5–7 September 2016, Istanbul, Turkey.
[176] P. Schittekat, J. Kinable, K. Sörense, M. Sevaux, F. Spieksma, J. Springael, A metaheuristic for the school bus routing problem with bus stop selection, European Journal of Operational Research 229 (2013) 518–528.
[177] P. Schittekat, J. Kinable, K. Sörensen, M. Sevaux, F. Spieksma, J. Springael, An efficient metaheuristic for the school bus routing problem, in: 13th Congress of the French National Society of Operations Research and Decision Science, ROADEF 2012, Angers, France, 2012, pp. 508–509.
[178] V. Schmid, Hybrid large neighborhood search for the bus rapid transit route design problem, European Journal of Operational Research 238 (2014) 427–437.
[179] M. Schneider, The vehicle-routing problem with time windows and driver-specific times, European Journal of Operational Research 250 (2016) 101–119.
[180] A.Z. Şevkli, B. Güler, A multi-phase oscillated variable neighbourhood search algorithm for a real-world open vehicle routing problem, Applied Soft Computing 58 (2017) 128–144.
[181] S.B. Sghaier, N.B. Guedria, R. Mraihi, Solving school bus routing problem with genetic algorithm, in: International Conference on Advanced Logistics and Transport, Sousse, Tunisia, 2013, pp. 7–12.
[182] S. Shao, J. Bi, Electric vehicle routing problem with simultaneously delivery and pick-up, Beijing Jiaotong Daxue Xuebao/Journal of Beijing Jiaotong University 41 (2017) 15–21.
[183] S. Shao, W. Guan, J. Bi, Electric vehicle-routing problem with charging demands and energy consumption, IET Intelligent Transport Systems 12 (2018) 202–212.
[184] S. Shao, W. Guan, B. Ran, Z. He, J. Bi, Electric vehicle routing problem with charging time and variable travel time, Mathematical Problems in Engineering 2017 (2017).
[185] J. Shi, J. Zhang, K. Wang, X. Fang, Particle swarm optimization for split delivery vehicle routing problem, Asia-Pacific Journal of Operational Research 35 (2018).
[186] C.M. Silva, J.F.M. Sarubbi, D.F. Silva, M.F. Porto, N.T.R. Nunes, A mixed load solution for the rural school bus routing problem, in: 2015 IEEE 18th International Conference on Intelligent Transportation Systems, Canary Islands, Spain, 2015, pp. 1940–1945.
[187] V.S. de Siqueira, F.J.E.L. e Silva, E.N. da Silva, R.V.S. da Silva, M.L. Rocha, Implementation of the metaheuristic grasp applied to the school bus routing problem, International Journal of e-Education, e-Business, e-Management and e-Learning 6 (2016) 137–145.
[188] A.J.V. Skriver, A classification of bicriteria shortest path (BSP) algorithms, Asia-Pacific Journal of Operational Research 17 (2000) 199–212.

[189] N. Smiti, M. Dhiaf, B. Jarboui, S. Hanafi, Skewed general variable neighborhood search for the cumulative capacitated vehicle routing problem, International Transactions in Operational Research (2018).

[190] H. Soleimani, Y. Chaharlang, H. Ghaderi, Collection and distribution of returned-remanufactured products in a vehicle routing problem with pickup and delivery considering sustainable and green criteria, Journal of Cleaner Production 172 (2018) 960–970.

[191] F.M. De Souza Lima, D.S.D. Pereira, S.V. da Conceição, R.S. de Camargo, A multi-objective capacitated rural school bus routing problem with heterogeneous fleet and mixed loads, 4OR 15 (2017) 359–386.

[192] F.M. Souza Lima, D.S. Pereira, S.V. Conceição, N.T. Ramos Nunes, A mixed load capacitated rural school bus routing problem with heterogeneous fleet: algorithms for the Brazilian context, Expert Systems with Applications 56 (2016) 320–334.

[193] K. Sundar, S. Rathinam, Algorithms for routing an unmanned aerial vehicle in the presence of refueling depots, IEEE Transactions on Automation Science and Engineering 11 (2014) 287–294.

[194] J. Sze, S. Salhi, N. Wassan, The cumulative capacitated vehicle routing problem with min-sum and min-max objectives: an effective hybridisation of adaptive variable neighbourhood search and large neighbourhood search, Transportation Research Part B: Methodological 101 (2017) 162–184.

[195] L. Talarico, K. Sörensen, J. Springael, The k-dissimilar vehicle routing problem, European Journal of Operational Research 244 (2015) 129–140.

[196] L. Tan, F. Lin, H. Wang, Adaptive comprehensive learning bacterial foraging optimization and its application on vehicle routing problem with time windows, Neurocomputing 151 (2015) 1208–1215.

[197] H.B. Ticha, N. Absi, D. Feillet, A. Quilliot, Empirical analysis for the VRPTW with a multigraph representation for the road network, Computers & Operations Research 88 (2017) 103–116.

[198] E.M. Toro O., A.H. Escobar Z., M. Granada E., Literature review on the vehicle routing problem in the green transportation context, Luna Azul (2016) 362–387.

[199] K. Udin, R. Gui, A. Rahmawan, Green vehicle routing problem with heterogeneous fleet and time windows, in: 6th International Conference on Software and Computer Applications, ICSCA'17, Bangkok, Thailand, 2017, pp. 223–227.

[200] M. Ulmer, N. Soeffker, D. Mattfeld, Value function approximation for dynamic multi-period vehicle routing, European Journal of Operational Research 269 (2018) 883–899.

[201] Ö. Ünsal, T. Yigit, Using the genetic algorithm for the optimization of dynamic school bus routing problem, BRAIN. Broad Research in Artificial Intelligence and Neuroscience 9 (2018) 6–21.

[202] Ö. Ünsal, T. Yiğit, C. Altıntaş, Optimization of dynamic school bus routing problem by using metaheuristic and clustering methods, in: 11th International Conference on Practice and Theory of Automated Timetabling, PATAT-2016, Udine, Italy, 2016, pp. 555–559.

[203] K. Van Moffaert, M. Mihaylov, B. Van Vreckem, A. Nowe, A learning approach to the school bus routing problem, in: BNAIC 2011 - 23rd Benelux Conference on Artificial Intelligence, Ghent, Belgium, 2011, pp. 280–288.

[204] P. Vaz Penna, H. Afsar, C. Prins, C. Prodhon, A hybrid iterative local search algorithm for the electric fleet size and mix vehicle routing problem with time windows and recharging stations, IFAC-PapersOnLine 49 (2016) 955–960.

[205] M. Veenstra, M. Cherkesly, G. Desaulniers, G. Laporte, The pickup and delivery problem with time windows and handling operations, Computers & Operations Research 77 (2017) 127–140.

[206] A. Veneti, C. Konstantopoulos, G. Pantziou, Continuous and discrete time label setting algorithms for the time dependent bi-criteria shortest path problem, in: 14th INFORMS Computing Society Conference, Richmond, Virginia, USA, 2015, pp. 62–73.

[207] M. Voorneveld, Characterization of Pareto dominance, Operations Research Letters 31 (2003) 7–11.
[208] J. Wang, X. Huang, Routing school bus for better student learning, in: 25th International Conference on Geoinformatics, Buffalo, NY, USA, 2017, pp. 1–7.
[209] X. Wang, C. Yuen, N.U. Hassan, N. An, W. Wu, Electric vehicle charging station placement for urban public bus systems, IEEE Transactions on Intelligent Transportation Systems 18 (2017) 128–139.
[210] Z. Wang, A. Shafahi, A. Haghani, Scda: School compatibility decomposition algorithm for solving the multi-school bus routing and scheduling problem, 2017, ArXiv e-prints.
[211] J. Widuch, A label correcting algorithm for the bus routing problem, Fundamenta Informaticae 118 (2012) 305–326.
[212] J. Widuch, A label correcting algorithm with storing partial solutions to solving the bus routing problem, Informatica 24 (2013) 461–484.
[213] J. Widuch, A relation of dominance for the bicriterion bus routing problem, International Journal of Applied Mathematics and Computer Science 27 (2017) 133–155.
[214] K. Worwa, A case study in school transportation logistics, Research in Logistics & Production 4 (2014) 45–54.
[215] W. Wu, R. Liu, W. Jin, Modelling bus bunching and holding control with vehicle overtaking and distributed passenger boarding behaviour, Transportation Research Part B: Methodological 104 (2017) 175–197.
[216] Y. Wu, Y. Wang, G. He, S. Zhao, An improved adaptive large neighborhood search algorithm for the heterogeneous fixed fleet vehicle routing problem, in: 8th IEEE International Conference on Software Engineering and Service Sciences, ICSESS 2017, Beijing, China, 2017, pp. 657–663.
[217] H. Yang, S. Yang, Y. Xu, E. Cao, M. Lai, Z. Dong, Electric vehicle route optimization considering time-of-use electricity price by learnable partheno-genetic algorithm, IEEE Transactions on Smart Grid 6 (2015) 657–666.
[218] J. Yang, H. Sun, Battery swap station location-routing problem with capacitated electric vehicles, Computers & Operations Research 55 (2015) 217–232.
[219] M. Yavuz, An iterated beam search algorithm for the green vehicle routing problem, Networks 69 (2017).
[220] T. Yiğit, Ö. Ünsal, Using the ant colony algorithm for real-time automatic route of school buses, The International Arab Journal of Information Technology 13 (2016) 559–565.
[221] V. Yu, P. Redi, Y. Agustina, O. Jimat Wibowo, A simulated annealing heuristic for the hybrid vehicle routing problem, Applied Soft Computing 53 (2017) 119–132.
[222] D. Zhang, S. Cai, F. Ye, Y.W. Si, T.T. Nguyen, A hybrid algorithm for a vehicle routing problem with realistic constraints, Information Sciences 394–395 (2017) 167–182.
[223] J. Zhang, L. Jia, S. Niu, F. Zhang, L. Tong, X. Zhou, A space-time network-based modeling framework for dynamic unmanned aerial vehicle routing in traffic incident monitoring applications, Sensors 15 (2015) 13874–13898.
[224] J. Zhang, Y. Li, School bus problem and its algorithm, IERI Procedia 2 (2012) 8–11.
[225] S. Zhang, Y. Gajpal, S. Appadoo, A meta-heuristic for capacitated green vehicle routing problem, Annals of Operations Research (2017) 1–19.
[226] G. Zhenfeng, L. Yang, J. Xiaodan, G. Sheng, The electric vehicle routing problem with time windows using genetic algorithm, in: IEEE 2nd Advanced Information Technology, Electronic and Automation Control Conference, IAEAC 2017, Chongqing, China, 2017, pp. 635–639.
[227] M. Ziebuhr, T. Buer, H. Kopfer, A column generation-based heuristic for a green vehicle routing problem with an unlimited heterogeneous fleet, in: 2017 IEEE Symposium Series on Computational Intelligence, SSCI, Honolulu, Hawaii, USA, 2017, pp. 1–8.
[228] X. Zuo, C. Zhu, C. Huang, Y. Xiao, Using AMPL/CPLEX to model and solve the electric vehicle routing problem (EVRP) with heterogeneous mixed fleet, in: The 29th Chinese Control and Decision Conference, CCDC 2017, Chongqing, China, 2017, pp. 4666–4670.

Chapter 2

On a road to optimal fleet routing algorithms: a gentle introduction to the state-of-the-art

Paweł Gora, Dominika Bankiewicz, Katarzyna Karnas, Wojciech Kaźmierczak, Michał Kutwin, Przemysław Perkowski, Szymon Płotka, Anna Szczurek, and Damian Zięba
Faculty of Mathematics, Informatics and Mechanics, University of Warsaw, Warsaw, Poland

2.1 Introduction

Finding optimal routes in a graph is an important computational problem, both from the theoretical perspective and real-world applications. Over the years, many variants of routing problems have emerged, and many algorithms solving such problems have been developed. Due to abundant and still growing literature on routing problems, it is not easy to stay up to date with the newest and most efficient methods. Therefore, there is a need for metaresearch works which summarize and compare existing topics and solutions, and many such works have been already published, but in most cases the scope of such review works is limited to particular types of problems, for example, Vehicle Routing Problem or Traveling Salesman Problem.

In this work, our goal is to review the most important routing problems and to give a concise and self-contained introduction to these research topics. We start from the Optimal Route Choice problem (Section 2.2), in which the goal is to find the optimal route (and sometimes also times of departure) for a given vehicle (or set of vehicles) traveling between a given origin–destination pair. Then we focus on its extension, the Traveling Salesman Problem (Section 2.3), in which the goal for a vehicle is to visit many destinations at minimal cost. Finally, we describe the Vehicle Routing Problem (Section 2.4), which can be seen as an extension of the Traveling Salesman Problem to the case of a fleet of vehicles. Thanks to such a structure, we can observe and understand how different problems and algorithms are linked. We hope that it may help researchers and experts working in the industry understand basic concepts of the fleet routing

Smart Delivery Systems. https://doi.org/10.1016/B978-0-12-815715-2.00014-2

problems, apply best algorithms in practice, and possibly even develop better algorithms.

2.2 Optimal Route Choice problem

2.2.1 Introduction

In this section, we present an overview of the problem of choosing an optimal route in the road network. We divide this problem into two main subproblems, Shortest Path problem (SP) and Traffic Assignment problem (TA), which is an extension of SP. SP aims at finding a path between two nodes in a graph such that the sum of weights of its constituent edges is minimized. TA aims at assigning routes to many cars in an optimal way, where optimality may be considered from a user perspective or traffic system perspective. In general, SP and TA assume the deterministic behavior of network participants. However, such an assumption is often far from reality. Therefore, we begin the chapter with the brief description of Discrete Choice Models which allow considering the stochastic components of the user's decision-making process.

2.2.2 Discrete choice models

Discrete choice models are a widely used tool for modeling short-term travel behavior. They are a complementary tool to the algorithms used for finding the shortest path on a particular graph. More specifically, in the context of ORC problem, Discrete Choice Models allow to model the behavior of network participants, taking into account the uncertainty related with user's choice of an optimal path. Therefore, these models are applied both in the case of SP and TA, particularly in stochastic variants, in order to analyze the demand on a particular road network [28].

Short-term travel decision concerns mainly the choice of the following factors: destination, travel mode, departure time, and route. These short-term decisions depend on the long-term travel decisions such as vehicle ownership and origin locations. Discrete choice methods are based mainly on the random utility theory. In such models the convention distinguishing between four types of assumptions concerning:

1. *Decision-maker: an individual*, that is, a person or a group of people (e.g., household, organization, company) and his/her characteristics.
2. *Alternatives: possible choices* available for the decision-maker; the set of considered alternatives is called the *choice set*.
3. *Attributes: a set of attributes* characterizing each alternative in the choice set; some attributes may be generic to all alternatives, but some may be specific to particular alternative.
4. *Decision rule: decision-making criteria* for the chosen alternative, such as the shortest path, fastest path, path with fewest stop signs/traffic lights, path with least congestion, path with no left turns, and so on.

The most popular models belong either to the *logit* family (e.g., Nested logit or Generalized Extreme Value model) or *probit* family (e.g., Multinomial probit), but there are also models attempting to bridge the gap between those two families (e.g., Generalized Factor Analytical Representation and Hybrid logit). The main difference between the logit and probit models is its capability of capturing all correlations among available alternatives, whereas an advantage of the logit model is its tractability [28].

In general, logit or probit regression models are applied when our goal is estimating the probability of an outcome as a function of independent variables. In the simplest (binary) case, such models can be formulated as

$$P(y=1|x) = F(\beta_0 + \beta_1 x_1 + \cdots + \beta_k x_k), \tag{2.1}$$

where F is a function which takes values strictly between zero and one: $0 < F(z) < 1$. In the logit model, F is the logistic function, that is, the cumulative distribution function for a standard logistic random variable:

$$F(z) = \frac{\exp(z)}{1+\exp(z)}, \tag{2.2}$$

whereas in the probit model, F is the is the standard normal cumulative distribution function:

$$F(z) = \int_{-\infty}^{z} \phi(v)dv, \tag{2.3}$$

where $\phi(z)$ is the standard normal density

$$\phi(z) = (2\pi)^{-1/2} \exp(\frac{-z^2}{2}). \tag{2.4}$$

In all cases, $F(z)$ is between zero and one for all real numbers z [249].

Most of discrete choice models are based on the random utility theory. Similarly to the general economic consumer theory, random utility theory assumes that the decision-maker has a perfect discriminatory power (capability of assigning different cost to each unit/segment of the road). The decision-maker is also assumed not to have complete information, so some portion of uncertainty must be also taken into account. Therefore, to reflect this uncertainty, the utility is modeled as a random variable, that is, the utility that the decision-maker n associates with the alternative i in the choice set C_n is given by

$$U_{in} = V_{in} + \varepsilon_{in}, \tag{2.5}$$

where V_{in} is the deterministic part of the utility U_{in}, and ε_{in} is a random term representing uncertainty. The decision-maker chooses the alternative with the highest utility. Therefore the probability that decision-maker n chooses the alternative i from the choice set C_n can be written as

$$P(i|C_n) = P[U_{in} \geq U_{jn} \ \forall j \in C_n] = P[U_{in} = \max_{j \in C_n} U_{jn}] \tag{2.6}$$

In other words, our choice set C_n is the set of routes (alternatives) to be chosen from by the road user. V_{in} is the function of length of a particular route which can be computed with the use of appropriate algorithms. In TA, we assume that we choose the shortest path (or to be more precise, the least costly path), taking into account the decisions of other road participants - the lower the path length the higher the value of V_{in}. However, we need to consider the stochastic nature of road participants' decision-making process because in reality it is never fully deterministic. Therefore, we include a stochastic component ε_{in} in the utility function, in order to take into account the uncertainty related to the travel decision.

2.2.3 Shortest Path problem

The SHORTEST PATH problem (SP) is the problem of finding a path between two nodes of the graph, where the sum of the weights of its constituent edges is minimized. SP has numerous successful applications in road networks, that is, finding the shortest route between particular origins and destinations, where the whole road network is represented as a graph.

One of the commonly known algorithms for finding the shortest path was developed by Dijkstra [64]. It is worth noting that this algorithm is very similar to Prim's algorithm [203] for finding the Minimum Spanning Tree. Therefore it is also referred to as the Prim–Dijkstra algorithm. Dijkstra's algorithm was originally developed to solve the problem referred to as the SINGLE SOURCE SHORTEST PATH problem (SSSP). This variation of SP focuses on finding the shortest path between the particular pair of origin and destination. Dijkstra's algorithm is a greedy algorithm that selects the closest vertex that has not yet been processed and performs this relaxation process on all of its outgoing edges. Moreover, it can only be applied to the graph with nonnegative edge weights [113]. Dijkstra's algorithm was further extended by other researchers to the form of exact algorithms [62,82] as well as of heuristics, such as a very popular A* algorithm [111], where the information about the estimated distance between nodes is provided. A comprehensive review of heuristic algorithms for solving the SP on the road network is provided by Fu et al. [94].

The other well-known algorithm for solving SSSP is the Bellman–Ford algorithm [26,91], which also has its modifications [105,20]. Contrary to Dijkstra's algorithm, the Bellman–Ford algorithm can be applied to graphs with negative edge weights. The algorithm relaxes all the edges (E) and does this $V - 1$ times, where V is the number of vertices in the graph. In each of these repetitions, the number of vertices with correctly calculated distances grows, from which it follows that eventually all vertices will have their correct distances. Moreover, the time complexity of the Bellman–Ford algorithm is much higher than in the case of Dijkstra's algorithm, that is, the time complexity for Dijkstra's algorithm is $O((V + E) \log V)$, and for the Bellman–Ford algorithm, it is $O(VE)$ [113]. A review and the presentation of the pseudocode for the most commonly known shortest path algorithms is provided by Magzhan and Jani [158].

Apart from SSSP, the other commonly known SP variant is the ALL-PAIRS SHORTEST PATH problem (APSP). Algorithms solving APSP find the shortest path between all nodes in a graph, such as the Floyd–Warshall algorithm [90, 247], Hidden Path algorithm [93,128], Uniform Path algorithm [59], or Propagation algorithm [36]. The Floyd–Warshall algorithm enables finding shortest routes in a weighted graph with positive or negative edge weights. Hidden Path algorithm is a modification of Dijkstra's algorithm, where the single-source algorithm is repeated using Fibonacci heaps, it runs in parallel from all vertex sources in the graph, and then reuses the computations performed by other vertices. Uniform Path algorithm is a refined variant of Hidden Path algorithm. Finally, Propagation algorithm is another modification of Dijkstra's algorithm [36].

2.2.4 Traffic Assignment problem

TRAFFIC ASSIGNMENT problem (TA) was originally formulated by Beckmann et al. [24]. The main goal is finding optimal routes between particular origins and destinations, that is, such routes where travelers incur the minimal costs. Since 1956, many methods to solve TA have been presented [196,205], all of them are iterative, that is, they start by considering some initial assignment, calculate the cost associated with it, and then iteratively modify both the assignment and the cost.

TA can be divided into two main problems, User Optimum traffic assignment (UO), also referred to as User Equilibrium traffic assignment (UE), and System Optimum traffic assignment (SO). In UO the aim is to minimize the travel cost of each traveler individually, whereas in SO the aim is minimizing the total travel cost of all travelers in the system. Each of these problems has two main variants, Static Traffic Assignment (STA) and Dynamic Traffic Assignment (DTA). In STA, it is assumed that the number of vehicles on each route between the origin and destination does not change. In DTA, it is assumed that the number of vehicles varies over time within the particular period of interest, which is a more realistic case. Furthermore, both UO and SO problems can be considered to be either deterministic or stochastic. In the deterministic variant, it is assumed that each traveler has a perfect knowledge about conditions on the road, whereas in the stochastic variant, randomness of the demand or supply is assumed.

TA is a subject of many papers describing not only its main types but also including the methods used for analyzing and solving it [197,248,198,155,226, 217]. In this section, we attempt to structure the gathered knowledge.

2.2.4.1 Deterministic traffic assignment

Road users tend to choose the shortest possible route on their way in terms of distance, bearing the least possible cost of travel from their origin to destination (O-D). However, their travel time and cost also depend on the decisions of other travelers. If road users know travel times and are reasonable, they can select the fastest/cheapest route, but their joint decisions may influence times of travel, so in the next iterations of making decisions, they can modify their selected routes (or times of departure) and change travel times and costs again. After some number of iterations, the system may converge to the state of equilibrium, in which each selected route has a minimum time (or cost) between the user's origin and destination. This condition is known as *user equilibrium* or *user optimum* and is a solution for the User Optimum traffic assignment problem (UO), which is a type of TA. In equilibrium, none of the travelers can find a faster route, and the demand for the network and road segments does not change, because no traveler has the motivation to choose different routes [46].

There are two types of user equilibrium/optimum, static and dynamic, which are solutions of the Static User Optimum problem (SUO) and the Dynamic User Optimum problem (DUO), respectively. In the former problem, we assume that the number of cars traveling between a given O-D pair is constant in time, and in the latter the number of such travels may change, usually on a daily basis.

Another type of DTA is the Dynamic System Optimum problem (DSO), in which the goal is to find routes maximizing the overall benefit of the system (instead of individual benefits) under time-dependent traffic conditions. DSO was mathematically formulated for the first time in 1978 by Merchant and Nemhauser [166].

2.2.4.1.1 Wardrop principles

User and system optimums are the core concepts occurring in Wardrop principles.

The first Wardrop principle means that in the user equilibrium, travel times on all used routes for a given O-D pair are equal and are not greater than those that would be felt by a single vehicle on any unused route. In other words, the travel times of all routes used for the same O-D pair are equal and minimal.

Alternatively, the first Wardrop principle can be presented as minimizing the following function:

$$\min\ S(x) = \sum_{n} \int_{0}^{x_n} t_n(w)dw, \tag{2.7}$$

where n is a route between origin–destination (O-D) pair, $t_n(w)dw$ is Highway Performance Function (HPF), and w is a flow.

The second Wardrop principle means that during equilibrium (system optimum), the total travel time is minimized [245]. Each traveler choosing the route behaves in such a way as to minimize the total travel time of all travelers. This rule is useful when planning high traffic, in which traffic management techniques such as signal timing, lane allocation, and tolls are used to discourage or encourage traffic so that the total travel time is minimized. The rule is useful for the logistic route flow, which is centrally controlled by the logistics manager.

The second Wardrop principle can be presented as minimizing the following function:

$$S(x) = \sum_{n} x_n t_n(x_n), \tag{2.8}$$

where n is a route for a given O-D pair, t_n is the travel time for a specific route, and x_n is a flow on a specific route.

2.2.4.2 Stochastic extensions for Traffic Assignment problem

The main difference between stochastic and deterministic traffic assignments (both in UO or SO) is that in the deterministic variant the actual travel time is minimized, whereas in the stochastic variant the perceived travel time is minimized [159].

A stochastic extension of TA enables to consider more realistic cases than deterministic ones. Stochasticity is captured in two dimensions, traveler's perception of the travel cost due to imperfect knowledge of traffic condition (demand) and uncertain travel times due to random events (supply) [226].

2.2.4.3 Solution methods for Traffic Assignment problem

Computing user or system optimum is important, because such traffic states are crucial from the perspective of various applications, for example, in case of the future traffic management systems, finding optimal configurations of routes, which can be transferred to connected and autonomous vehicles [109]. Nowadays, the concept of user or system optimum is less useful, because such traffic states are usually not achieved in the real world [204]; however, DTA models provide realistic traffic data for estimations of environmental impacts and so have been adopted in transportation planning and traffic management models concerning environmental sustainability [244].

Usually, the user optimum is sough using iterative algorithms. To find a user or system equilibrium (dynamic or static, deterministic or stochastic), several assumptions must be made:

- the way of modeling behaviors related to route selection (described in Section 2.2.2),
- the way of representing traffic flow and conditions (described further).

The approaches for solving the traffic assignment problem may be divided into two broad groups of methodologies, analytic methods and simulation-based methods. Furthermore, analytic methods may be categorized based on the type of the formulation of the problem [198] or the level of aggregation in which they store the solution between iterations [22]. In this section, we focus on the latter type of categorization of analytic methods for deterministic TA, since they are often compared between each other in terms of computational performance, that is, the link-based, the route-based (also referred to as path-based), and the origin-based approaches. Moreover, we also refer to the simulation-based methods used for solving stochastic traffic assignment problem.

The link-based approach is the most aggregated one, where only total link flows, aggregated over all origin-destination pairs, are stored between iterations. Due to the relatively low memory requirements, this approach had been highly preferred in the early years of development of TA solution methods. The most widely used link-based method is the Frank–Wolfe (FW) algorithm [92], which is a general nonlinear optimization. The FW algorithm is known for its slow convergence rate. Therefore numerous modifications enhancing its convergence rate have been proposed, especially the PARTAN technique [96,147,156,89,8]. The Restricted Simplicial Decomposition (RSD) algorithm [112] is the other very common link-based method. The similar approach is proposed by Larsson et al. [146], where a similar multidimensional nonlinear problem, as in the case of RSD method, is solved.

The route-based (or path-based) approach is, on the other hand, the most disaggregated one, where all used routes and the flow on each route are stored between iterations. There are two most commonly known route-based methods, Disaggregated Simplicial Decomposition (DSD) algorithm [145] and Gradient Projection (GP) algorithm [121]. Route-based algorithms have been proven to achieve more accurate solutions than link-based methods, but they come with the cost of large memory requirements.

The alternative approach are origin-based solution methods. The best known algorithm of this type was developed by Bar-Gera [22]. In the Bar-Gera algorithm, link flows representing the solution are aggregated over all destinations. The solutions obtained from using this algorithm are highly accurate and detailed, similarly as in the case of route-based algorithms. Moreover, the amount of memory required by the origin-based algorithm is relatively small in comparison with the equivalent route-based method. The other commonly known *origin-based* (also referred to as *bush-based*) algorithm was proposed by Dial [63], commonly known as algorithm B. The main improvement introduced by this algorithm is that it does not require the storage of paths and the enumeration.

Methods for solving the stochastic TA can be classified, similarly to the deterministic case, into analytic and simulation-based methods [198]. Duell et al. [71] provide a comprehensive review of methods used for solving stochastic TA. They take a simulation-based method and develop a so-called scenario-based stochastic time-dependent shortest path algorithm focusing on randomness in the demand side. All in all, authors stress out the necessity for wide range of further developments in the field of stochastic traffic assignment.

2.2.4.3.1 Comparison of performance between solution methods

In the literature on TA, we can find numerous studies focused on comparing the performance of different solution algorithms applied to this problem. Early studies were mainly focused on comparing the most commonly known link-based Frank–Wolfe (FW) algorithm with route-based algorithms [230,42], whereas Lee et al. [148] extend such studies by comparing two link-based algorithms with three route-based algorithms. The results suggested that route-based algorithms, in general, outperform link-based ones. Therefore Chen et al. [40] compare only the performance of route-based algorithms between each other. However, Perederieieva et al. [199] provide a comprehensive summary of results from existing computational studies of performance of TA solution algorithms. A comparison of all three types of algorithms shows that each type has its own advantages and disadvantages.

Tatineni et al. [230] compare and evaluate the performance of FW and DSD algorithms on a large network in Chicago area (referred to as ADVANCE network) while solving the deterministic SUO problem. Their results indicate the superiority of DSD over FW. More specifically, they find that DSD is able to achieve a much more accurate solution than FW and does it in much fewer iterations. Moreover, even though DSD requires more CPU time per one iteration than FW, it takes less time to achieve a more accurate solution. Furthermore, they find that DSD requires fewer routes and links loaded to achieve a better solution.

Chen et al. [42] compare the performance of FW with two modifications of DSD, that is, DSD1 [145] and DSD2 [242], and GP on five test networks while solving the deterministic DUO problem. Their results indicate that GP is the fastest to achieve the optimum solution whereas FW is the slowest. Moreover, FW and DSD require the lowest amount of memory. Therefore, they recommend either GP or DSD1 for solving DUO problem.

Lee et al. [148] compare the performance of five algorithms, that is, FW, PARTAN, GP, RSD, and DSD using mid- to large-scale randomly generated networks. They find that GP is faster in approaching the neighborhood of the optimal solution than the other compared algorithms when the required level of accuracy is not stringent (low accuracy level). According to authors, such efficiency of GP is achieved mainly because it avoids performing expensive line searches and reduces the number of stored paths by the projection procedure. However, when the high accuracy level is required, GP is inappropriate because of its linear convergence rate. On the other hand, RSD outperforms GP especially in the case where high accuracy level is required, particularly in terms of the convergence speed. Moreover, similarly to other link-based algorithms (FW and PARTAN), the zone-node ratio, network size, or congestion level increase, but the performance of RSD is not affected negatively that much.

Perederieieva et al. [199] provide an extensive computational study, where they compare the performance of three link-based algorithms, four path-based algorithms, and one origin-based algorithm for deterministic SUO. For their study, they use realistic networks of different scales. Their results indicate that link-based algorithms perform relatively well in comparison to other algorithms when the required level of accuracy is low. Path-based algorithms were not able to reach the required level of accuracy on the large-scale network due to their high memory requirements. The origin-based algorithm B [63] efficiently achieved the highest level of accuracy while consuming a moderate amount of memory.

All in all, the results present advantages and disadvantages of considered algorithms, depending on the requirements of a particular solution (accuracy level, available resources, scale of the network).

2.3 Traveling Salesman Problem

2.3.1 Introduction

The Traveling Salesman Problem (TSP) is a famous classical optimization problem, which has applications in various disciplines such as physics, vehicle routing, logistics, or computer science. In particular, TSP plays a starring role in the development of algorithms; it is commonly used as a test case for almost every new discrete optimization algorithm.

The first formal definition of TSP was formulated in the 1800s by W. R. Hamilton and T. Kirkman, who defined TSP as a problem of finding a circuit that passes through each vertex of a graph/polyhedron once and only once [131]. However it is rather Euler, thanks to his works (problem of seven bridges of Köningsberg, [79] and the Knight's tour problem [80]), who can be considered as a father of TSP. A contemporary definition of the problem comes from 1930 (Karl Menger) and was first known as a Messenger Problem. A detailed history and the growth of the TSP as a topic of study can be found in [219].

After 1950, the TSP was a subject of intense research, which resulted in the development of many exact and approximate algorithms to solve it [11]. Currently, the most efficient method for solving large instances of the TSP is the *Concorde* solver implemented by Applegate et al. [6], which plays a role of a baseline for newly invented algorithms. However, the center of gravity of the research was moved to more realistic problems like the asymmetric TSP (Section 2.3.2.1), constrained TSP (Section 2.3.2.2), or Vehicle Routing Problem (Section 2.4).

In this section, we attempt to structure the gathered knowledge on the TSP and its most important generalizations. They are briefly described in Section 2.3.2. Next, we give an overview of exact and approximate solutions and, finally, compare their complexity and performance.

2.3.2 TSP and its generalizations

Originally, TSP takes an undirected graph $G = (V, E)$, where V denotes a set of vertices $V = \{1, \ldots, n\}$, and E is the set of weighted edges $E = \{d_{ij}\}$, and returns the shortest Hamiltonian cycle $\mathcal{H}$, that is, a cycle that passes through every vertex exactly once. To provide the existence of $\mathcal{H}$, we usually require $G = (V, E)$ to be complete. In most papers, it is also assumed that d_{ij} satisfy the triangle inequality, which is a natural restriction for real-life problems. In particular, if $G(V, E)$ belongs to a two-dimensional space (e.g., in vehicle routing problems), then the Euclidean metric is distinguished among other metrics, and TSP is called the *Euclidean TSP*.

An equivalent definition of TSP can be expressed as an Integer Program. The first such definition was formulated by Dantzig et al. [58]. Define

$$x_{ij} = \begin{cases} 1 & \text{path from node } i \text{ to node } j, \\ 0 & \text{otherwise,} \end{cases} \tag{2.9}$$

$$x_{ij} \in \{0, 1\}, \quad i, j = 1, \dots, n, \tag{2.10}$$

where d_{ij} is the "distance" between the nodes i and j. Then TSP is given by

$$\min \sum_{i=1}^{n} \sum_{i \neq j, j=1}^{n} d_{ij} x_{ij} \tag{2.11}$$

with the constraints

$$\sum_{i=1, i \neq j}^{n} x_{ij} = 1, \quad j = 1, \dots, n, \tag{2.12}$$

$$\sum_{j=1, i \neq j}^{n} x_{ij} = 1, \quad i = 1, \dots, n, \tag{2.13}$$

$$\sum_{i \in Q, j \in Q} x_{ij} \leq |Q| - 1, \quad \forall Q \subseteq \{2, \dots, n\}. \tag{2.14}$$

Constraint (2.12) informs that each city is arrived at from exactly one other city, whereas (2.13) requires that from each city there is a departure to exactly one other city. The last constraint, called a *Subtour Elimination Constraint* (SEC), ensures that there are no subtours among the nonstarting vertices, so the solution returned is a single tour and not the union of smaller tours.

2.3.2.1 Asymmetric TSP

Recall that d_{ij} is the distance between nodes (cities) i and j. If $d_{ij} = d_{ji}$, then the problem is *symmetric* (TSP), otherwise it is *asymmetric* (ATSP). In any case, ATSP can be translated to the classical TSP [126]. A practical application of the ATSP is street-level route optimization (which is usually asymmetric due to one-way streets, slip-roads, motorways, etc.). It is important that ATSP can be transformed to the min-sum Assignment Problem, which always has an integer-optimal solution [209].

2.3.2.2 Precedence-constrained TSP

In contrast to the original TSP and ATSP, real-life instances of these problems are always restricted in some way. Among the constrained variables, we can distinguish time, cost, distance, and precedence of visited nodes. In this section, we focus only on TSP with precedence constraints as having wide applications and significantly different from the problems discussed previously.

The precedence-constrained TSP (PCTSP) can be defined as the symmetric or asymmetric TSP such that the salesman should start from a prescribed node and each admissible tour must satisfy k precedence relations denoted by $i_r < j_r$, $r = 1, 2, \ldots, k$. Given a directed complete graph $G(V, E)$, a distance d_{ij} on each edge, and precedence constraints $<$ on the node set V, a solution of the PCTSP is a minimum distance tour that starts from the node $n_0 \in V$, visits all the nodes in $V - n_0$, and returns n_0 such that a node i is visited before node j when $i < j$. Possible applications of PCTSP include public transport scheduling and circuit board assembly [72].

2.3.2.3 TSP with time windows

TSP with time windows (TSPTW) is a time-dependent variant of the TSP defined as follows: given an undirected complete graph $G = (V, E)$, where $V = \{0, 1, \ldots, n\}$ is a set of nodes representing the depot (node 0) and n customers, a TSPTW solution is a sequence visiting each node of G exactly once, starting and ending at the depot, starting service of each customer i within a given time window $[e_i, l_i]$, and minimizing the given objective function (e.g., cost, distance, or time). Waiting times are generally permitted, that is, customer i may be reached before the start e_i of its time window, but the service cannot start until e_i.

2.3.3 Exact methods

Naive method for solving TSP relies on comparing all the possible routes and choosing the shortest one. Complexity of this method is $O((n-1)!)$, and hence the naive method becomes inefficient even for a moderately large instance of TSP. Thus a great effort has been put in search for other solutions.

Rapid development of exact solutions took place after formulating the Integer Program [58] and *LP-relaxation*,[1]

that is, transformation of the Integer program to the corresponding Linear Program (LP) by replacing (2.10) with $0 \leq x_{ij} \leq 1$. It is important that LP can be solved in a polynomial time [124].

According to Matai et al. [164] and Junger et al. [127], exact methods for TSP can be divided into the following classes:

1. Dynamic Programming, for example, the Held–Karp algorithm.

1. In general, a relaxation of an optimization problem P is another optimization problem R whose set of feasible solutions properly contains all feasible solutions of P [127].

2. Exact methods for solving Integer Problems, for example, the simplex algorithm.
3. Algorithms for solving LP with a later return to integrality:
   a. Branch-and-bound algorithms.
   b. Branch-and-cut algorithms.

In this section, we describe the most commonly used and historically significant algorithms.

2.3.3.1 *Held–Karp algorithm*

The Held–Karp algorithm (HK) is a dynamic programming technique [27,114], which uses the property that every subpath of a path of minimum distance is itself of the minimum distance. They showed that finding a proper spanning tree is equivalent to solving a linear program for the TSP with only node-degree constraints and subtour elimination constraints [129,220].

The Held–Karp algorithm can be evaluated exactly by Linear Programming techniques [124], but its exact implementations become impractical for large instances of TSP. Therefore an iterative estimation of HK solution is preferred more than the exact method [114,236,220]. The best approximated HK solution is called the *Held–Karp lower bound,* and the difference between the lower bound and the exact solution is the *Held–Karp gap.*

2.3.3.2 *Branch-and-bound algorithm*

The first formulation of the branch-and-bound algorithm (B&B) was provided by Land and Doig [140]. Its idea is finding a value x that maximizes or minimizes the value of a real-valued function $f(x)$, called an *objective function,* among some set S of candidate solutions. Next, the algorithm recursively splits the search space into smaller spaces and minimizes $f(x)$ on them (*branching*). The algorithm keeps track of bounds on the minimum that is trying to find and uses these bounds to the search space, eliminating candidate solutions that contain no optimal solution.

2.3.3.3 *Branch-and-cut algorithm*

The degree equations (2.12), (2.13), nonnegativity inequalities, and SECs of LP define the polyhedron, called the Subtour Elimination Polytope (SEP). The integer solutions for TSP form a convex hull in SEP, called the symmetric traveling salesman polytope (STSP). Therefore analysis of the SEP corresponding to a given TSP allows us to find approximate and exact solutions [15,14,127].

The idea of the branch-and-cut algorithm (B&C) is solving the LP corresponding to a given TSP and then gradually search for the STSP, which can be done by choosing a proper cutting plane in SEP. To this end, additional constraints are required; for details, see [192,175,191,127,174].

2.3.3.4 Exact algorithms for TSP and ATSP

All the exact methods presented in Section 2.3.3 were originally formulated for solving TSP.

Most of papers about the Held–Karp method are concentrated on approximate solutions and their mathematical analysis [129,124]. Valenzuela and Jones [236] propose novel iterative schemes for the HK algorithm applied for the large Euclidean TSP. Shmoys and Williamson [220] proves the monotonicity for the Held–Karp gap, which is an interesting result for general LP theory.

The first formulation of B&B for symmetric TSP was presented by Land and Doig [140], whose work was an extension of Eastman's thesis for ATSP [73]. The method was used by Miliotis [168] to reach integrality in the first fully automated LP algorithm for solving TSP [164]. Performance of the algorithm for various types of LP-relaxation was studied by Balas and Toth [15]. However, much more work was devoted to the B&C method, especially to the problem of constructing proper inequalities defining cutting planes in SEP [192,175,191, 127,174].

Among exact methods for ATSP, two classes can be distinguished. The first one consists of the algorithms solving the Assignment Problem (AP), to which ATSP can be easily transformed, and its relaxed (LP) version can be solved in polynomial time [209]. The solution algorithms include:

1. Exact algorithms for integer AP [39,37].
2. Algorithms based on AP-relaxation [73,154].

The second class of ATSP solutions includes the algorithms applied directly to ATSP. In most cases, they are modified versions of the algorithms for solving symmetric TSP. Roberti and Toth [209] proposed an additive branch-and-bound approach, which can be briefly described as follows: given b possible bounding procedures $b_1, \ldots, b_b$ for ATSP, the additive approach generates a sequence of ATSP instances obtained by applying $b_1, \ldots, b_b$. A branch-and-cut method for ATSP, equipped with a novel variable-pricing technique, was presented by Fischetti and Toth [85].

Gharan and Saberi [102] presented a solution for ATSP on bounded-genus graphs. The main step of the algorithm is finding a "thin" tree in the fractional solution of the Held–Karp linear programming relaxation. A similar approach was published by Svensson et al. [225]. Their algorithm returns for a given ATSP a tour of value at most $c \cdot hk$, where hk denotes the Held–Karp lower bound, and c is a real constant. The algorithm, however, does not optimize c. Another example of a Held–Karp-based algorithm was announced by Fages and Lorca [81], who analyzed connections between the graph structure and time complexity or consistency level and showed that the graph structure had a serious impact on resolution.

2.3.3.4.1 Exact methods for PCTSP and TSPTW

PCTSP became a subject of intensive research not only because of the possible applications, but also because interesting mathematical structures behind it. An exhaustive theoretical analysis of the asymmetric PCTSP was presented in [14]. Sarin et al. [214] announced novel polynomial formulations for the constrained and unconstrained ATSP and used relaxed Dantzig–Fulkerson–Johnson subtour constraints to find the solution.

Most of algorithms for solving PCTSP are based on exact algorithms for solving TSP with some modifications, for example:

1. The B&B method such that branching rules preserve precedence relation [136].
2. Modified Lagrangean approach based on Balas and Christofides [13] and Kubo and Kasugai [136].
3. B&C algorithm [10].

Although TSPTW is particularly difficult to solve and time-consuming, there are few already implemented exact algorithms for this problem. Albiach et al. [3] announced an exact solution through a proper SEP-graph transformation for asymmetric TSPTW. Mingozzi et al. [169] presented a dynamic programming approach using a generalized version of the State Space Relaxation [49]. The B&B approach was used in turn by Pesant et al. [200], who applied an exact B&B optimization without any restrictive assumption on the time windows. Their algorithm outperformed LP solutions, but only for some classes of TSPTW instances.

2.3.4 Approximate solutions

We have already discussed the exact algorithms, the methods that, for every finite-size problem, guarantee to find the optimal solution in a bounded time [33]. However, in a complicated optimal fleet routing problem the exact algorithms may not be effective by leading to a dramatic increase in computation time and cost. Consequently, the contemporary research concentrates on approximated algorithms that are capable of finding high-quality solutions in a limited time, which can be applied to a real-life instances characterized by large vehicle fleets and huge number of cities. This is the main reason why approaching methods, *heuristics* and *metaheuristics*, instead of the exact approaches, are commonly used in practise in optimal fleet routing NP-hard problems [53]. Those algorithms are attractive, since based on the input information (usually incomplete or imperfect), they provide an approximate solution using far fewer computational resources than optimal algorithms [182,33,77].

2.3.4.1 Types of approximate algorithms

Initially, classical heuristics were focused on obtaining a feasible solution as quickly as possible and then applying it to the post-optimization procedure.

However, over the last 40 years, more emphasis is put on the development of metaheuristics, higher-level heuristics-based procedures that are used to select, find, or generate a heuristic providing a sufficiently good answer to an optimization problem [33,190]. They are methods trying to combine basic heuristic algorithms in higher-level frameworks to explore a search space in an efficient and effective way [33]. In general, metaheuristics can be classified into different groups based on their characteristics [190,33,67,53]). However, in this section, we use the classification provided by Junger et al. [127], La Maire and Mladenov [138], and Laporte [141], who divided the heuristics and metaheuristics for solving TSP as follows:

1. Constructive heuristics.
2. Improvement heuristics.
3. Nature inspired algorithms.

2.3.4.1.1 Constructive heuristics

Constructive heuristics gradually build a feasible solution while keeping an eye on solution cost, but they do not contain an improvement phase per se. Examples of constructive heuristics that are particularly useful in solving TSP are the following:

1. Insertion heuristics.
2. Local searches.
3. Heuristics based on constructing spanning trees.

2.3.4.1.1.1 Insertion heuristics Insertion heuristics start with a subset S of all nodes and then construct the whole tour step by step by inserting new nodes to S [127]. In the method of nearest insertion, a node $k \notin S$ that is closest to any of the subtour nodes is added to the tour. An opposite approach is the method of farthest insertion, where the node that is farthest to any of the subtour nodes has to be added. Both of these methods have the worst-case complexity $O(n^2)$. Another commonly used insertion heuristic is the method of cheapest insertion, having the complexity $O(n^2 \log n)$ [127].

Insertion heuristics are usually used only in combination with other methods due to their low performance [127].

2.3.4.1.1.2 Local searches The local search heuristics gradually construct a tour by repeatedly selecting the shortest edge and adding it to the tour as long as it does not create a cycle with less than N edges or increases the degree of any node to more than 2. Its complexity is $O(n^2 \log_2(n))$ [127,164].

One of the most commonly used metaheuristics based on local search is Tabu Search [117,74,153]. It can be defined as a local search equipped with short-term memory structures that describe the visited solutions or/and user-provided sets of rules. Potential solutions that have been previously visited within a certain period or that violate a user-provided rule are added to the list of forbidden solutions (tabu list), which implies that the algorithm does not consider that possibility repeatedly.

2.3.4.1.1.3 Heuristics based on constructing spanning trees Christofides algorithm [47] is one of the best approximation algorithms for the metric TSP, which achieves an approximation ratio of $3/2$. It returns a Hamiltonian cycle of a graph G. Its main idea is based on the iterative constructing of minimum spanning trees for a given input and computing minimum weight matching on the nodes that have an odd degree in these trees. Analysis of the algorithm and its comparison to the local search was presented, for example, by Nallaperuma et al. [181].

2.3.4.1.2 Improvement heuristics

Improvement methods attempt to upgrade any feasible solution by performing a sequence of edge or vertex exchanges for a Hamiltonian cycle of a graph $G(V, E)$. Improvement can be obtained using *k-opt moves*. One of the improvement heuristics, the Lin–Kernighan algorithm, is one of the best and most frequently used heuristics for the TSP [123].

2.3.4.1.2.1 k-opt heuristics Improvement can be obtained using k-opt moves [127]. K-opt move consists of eliminating k edges and reconnecting the resulting paths in a different way to obtain a new cycle. Checking whether an improving k-opt move exists takes time $O(n^k)$ and is therefore only applicable for small problems (K-opt heuristic is also very sensitive with respect to the sequence in which moves are performed). However, it is possible to design restricted searches for higher values of k by restricting the number of nodes to check in each iteration (see discussion in [162]).

2.3.4.1.2.2 Lin–Kernighan heuristics The Lin–Kernighan heuristic [152] is known to be one of the best heuristics for the TSP. Its idea builds upon 2-opt and 3-opt moves with node insertion. The number of edges to switch is decided at each step of the algorithm. Results of a very successful test of this approach were published in 2000 [116]. They showed that the Lin–Kernighan heuristic was able to find the solution for 13509-city TSP.

2.3.4.1.3 Nature inspired algorithms

Most of the optimization algorithms nowadays are nature-inspired: the real world has been the inspiration for researchers in many ways [190,87]. The majority of nature-inspired methods are based on some successful characteristics of a biological system, but many algorithms have been developed by drawing the inspiration from physical and chemical systems as well [33,87].

A number of other nature-inspired algorithms applied to TSP can be found in the literature; see, for example, the bumble bee algorithm [149], the ant colony algorithm [66], or the water wave optimization [252]. It is difficult to compare and clearly identify the best algorithm, because its performance highly depends on the specific TSP and the hyperparameters of the metaheuristic. All of these algorithms, however, had better performance and runtime, comparing to genetic algorithms or local search techniques and seemed to be an interesting direction for optimization algorithms.

Among nature inspired algorithms for solving TSP and its variants, we can distinguish artificial neural networks and some metaheuristics.

2.3.4.1.3.1 Artificial neural networks *Artificial neural networks* (ANN) are computing systems vaguely inspired by the biological neural networks [118, 134], which learn to solve computational problems by considering examples without being programmed with any task-specific rules. An ANN can be described as a collection of connected units or nodes called *artificial neurons*, which are organized in layers. The learning process includes forward propagation, where the signal goes through the ANN and gives a result and backpropagation, where the weights of the network are updated starting from the last layer. Forward propagation and backpropagation are repeated several times until the difference between the final result and the ground truth becomes small enough. ANNs are called *deep* if they consist of at least two layers and have a certain level of complexity. Deep neural networks have been used to solve TSP, for example, by La Maire and Mladenov [138] and Miki et al. [167].

2.3.4.1.3.2 Metaheuristics As explained in Section 2.3.4.1, metaheuristics are higher-level heuristics-based procedures that are used to select, find, or generate a heuristic providing a sufficiently good answer to an optimization problem [33,190]. Metaheuristics may be based, for example, on local search (e.g., tabu search, a higher-level heuristics-based procedures that are used to select, find, or generate a heuristic providing a sufficiently good answer to an optimization problem [33,190]), but most of them are inspired by the nature. Among them, we can distinguish, for example, simulated annealing [33], genetic algorithms [181,29,83], water wave algorithm [252], particle swarm optimization [137], or ant colony algorithm [224]. They are powerful algorithms commonly used in solving fleet optimization problems nowadays. A possibility to accept worse solutions is a fundamental property of metaheuristics enabling a more extensive exploration of the search space to find the global optimal solution [77].

2.3.4.1.3.2.1 Genetic algorithms *Genetic algorithms* are inspired by the biological evolution, the process of natural selection based on the concept of Darwin's evolution theory. The general algorithm looks as follows: in the first step, the population as a group of individuals is generated randomly and encoded into genotypes. Individuals are evaluated using a fitness function (which has to be optimized), best of them are selected for the further reproduction, and then they exchange a genetic material (part of its structure) in the crossover procedure and produce the next population in which genotypes of individuals may be additionally mutated. The process continues until a suitable solution is found or a certain number of generations have passed, depending on the stopping criterion [33,67,87].

2.3.4.1.3.2.2 Simulated annealing *Simulated Annealing* (SA) is considered to be the oldest metaheuristic. It was introduced in 1983, and its origins are in the statistical mechanics (Metropolis algorithm) [33]. At each step the SA algorithm randomly selects a state (solution) within the current state neighborhood, examines its quality, and then decides whether to move to this new state or to stay in the current one. This move decision is made based on the *acceptance probability function* that depends on the energy of considered states and the control parameter called *temperature* [67,77,33].

2.3.4.1.3.2.3 Algorithms based on social behavior of animals Particle Swarm algorithm (PS), Ant Colony algorithm (AC), Bat algorithm, and Bumble Bee algorithm (BB) are based on social behavior of animals. They are inspired by the observation that animals are able to solve complex problems in a natural way [67], for example, the problem of choice at the time of finding food sources. The whole process eventually leads to the choice of finding the best (shortest) path between the nest and a source of food without the individuals having a global vision of the path. Such methods are able to solve a computational problem of finding specific paths through graphs.

2.3.4.1.3.2.4 Water wave algorithm Water Wave algorithm (WW) is based on the observation that water waves, which can be described as planar waves, lose their energy in time, which makes their amplitude decrease [256]. The algorithm starts from an initial population of solutions and applies a propagation, breaking, and refraction operator on each solution respectively according to the appropriate equations for plane waves.

2.3.4.1.3.2.5 Other nature-inspired metaheuristics The literature on nature-inspired metaheuristics for symmetric TSP is very rich and diverse. As a curiosity, it is worth mentioning the solution of TSP performed by simple living organisms. Jones and Adamatzky [125] presented results of 20-city TSP obtained by *Physarum polycephalum*, which achieved the approximation ratio 1.1 comparing to the Concorde solver. Also, recent results suggest that unicellular amoeboid organisms may possess an interesting ability of solving TSP very fast [257].

2.3.4.2 Approximate algorithms for TSP and ATSP

An overview of approximation algorithms for ATSP can be found in [104], where the following heuristics are compared: the local search algorithm, Random Insertion algorithm, modified Karp-Steele patching heuristic [130], Recursive Path Contraction algorithm [255] and Contract-or-patch heuristic [107], which had the best performance among all the presented algorithms.

As for nature-inspired algorithms for solving ATSP, it is worth mentioning the improved version of discrete Bat algorithm proposed by Osaba et al. [189]. The algorithm was compared with SA and GA and outperformed both of these methods. Another very recent approach is a modified AC with Short-Term Memory [218]. Other examples of nature-inspired metaheuristics can be found in [97,177].

2.3.4.3 Approximate algorithms for PCTSP and TSPTW

The literature on approximate methods for PCTSP is relatively limited comparing to other variants of TSP. Among nature-inspired heuristics, we can mention a Hybridized AC system combined with several local search procedures [206] and genetic algorithm [206]. A heuristic based on restricted 3-opt moves was used in the problem of printed circuit board assembly [72,188].

On the other hand, there is a wide variety of metaheuristics for solving TSPTW. In this section, we present a few interesting applications of metaheuristics to TSPTW.

The algorithm announced by Calvo [38] consists of construction heuristics and local search procedures. Its performance strongly depends on the choice of the objective function and varies from 15% to 70% of proper solutions. A branch-and-cut approach with many auxiliary heuristics (sorting, swapping, node insertion, edge exchange, etc.) was described by Ascheuer et al. [9].

Problems extending TSPTW with additional constraints or many salesmen become more and more popular. An interesting example is a TSPTW with precedence constraints in Air Travel, for which Saradatta and Pongchairerks [213] proposed a heuristic based on the nearest-neighbor algorithm, local search, and k-opt moves. In turn, one of the first real cases of the clustered TSPTW, a problem of optimizing parcel deliveries in London, was described by Nguyêñ et al. [185]. The authors transformed the problem to the corresponding LP and used commercial Mixed-Integer LP optimizers[2] to find the solution.

Among the nature-inspired heuristics, AC seems to be particularly popular for solving TSPTW. Cheng and Mao [43] proposed a modified ant colony algorithm, where the time-window constraint is included by incorporating additional local heuristics into the state transition rule. The performance of this approach was evaluated by assessing their distance from the lower bounds of the corresponding problem instances. The results have shown that the algorithm either finds the optimum solution or returns a result having a high likelihood of being the optimal solutions.

2. The model was implemented in IBM ILOG CPLEX Optimization Studio V12.70.

2.3.5 Quantum algorithms

Quantum computers enable natural parallelization, hence they provide the better time-scaling for many algorithms. However, problems with building and controlling quantum computers and non-intuitiveness of quantum mechanics resulted in the fact that only a little work has been done on quantum algorithms for the TSP.

Before we present the algorithms, we need to introduce the most fundamental concepts from quantum computing.

2.3.5.1 *Qubit*

The quantum analogue of a bit is called a *qubit*. Physically, a qubit is a particle having two degrees of freedom. Qubit states can be represented as normalized vectors in 3-dimensional real space and their time evolution is driven by the energy operator of the system, H. The evolution equation is called the Schrödinger equation:

$$\frac{d}{dt}|\phi\rangle = H|\phi\rangle. \tag{2.15}$$

A problem to be solved is encoded as some initial state of qubits, $|\phi(0)\rangle$, then a number of unitary operations is performed on $|\phi(0)\rangle$. The final state $|\phi\rangle$ corresponds to the solution.

2.3.5.2 *Quantum annealing*

Quantum annealing [195,240] can be thought as a quantum analogue of SA. Given a quantum system, the computing procedure starts from a quantum-mechanical superposition of all possible states. Then the system evolves very slowly and changes external field according to Eq. (2.15) and finally attains its ground state, that is, the state of minimal energy. The ground state corresponds to the solution of the optimization problem.

2.3.5.3 *Algorithms*

One of the first papers concerning the quantum TSP was by Mandra et al. [161], who developed an algorithm for finding a Hamiltonian cycle for a degree-k graph in time $O(2^{(k-2)n/4})$. The lowest bounds for the algorithm are $O(1.189^n)$ for $k = 3$ and $O(1.414^n)$ for $k = 4$. However, for $k \geq 5$, the algorithm becomes slower than classical algorithms.

A faster approach was announced by Moylett et al. [173], which is based on Montanaro's paper on quantum backtracking [172] and the algorithm of Xiao and Nagamochi [176,253]. The authors applied the backtracking to the quantum analogue of Xiao and Nagamochi method and constructed an algorithm of complexity $O(C(k)^n \log L \log\log L)$, where $C(k)$ is a real factor depending on the degree k of the graph, and L is the maximal edge cost. They showed that this method does not enable getting a quantum speedup over the fastest classical algorithms for degree-k graphs, where $k \geq 8$.

A completely different approach is based on quantum annealing. Martonak et al. [163] compared SA with quantum annealing. The quantum model was mapped onto the classical model and referenced to the Lin–Kernighan algorithm. It was shown that quantum annealing worked more efficiently and reduced the error at a much steeper rate than SA. Experimental results on QA performed in a nuclear-magnetic-resonance quantum simulator were presented for the TSP by Chen et al. [41].[3] Another metaheuristics that have been already implemented on quantum simulators are genetic algorithms [228], the particle swarm algorithm [243], and the clonal selection algorithm [56]. We expect that availability of working quantum computers will result in even faster development of further quantum algorithms.

As it was pointed by Bang et al. [19], quantum annealing does not provide a proper solution to the large number of cities since finding a ground state becomes extremely difficult for large systems. One of the possible methods to avoid this problem is based on Grover's search [186] with oracle[4] [19]. The algorithm works in time $O(1/\sqrt{f})$, where f is the probability of finding a solution for asymptotically large number n of cities,

$$f \simeq e^{-3n/2-1/2 \ln n}. \tag{2.16}$$

It is interesting that first experimental results for quantum TSP have also been already published. Srinivasan et al. [223] announced a unique algorithm for the four-cities TSP, which was run on an IBM quantum simulator for a custom qubit topology.

2.3.6 Computational complexity

The Euclidean TSP and its variations are known to be NP-hard as the problem of finding the Hamilton cycle is NP-hard [129]. Recall that a computational problem L is NP-hard if every other NP-hard problem can be reduced in polynomial time to L. In particular, finding a polynomial algorithm to solve any NP-hard problem means the equality of P and NP classes, which has been neither proven nor refuted yet. Euclidean TSP is also NP-complete [194], that is, its solutions can be verified in polynomial time.

For many years, the algorithm with the best proven worst-case bounds for the general TSP was the Held–Karp algorithm [115], which runs in $O(n^2 2^n)$ time and uses $O(n2^n)$ space. This result was improved by Björklund et al. [30], who showed that modifications to the Held-Karp algorithm could yield a runtime

3. There is also possibility to run quantum annealing algorithm on a D-Wave quantum computer; see the open-source project https://github.com/BOHRTECHNOLOGY/quantum_tsp.
4. An oracle O is a "black box" operation defined as a classical function $f : \{0, 1\}^n \to \{0, 1\}^m$ that is used as input to another quantum algorithm.

$O((2^{k+1}-2k-2)^{n/(k+1)} poly(n))$, where k was the largest degree of any vertex in the graph. This bound is strictly less than $O(2^n)$ for all fixed k.

A recent line of research has produced a sequence of nongeneral algorithms, which improve on the $O(2^n)$ runtime of the general Held–Karp algorithm in this setting [173]. First, Eppstein [78] presented algorithms that solve the TSP on degree-3 graphs in time $O(2^{n/3}) \simeq O(1.260^n)$ and on degree-4 graphs in time $O\left(\left(\frac{27}{4}\right)^{n/3}\right) \simeq O(1.890^n)$. The algorithms are based on the standard classical technique of backtracking.[5] The best classical runtimes known for algorithms based on this general approach are $O(1.232^n)$ for degree-3 graphs and $O(1.692^n)$ for degree-4 graphs, in each case due to Xiao and Nagamochi [253, 176]. All of these algorithms use polynomial space in n. In the case where edge costs in the graph are bounded by some upper bound L, there is a randomized algorithm solving the TSP on arbitrary graphs in $O(1.657^n L)$ time and polynomial space, which is an improvement on the runtime of the Xiao–Nagamochi algorithm for degree-4 graphs when L is subexponential in n [30].

Quantum computers seemed to be a way to speed up the algorithms solving TSP, but the results obtained so far have not shown a significant speedup. It has been proven that quantum TSP was BQP-hard [186], that is, it can be solved by an universal quantum computer in polynomial time with the probability of error bounded by $1/3$. However, this definition assumes that a fully universal quantum computer exists (which is not true now). Moreover, it is not yet known whether $NP \subset BQP$, and the informal consensus is that this containment is in fact very unlikely [173]. Hence *quantum computers do not provide any known superpolynomial advantage for solving NP-complete problems* [173].

2.4 Vehicle Routing Problem

2.4.1 Introduction

VEHICLE ROUTING PROBLEM (VRP) is a generic name given to a whole class of problems concerning the optimal design of routes to be used by a fleet of vehicles to serve a set of customers [18]. VRP is a generalization of the TSP problem widely described in the previous section. It was first introduced in Dantizg and Ramser [57] under the name *Truck Dispatching Problem*. The goal was to plan routes for a fleet of gasoline delivery trucks between a bulk terminal and a large number of service stations supplied by the terminal. It was presented as follows: we have a homogeneous fleet of vehicles initially stationing at the terminal and having some limited capacity. Each service station demands a given amount of gasoline. The goal is to assign trucks to routes so that all the demands are satisfied and capacity of trucks is not exceeded. This class of problems is called

5. Backtracking is an recursive algorithm which solves problems by trying to build a solution incrementally and removing the solutions that fail to satisfy the constraints of the problem at any point of time.

CAPACITATED VEHICLE ROUTING PROBLEM and is the most studied variant of VRP. We focused closer on it in Section 2.4.3.

Toth and Vigo [235] wrote: *"The* VEHICLE ROUTING PROBLEM *is definitely one of the most studied among the combinatorial optimization problems."* This statement is hard to contest and has not lost its genuineness over years. According to GoogleScholar [108], more than 5400 research papers with phrase "Vehicle routing problem" in the title were published since 1959, and more than 500 of them only in 2018. In particular, it is supposed that raising awareness of climate changes has significantly contributed to the increase in the number of articles due to taking into consideration the environmental footprint [133,157].

The taxonomy of VRP is very complex, and we devoted the whole Section 2.4.2 to its overview. Since the number of VRP variants is large, we decided to focus only on three variants, which seem to have the most important industrial applications: Capacitated Vehicle Routing Problem (CVRP), Vehicle Routing Problem With Time Windows (VRPWTW or VRPTW), and (general) Pickup and Delivery problem (GPDP).

In CVRP (Section 2.4.3), we assume that vehicles have a given maximum capacity (usually, the same for all vehicles).

In VRPTW (Section 2.4.4), service at any customer starts within a given time interval called a *time window*.

In GPDP (Section 2.4.5), all locations are divided into two types, origins and destinations. Each vehicle in a fleet has a given capacity, start location, and end location. Each load has to be transported by one vehicle from its set of origins to its set of destinations without any transshipment at other locations [216].

2.4.2 Taxonomy

There exist many variants of VRP, and to unify classification of its literature, many taxonomies have been developed [241,34,75]. In our paper, we adapt the taxonomy proposed by Eksioglu et al. [75], which highlights five main characteristics of VRPs: the type of study, scenario characteristics, problem physical characteristics, information characteristics, and data characteristics. For each category, several subcategories are presented.

Table 2.1 summarizes the previous approaches to classify the studies on VRP and enables us to have a reference point.

2.4.2.1 Variants of CVRP

Since most of the variants of VRP (and all variants with reasonable real-world applications) assume a limited capacity of vehicles, they are in fact also variants of CVRP. Due to its importance, CVRP has been intensively studied for over 60 years, and many its variants, mostly driven by business requirements, have been identified. In this subsection, we present the most popular ones.

TABLE 2.1 Table presenting a taxonomy from [75].

- **1.** Type of study
  - **1.1.** Theory
  - **1.2.** Applied methods
    - 1.2.1. Exact methods
    - 1.2.2. Heuristics
    - 1.2.3. Simulation
    - 1.2.4. Real time solution methods
  - **1.3.** Implementation documented
  - **1.4.** Survey, review or metaresearch
- **2.** Scenario characteristics
  - **2.1.** Number of stops on route
    - 2.1.1. Known (deterministic)
    - 2.1.2. Partially known, partially probabilistic
  - **2.2.** Load split constraint
    - 2.2.1. Splitting allowed
    - 2.2.2. Splitting not allowed
  - **2.3.** Customer service demand quantity
    - 2.3.1. Deterministic
    - 2.3.2. Stochastic
    - 2.3.3. Unknown
  - **2.4.** Request time of new customers
    - 2.4.1. Deterministic
    - 2.4.2. Stochastic
    - 2.4.3. Unknown
  - **2.5.** On site service/ waiting times
    - 2.5.1. Deterministic
    - 2.5.2. Time dependent
    - 2.5.3. Vehicle time dependent
    - 2.5.4. Stochastic
    - 2.5.5. Unknown
  - **2.6.** Time window structure
    - 2.6.1. Soft time windows
    - 2.6.2. Strict time windows
    - 2.6.3. Mix or both
  - **2.7.** Time horizon
    - 2.7.1. Single period
    - 2.7.2. Multi period
  - **2.8.** Backhauls
    - 2.8.1. Nodes request simultaneous pickup and deliveries
    - 2.8.2. Nodes request either linear or backhaul service, but not both
  - **2.9.** Node/Arc covering constraints
    - 2.9.1. Precedence and coupling constraints
    - 2.9.2. Subset covering constraints
    - 2.9.3. Re course allowed
- **3.** Problem physical characteristics
  - **3.1.** Transportation network design
    - 3.1.1. Directed network
    - 3.1.2. Undirected network
  - **3.2.** Location or addresses (customers)
    - 3.2.1. Customers on nodes
    - 3.2.2. Arc routing instances
  - **3.3.** Geographical location of customers
    - 3.3.1. Urban (scattered with a pattern)
    - 3.3.2. Rural (randomly scattered)
    - 3.3.3. Mixed
  - **3.4.** Number of points of origin
    - 3.4.1. Single origin
    - 3.4.2. Multiple origin
  - **3.5.** Number of points loading/unloading facilities (depot)
    - 3.5.1. Single depot
    - 3.5.2. Multiple depot
  - **3.6.** Time window type
    - 3.6.1. Restriction on customers
    - 3.6.2. Restriction on roads
    - 3.6.3. Restriction on depots/hubs
    - 3.6.4. Restriction on drivers/vehicles
  - **3.7.** Number of vehicles
    - 3.7.1. Exactly n vehicles (*TSP* in this segment)
    - 3.7.2. Up to n vehicles
    - 3.7.3. Unlimited number of vehicles
  - **3.8.** Capacity consideration
    - 3.8.1. Capacitated vehicles
    - 3.8.2. Uncapacitated vehicles
  - **3.9.** Vehicle homogeneity (Capacity)
    - 3.9.1. Similar vehicles
    - 3.9.2. Load-specific vehicles
    - 3.9.3. Heterogeneous vehicles
    - 3.9.4. Customer-specific vehicles
  - **3.10.** Travel time
    - 3.10.1. Deterministic
    - 3.10.2. Function dependent (a function of current time)
    - 3.10.3. Stochastic
    - 3.10.4. Unknown
  - **3.11.** Transportation cost
    - 3.11.1. Travel time dependent
    - 3.11.2. Distance dependent
    - 3.11.3. Vehicle dependent
    - 3.11.4. Operation dependent
    - 3.11.5. Function of lateness
    - 3.11.6. Implied hazard/risk related
- **4.** Information characteristics
  - **4.1.** Evolution of information
    - 4.1.1. Static
    - 4.1.2. Partially dynamic
  - **4.2.** Quality of information
    - 4.2.1. Known (Deterministic)
    - 4.2.2. Stochastic
    - 4.2.3. Forecast
    - 4.2.4. Unknown (Real-time)
  - **4.3.** Availability of information
    - 4.3.1. Local
    - 4.3.2. Global
  - **4.4.** Processing of information
    - 4.4.1. Centralized
    - 4.4.2. Decentralized
- **5.** Data characteristics
  - **5.1.** Data used
    - 5.1.1. Real world data
    - 5.1.2. Synthetic data
    - 5.1.3. Both real and synthetic data
  - **5.2.** No data used

2.4.2.1.1 Heterogeneous fleet

When describing CVRP (Section 2.4.1), we assumed that the fleet of vehicles is homogeneous; however, in the real world the vehicles are rarely identical. Such a problem is known as the Mixed Fleet VRP or as the Heterogeneous Fleet VRP [16]. Vehicles can differ in many ways; for example, Drexl [68] highlights the following dimensions:

- **Cost for driver**: additional payment for driver might be included when time or distance is exceeded.
- **Tolls**: some vehicles may require paying a toll for using a road.
- **Capacity**: vehicles may differ in capacity dimensions such as weights, volume, loading meters, or number of pallet places.
- **Location**: locations or depots at the beginning and at the end of the planning horizon.
- **Vehicle class**: there might be different classes of vehicles (e.g., lorry, train, ship) to be considered. Apart from that, there are different types of vehicles (e.g., swap-body vehicle or tank vehicle).

2.4.2.1.2 Multicompartment vehicles

Another type of CVRP is MULTI-COMPARTMENT VEHICLE ROUTING PROBLEM (MC-VRP) when products must be transported in separated vehicle compartments [165]; for example, we may have delivery vehicles that have refrigerated and nonrefrigerated compartments for foodstuffs and tankers, which distribute different types of petroleum products, as it was introduced in the first paper concerning MC-VRP [48]. There are several real-world use cases where solving such problem is crucial, for example, fuel distribution in Poland [211], collection of olive oil in Tunisia [139], and recyclable waste from households [207].

2.4.2.1.3 Pollution-Routing Problem

The Pollution-Routing Problem (PRP) was first introduced by Bektas and Laporte [25] as an extension of VRP. Apart from the constraints that a set of customers must be served within a specified time window, they took into consideration pollution side-effect such as Greenhouse Gas (GHG) and, in particular, CO_2 emission. The authors suggested minimizing the total cost function composed of labor, fuel, and emission costs expressed as functions of load, speed, and other parameters. This problem was extended by Koc et al. [132], who investigated the impact of heterogeneous fleet on environment pollution. A complex review of recent studies might be found in [60,150].

2.4.2.1.4 Split deliveries VRP

The split delivery vehicle routing problem (SDVRP) was formulated by Dror and Trudeau [69], who analyzed the possibility of splitting the delivery—the assumption of the classical VRP that customers are visited only once is relaxed, which may lead to a significant reduction in the total distance and the total number of required vehicles. Dror and Trudeau [70] extended their analysis and compared gains from splitting the delivery in hundreds of problems. In recent years, many researches on SDVRP were carried out. Tavakkoli-Moghaddam et al. [231] studied the split delivery problem with a heterogeneous fleet of vehicles. Mitra [170] researched the problem with both linehauling (satisfying a demand for finished goods) and backhauling (items to be returned to the depot). In addition, deliveries and pickups can occur in any sequence and a customer may be visited more than once by the same vehicle. Thangiah et al. [232] considered time windows and real-time events such as deletion, insertion, and modification of a shipment. A detailed survey of SDVRP might be found in [110] or [7].

2.4.2.1.5 Time-Dependent VRP

Another variant of VRP is Time-Dependent VRP (TDVRP) proposed by Ichoua et al. [120], in which the time of travel between two locations depends not only on the distance but also on other factors including time of a day (models **usually** assume constant time). Owing to this constraint, the problem is more similar to real-world conditions. There are several papers presenting solutions for real-world networks in Padua [65], Torino [160], or Quebec [239]. Besides, TDVRP is also considered with time windows corresponding with customer requirements [84].

2.4.3 Capacitated Vehicle Routing Problem

2.4.3.1 Mathematical formulation

In Section 2.3, we focused on the TRAVELING SALESMAN PROBLEM (TSP), in which we have a set of nodes V and a set E of arcs between all the pairs of nodes annotated by distances between nodes. The problem is finding the shortest route starting at some node $x \in V$, visiting every node from V, and returning to x.

CAPACITATED VRP (CVRP) is a natural generalization of TSP. Again, we have $V = \{0, 1, \dots, n\}$ nodes, we denote node 0 as the *depot* and the remaining nodes $V' = \{1, \dots, n\}$ as *customers*. Each customer has an assigned demand, a non-negative amount of product d_i. The main difference from TSP is that instead of one vehicle, we operate a fleet of $F = \{1, \dots, f\}$ vehicles. In the standard formulation of $CVRP$ [57] the fleet is homogeneous, which means that all the vehicles are of the same type and have the same capacity Q ($d_i \leq Q$ for each customer i). A *route* defined as $R = \{v_0, v_1, \dots, v_k\}$ is an ordered list of customers $v_1, \dots, v_{k-1}$ ($k \leq n$ and $v_i \in V'$ for $i \in \{1, \dots, k-1\}$) starting and ending in a node $v_0 = v_k = 0$ (depot).

The goal is to deliver some goods from depot to each customer in the optimal way, that is, to find a set of routes S such that every customer v belongs to

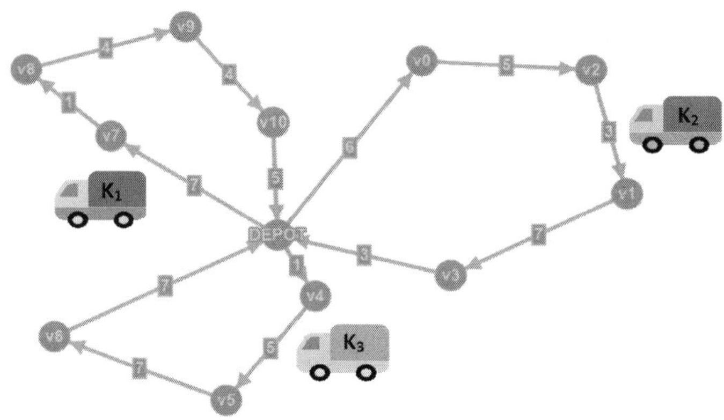

FIGURE 2.1 Example of a solution of CVRP. The solution is composed of three routes for three different vehicles, which are labeled as K_1, K_2, K_3. Nodes $v(i)$ (for $i \in \{0 \ldots 10\}$) symbolize the customers locations. The number on each edge is a cost of a given travel connection.

exactly one route $R \in S$, the number of routes is less than or equal to the size of a fleet f, and the capacity of vehicles is never exceeded (the sum of demands of all customers belonging to one route is less than or equal to Q). We want to achieve our goal at the lowest possible cost. The optimized *cost* of some feasible solution S may differ throughout specific applications. The most common cost, originating from TSP, is the distance traveled by all vehicles [57]. Vehicles and drivers usually also generate some fixed costs, and we may want to minimize some function $\sigma(d, f)$, where d is the total distance, and f is the total number of vehicles in the fleet.

Fig. 2.1 presents an example solution of a CVRP instance.

2.4.3.2 Exact solutions

VRP is NP-hard as TSP is (any instance of TSP can be easily transformed in polynomial time to a VRP instance), but VRPs are in general more difficult to solve than TSPs of the same size, as the number of possible solutions is much larger for large fleets. Since 1960, many papers introducing various methods to solve CVRP have been published, and it was quickly noticed that the exact algorithms running in exponential time could not solve big instances of CVRP fast enough. However, some relatively slow algorithms (presented in this section) are sufficient to solve particular real-live scenarios when we consider constant increase in processing speed.

Laporte and Nobert [142] summarized the achievements of the 1960s, 1970s, and 1980s by dividing all exact approaches into the following groups:

1. Direct tree search methods.
2. Dynamic programming (DP).
3. Integer Linear Programming (ILP):
   a. Vehicle flow formulations.

b. Commodity flow formulations.
c. Set partitioning formulations.

This section briefly describes the most important ones.

2.4.3.2.1 Dynamic programming

Let us define the cost $c(R)$ as the sum of distances between consecutive nodes on route R and the cost function of CVRP solution as $\sum_{j=1}^{f} c(S_j)$, where $\{S_1, S_2, \ldots, S_f\}$ is a partition of V'. The classic dynamic approach is to compute function g defined as follows:

$$g_k(U) = \begin{cases} c(U), & k = 1, \\ \min\limits_{U' \subseteq U \subseteq V'} [g_{k-1}(U - U') + c(U')], & k > 1. \end{cases} \tag{2.17}$$

Then $g_f(V')$ stands for the solution cost. The main problem in this approach is an enormous number of states.

Christofides et al. [49] introduced the state-space relaxation method for relaxing the dynamic programming recursions to reduce the number of states and obtain valid lower bounds on the value of the optimal solutions.

2.4.3.2.2 Integer linear programming

Integer linear programming, compared to linear programming, requires an additional constraint that some or all the variables must be integer. In this approach, both the objective function and the constraints are linear (apart from integer constraints) [142]. The main advantage of this method is that we can easily extend formulas and add some constraints to enhance a given problem with additional conditions (e.g., time windows).

2.4.3.2.2.1 Vehicle flow formulations These formulations use binary variables to indicate whether a vehicle travels between two given cities in the optimal solution. We distinguish two families of the vehicle flow formulations: three-index formulations and two-index formulations. The example of three-index formulations proposed by Golden et al. [106] included the following variables:

$$x_{i,j,k} = \begin{cases} 1 & \text{if vehicle } k \text{ passed route from } i \text{ to } j, \\ 0 & \text{otherwise,} \end{cases} \tag{2.18}$$

$$c_{i,j,k} = \text{the cost of using vehicle } k \text{ from } i \text{ to } j, \tag{2.19}$$

and many conditions (that we do not list here) to achieve the required properties. In two-index formulas, we define the variables

$$x_{i,j} = \sum_{l=1}^{k} x_{i,j,k},$$

meaning that *there is a vehicle traveling **directly** from i to j*, and we assume that the cost $c_{i,j}$ of travel from i to j is equal for each vehicle k (see, e.g., [88]). Conditions for two-index formulations are obviously different from those for three-index formulations.

2.4.3.2.2.2 Set partitioning formulas Agarwal et al. [2] defined this problem as follows: the route of vehicle j is represented by a binary n-vector a_j, and $a_{i,j} = 1$ iff i $(i = 1, 2, \ldots, n)$ is visited on route a_j. Cost c_j is associated with traveling route a_j. We can apply an arbitrary algorithm to solve TSP to calculate c_j.

The goal is to minimize $\sum_j c_j \cdot x_j$ subject to $\sum_j a_j \cdot x_j = e$ and $x_j = 0$ or 1 for all j, where e is the n-vector of all ones, the binary variable x_j determines the inclusion of the route represented by the column a_j in the solution, and the constraints ensure that each stop is visited on exactly one route.

Agarwal et al. [2] proposed a column-generation algorithm. Let us define A as the matrix constructed from vectors a_j. The idea is to avoid operating on a whole matrix A and rather construct it by starting with an empty matrix and adding particular columns on the fly.

Many works tried to merge these methods to get better results (e.g., the branch-and-cut-and-price algorithm by Fukasawa et al. [95] and Pessoa et al. [201]). Most of the formulations have been aggregated by Baldacci et al. [18], where reader can find references to many more works.

2.4.3.3 Heuristics for CVRP

Similarly as in case of TSP (Section 2.3.4), there are several heuristics proposed to solve the CVRP and its variants in the literature. Toth and Vigo [234] gave a detailed survey on VRP, which describes both exact and heuristic methods for VRPs up to 2002, whereas Laporte and Semet [143] divided heuristics for the CVRP into three groups:

1. Constructive heuristics.
2. Two-phase heuristics.
3. Improvement methods.

2.4.3.3.1 Constructive heuristics

In case of constructive heuristics, two main techniques are used, *merging existing routes using a savings criterion* and *gradually assigning vertices to routes using an insertion cost.*

The base method applying savings criterion is the *Clarke–Wright algorithm (CW)* [50]. The idea is to start from an initial set of routes, compute savings of adding each link, and process savings in a descending order if only it is possible to add a new link related to the given saving. There are sequential and parallel versions of the algorithm [143]. One of its drawbacks is that it tends to produce good routes at the beginning but less interesting routes toward the end, including some circumferential routes. Therefore generalized savings including a route shape parameter were introduced [143,99,254]. The larger the parameter, the more the emphasis put on the distance between the vertices to be connected, and it was reported that values 0.4 and 1 yield good solutions, taking into account the number of routes and the total length of the solution [106]. Golden et al. [106] also presented a way to increase efficiency of the algorithm by sorting savings using a heap structure and superimposing a grid over the network. A similar approach, using heaps to limit storage requirements and thus obtaining more efficient updating operations, was presented in [183], whereas Paessens [193] proposed disregarding edges with a cost too high comparing to other edges. Desrochers and Verhoog [61] and Altinkemer and Gavish [4] introduced a modification of the standard savings approach by matching paths corresponding to solutions of TSP on a given subset. Similarly, Wark and Holt [246] merge clusters defined as ordered sets of vertices at their endpoints. After a merge is performed, only a few lines or columns of the savings matrix need to be updated. If all clusters are matched, then some of them are split with a given probability. The process thus grows a tree of sets of clusters from which the best solution is selected.

The presented merging approaches [61], [4], [246] are compared by Toth and Vigo [234]. The Wark and Holt heuristic is clearly the best of the three matching-based methods in terms of solution quality, but not necessarily in terms of the required time of computations. In general, use of a matching-based algorithm yields better results than the standard CW, but at the expense of much higher computation time.

In case of the heuristics gradually assigning vertices to vehicle routes using an insertion cost, Laporte and Semet [143] investigate two representative algorithms proposed by Mole and Jameson (MJ) [171] and Christofides, Mingozzi, and Toth (CMT) [48]. Both apply to problems with an unspecified number of vehicles [143]. The MJ algorithm expands one route at a time. For each unrouted vertex k, it computes the feasible insertion cost and optimizes the current route by means of a 3-opt procedure [151]. A similar optimization approach (3-opt procedure) is used in CMT along with sequential and parallel route construction procedures. The comparison presented there shows that this heuristics is superior to MJ and CW but requiring about twice the computing time (however, the values obtained with the CMT heuristic are in general far from the best known).

2.4.3.3.2 Two-phase heuristics

In case of two-phase heuristics the problem is decomposed into its two natural components, clustering of vertices into feasible routes and actual route construction, with possible feedback loops between the two stages. There are two groups of such methods: *cluster-first route-second* and *route-first cluster-second*. Among cluster-first methods, we can distinguish *elementary clustering methods*, *truncated branch-and-bound methods,* and *petal algorithms*. In case of elementary clustering methods, there are again three groups of methods:

sweep algorithms, *generalized-assignment-based algorithm*, and *location-based heuristic*.

The sweep algorithm applies to planar instances of VRP. Feasible clusters are initially formed by rotating a ray centered at the depot, and a vehicle route is then obtained for each cluster by solving TSP. Some implementations include a postoptimization phase, in which vertices are exchanged between adjacent clusters and routes are reoptimized [144]. Examples of such sweep algorithms are included in [250,251,103].

The generalized-assignment-based algorithm invented by Fisher and Jaikumar [86] assumes a fixed value of the number of vehicles K and divides the plane into K cones according to the customer weights. The seed vertices correspond to dummy customers located along the rays bisecting the cones. Once the clusters have been determined, the TSPs are solved optimally using a constraint relaxation-based approach [168].

The representative example of *location-based heuristics* is Bramel and Simchi-Levi Algorithm [35]. The authors suggest first locating K seeds, called concentrators, among the n customer locations to minimize the total distance of customers to their closest seed while ensuring that the total demand assigned to any concentrator does not exceed Q. Vehicle routes are then constructed by inserting at each step the customer assigned to that route seed having the least insertion cost. Since the calculation of the insertion cost may be time consuming, two approximations are considered: using a direct cost and the nearest-insertion cost. In case of the first approach, the authors showed that the algorithm is asymptotically optimal [144].

Truncated branch-and-bound algorithms gradually add new levels of a tree. First, they generate all possible routes constructed by adding at each level a selected unrouted customer to vertices from the previous level. Then they evaluate possible routes with that customer by solving an instance of TSP and adding the estimated (using the shortest spanning tree) cost of all other routes. Finally, the best such route is selected, and its vertices are excluded from the further search [144].

Petal algorithms are extensions of sweep algorithms. They generate several routes, called petals, and make a final selection by solving a set partitioning problem [144]. The algorithm produces better results in shorter time on benchmark datasets.

The *route-first* methods construct in the first phase a giant TSP tour, disregarding side constraints, and decompose this tour into feasible vehicle routes in a second phase [23]. We are not aware of any experiments showing advantages of these methods over other approaches.

2.4.3.3.3 Improvement heuristics

Improvement methods attempt to upgrade any feasible solution by performing a sequence of edge or vertex exchanges within or between vehicle routes. Such algorithms may operate on each vehicle route taken separately or on several routes at a time. In the first case, any improvement heuristic for TSP can be applied, and most of them can be described in terms of Lin's λ-opt mechanism [151]. Here λ edges are removed from the tour, and the λ remaining segments are reconnected in all possible ways. The first (or the best) profitable reconnection is implemented, and the procedure stops at a local minimum when no further improvements can be obtained.

In case of multiroute improvements. Thompson and Psaraftis [233] describe a general b-cyclic k-transfer scheme in which a circular permutation of b routes is considered, and k customers from each route are shifted to the next route of the cyclic permutation. Van Breedam [237] classified the improvement operations as string cross (two strings of vertices are exchanged by crossing two edges of two different routes), string exchange (two strings of at most k vertices are exchanged between two routes), string relocation (a string of at most k vertices (usually, $k \in \{1, 2\}$) is moved from one route to another), and string mix (the best move between string exchange and string relocation is selected). Overall, string exchange moves seem to be best.

2.4.3.3.4 Conclusions

Several presented heuristics can be easily adapted to other variants of the VRP and are easy to implement, so they are often used in practice. The Clarke–Wright algorithm is probably the most popular method; when followed by the BI version of 3-opt, it produces very fast solution values at most 7% worse than the best known results. Better performances are observed with some other algorithms (e.g., 2-petal algorithm), but time of coding and computations may be larger.

2.4.3.4 Metaheuristics for CVRP

Tabu Search (TS), Evolutionary Algorithms (EA), and Simulated Annealing (SA) are frequently applied to CVRP, especially as a component of a hybrid metaheuristic; see, for example, [101].

An enhanced version of the Artificial Bee Colony (ABC) metaheuristic for solving the CVRP was introduced by Szeto et al. [227]. ABC is a swarm-based heuristic, a fairly new approach at the time of writing the paper [227], and at that time, it had not yet been applied to solve the CVRP mainly because of the computational performance aspects. The authors modified the algorithm by introducing the control of the quality of the food sources and by reducing its restrictions to prevent the heuristic from searching inadequate areas of the solution space. The modified algorithm was tested and compared with other metaheuristics, and it proved to produce much better solutions than simple ABC with only a slight increase in computation time.

Berger and Barkaoui [29] applied a hybrid genetic metaheuristic to CVRP. The concept is based on two concurrently evolving populations of solutions to minimize the total traveled distance. The authors combined operators typical for GA with various concepts inspired from routing techniques and search strategies, which enabled both search guidance and the control over algorithm intensification and diversification. The proposed approach has been tested and compared with other most commonly used methods, and it seems to be very competitive and promising in terms of computational aspects.

Ngueveu et al. [184] proposed an efficient memetic metaheuristic for solving CVRP using nontrivial evaluations of cost variations in the local search. Authors introduced the first upper and lower bounding procedures: the lower bounds were derived from CVRP properties, whereas the upper bounds were obtained by a memetic algorithm and its nontrivial evaluations of local search moves cost. The research showed that the proposed method can be applied to CVRP with good results.

Ant colony systems (ACS) are also used to solve a Multi-Compartment Vehicle Routing Problem (MCVRP), an extension of the classical CVRP in which different products are transported in multiple compartments of one vehicle. For example, [1] used hybridized ASC to solve MCVRP. Their approach included a hybridized metaheuristic combining ASC with local search. On the other hand, Reed et al. [207] focused on the collection of recycling waste from households sorted into separate compartments for paper, glass, and so on. An implemented ACS algorithm resulted in high-quality solutions for two-compartment test problems.

As for nonnature-inspired heuristics, Rego and Roucairol [208] proposed a Parallel TS algorithm for CVRP with distance restrictions. Their approach uses compound moves in a neighborhood area that are generated by the ejection chain process. The authors employ different parallel methods to search the solution space more extensively. Another parallel TS algorithm for solving CVRP was proposed by Jin et al. [122]. Their approach uses several different neighborhood structures that can cooperate with each other and generate the collective power of multineighborhood system.

2.4.4 Vehicle Routing Problem with time windows

2.4.4.1 *Definition*

The vehicle routing problem with time windows (VRPTW) can be defined as follows: let $G = (V, E)$ be a directed graph with vertices V and edges E; V is composed of $C + 1$ vertices, where v_0 represents the depot (starting point), and $v_i \in \{1, \ldots, C\}$ represent the customers; E is a set of arcs, which represents connections (roads) between the travel points. Each arc (i, j) $(i \neq j)$ has a cost c_{ij} and time t_{ij}. It is assumed that $c_i \geq 0$ and $t_i > 0$ for all i.

Each customer i has a demand d_i, and its service must start in a specified time interval called *the time window* $[e_i, l_i]$. It is assumed that $e_i \geq 0$ and $l_i \geq 0$ for all i. The vehicle can arrive at customer i before the opening time e_i of time window, but it must wait to reach time e_i before the delivery could be performed. The vehicle must arrive at costumer i before time l_i.

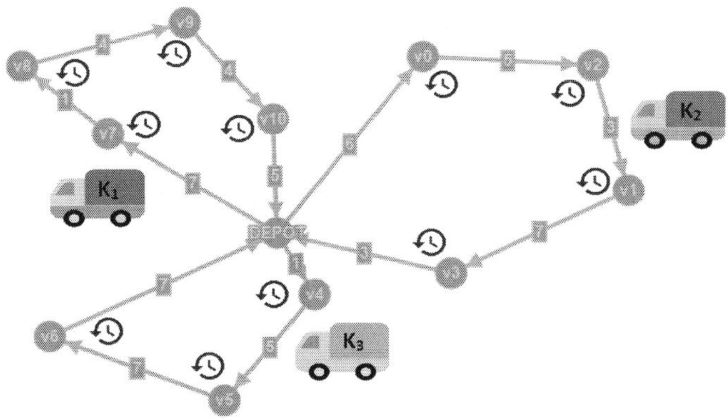

FIGURE 2.2 Example of a solution of VRPTW. The solution is composed of three routes for three different vehicles. The vehicles are labeled K_1, K_2, K_3. Node labeled DEPOT represents the starting point for all vehicles. Nodes with labels $v(i)$ (for $i \in \{0, \ldots, 10\}$) symbolize the costumers' locations. The number on each edge is a cost of a given travel connection. The clock icons symbolize the time windows constraint for each node.

Denote a set of vehicles by K. All vehicles $k \in K$ are homogeneous, that is, they have the same maximal capacity Q, which cannot be exceeded by any vehicle.

Finally, let f be a function defined as follows:

$$f(R) = \sum_{i=1}^{|K|} x_i^R, \tag{2.20}$$

where x_i^R is a cost of route R_i for vehicle i ($x_i = 0$ if vehicle i is not included) for a given solution R. The cost of a route is a function that depends on costs c_{ij}, travel distances d_{ij}, and travel times t_{ij} on edges. The goal of VRPTW is to minimize the goal function f and satisfies all constraints. Fig. 2.2 presents an example solution for VRPTW.

2.4.4.2 *Exact solutions*

The exact methods for all types of VRP are similar to the methods for CVRP and were briefly summarized in Section 2.4.3.2. We also can find examples of using these methods in time windows version of VRP in El-Sherbeny [77] and Baldacci et al. [17].

2.4.4.3 *Heuristics for VRPTW*

Since VRPTW is NP-hard problem [215], the exact algorithms are too time-consuming for its large instances, so multiple heuristics and metaheuristics have been developed.

El-Sherbeny [77] divides heuristics for the VRPTW into two groups:

1. Route-building heuristics.
2. Route-improving heuristics.

Cordeau et al. [52] adds one more type, composite heuristics.

In case of route-building heuristics, similarly as in case of CVRP (Section 2.4.3.3), we start from shorter routes (i.e., not visiting all required customers) and extend them by adding new points or combining shorter routes into longer routes. On the contrary, algorithms from the second group operate on sets of potential solutions (i.e., sets of routes that visit all customers) and gradually improve them to generate even better solutions.

2.4.4.3.1 Route-building heuristics

In the first paper on route-building heuristics for the VRPTW, Baker and Schaffer [12] extend the savings heuristic of [50]. The algorithm begins with all possible single-customer routes (depot-to-depot) and calculates in every iteration which two routes can be combined with the maximum saving. A time-oriented nearest-neighbor algorithm is developed by defining the savings as a combination of distance, time, and time until feasibility [77].

Van Landeghem [238] adapts the savings heuristic from Baker and Schaffer [12] and uses the time windows to get a measurement of how good a link between customers is in terms of timing.

A similar heuristic based on the savings algorithm is developed by Solomon [221], but the time aspect is not a part of the savings function: the arcs that can be used are limited by how large the waiting times become when they are used. Due to the existence of time windows, a route orientation should be taken into account, and a violation of the time windows should be checked when two routes are combined. In the first phase of the algorithm, each unrouted customer is assigned to its best feasible insertion position based on the minimum additional distance and time required. In the second step the method selects the customer to insert using a maximum savings concept. These heuristics have time complexity $O(n^2 \log n^2)$ [77]. A parallel version building several routes simultaneously was also developed by Potvin and Rousseau [202].

Another heuristic described by Solomon [221] is a time-oriented nearest-neighborhood heuristic. Every route is started by finding the unrouted customer closest to the depot with respect to the geographical and temporal distance. At every iteration the customer closest to the last customer added to the route is considered for insertion to the end of the presently generated route. When the search fails, a new route is started.

In general, the heuristics presented by Solomon [221] and Van Landeghem [238] return solutions fast, but their solutions are generally more than 10% from optimum.

In approach of Antes and Derigs [5], every unrouted customer requests and receives from every route a prize for insertion, defined in a similar way as in

[221]. Then the unrouted customers send a proposal to the route with the best offer, and each route accepts the best proposal among those customers with the fewest number of alternatives. If a certain threshold of routes is violated, then a certain number of customers is removed, and the process is initiated again. Generally, building several routes in parallel as in [5] and [202] gives better solutions than building the routes one by one.

2.4.4.3.2 Route improvement heuristics

In case of *route improvement* heuristics, the current solution is modified by performing local searches for better neighboring solutions. A neighborhood consists of solutions that can be reached from the present one by r-exchange, swapping a subset of r arcs between solutions, which is implemented only if it leads to an improved feasible solution (the procedure stops when no improving exchange is possible). In [212,51,12], small r (e.g., 2, 3) were tested, which led to large number of neighborhoods and effective but time-consuming methods. To accelerate the method, OR-opt exchanges considering only currently adjacent customers can be applied [188]. Solomon et al. [222] present a method accounting for the time orientation of a route, where at each iteration, up to three adjacent customers are shifted to a later position on the same route between two currently consecutive customers. Schedule shifts are also used to accelerate the screening of unfeasible solutions. Approach of Thompson and Psaraftis [233] attempts exchanges among a subset of routes that form a cyclic permutation. The authors implemented a method based on 2- and 3-cyclic 1-transfers.

2.4.4.3.3 Composite heuristics

Composite heuristics combine route building and improving heuristics. Kontoravdis and Bard [135] combine a greedy heuristic and randomization to produce in parallel initial routes, which are then improved through local search (certain routes may be eliminated). The authors also propose three lower bounds for the fleet size: two of them are based on bin packing structures generated by the capacity or time window constraints, and the third one is derived from the associated graph created by pairs of customers who have incompatible demands or time windows [52]. Russell [212] developed a procedure that embeds route improvement within the tour construction process. He proposes switching customers between routes and the elimination of routes during the construction process. Cordone and Calvo ([54], [38]) used similar ideas in the design of a composite heuristic, where local search is performed hierarchically. First, within a classical 2- and 3-opt exchange framework, he attempted to decrease the number of routes by moving a route into others, one customer at a time. Second, another heuristic was used to try to step away from a local optimum. This procedure resolves the problem with a partly modified objective function since the current solution may not be a local optimum for the related objective [52].

2.4.4.4 Metaheuristics for VRPTW

Simulated Annealing (SA), Evolutionary Algorithms (EA), and Tabu Search (TS) are commonly used in solving VRPTW [101,77], usually combined together and used as hybrid metaheuristics.

Three different SA methods to solve VRPTW were proposed by Chiang and Russell [44]. The first two of them assume different neighborhood structures, whereas the third one introduces a tabu list representing a hybrid approach. El-Sherbeny [76] proposed a multiobjective SA method focusing on three categories of objectives: vehicles, time, and duration. The first category includes the total number of vehicles and the number of covered/uncovered vehicles. The second category concerns time aspects such as the total duration of travel or working time, whereas the third category involves flexible duration of travel.

Genetic algorithms (GA) and ant colony systems (ACS) are very popular evolutionary algorithms used in solving VRPTW. GA were applied by Ombuki et al. [187], who interpreted VRPTW as a multiobjective problem with two objective dimensions: the number of vehicles and the total cost defined as distance, both of equal importance. Such an approach does not require to derive weights for scoring formula and consequently prevents the solution from the bias toward any dimension. A problem was solved using GA and the Pareto ranking technique and resulted in the set of good solutions considering both dimensions. A similar approach was proposed by Gambardella et al. [98], who introduced a multiple ACS consisting of two artificial ant colonies arranged hierarchically and designed to successively optimize a multiple objective function. Each colony included in the system was responsible for minimizing different objectives, the number of vehicles and the total traveled distances, assuming the cooperation between colonies.

Chiang and Russell [45] developed a reactive TS for solving VRPTW, which dynamically adjusts the size of the tabu list, based on the current search process, to avoid cycles and excessively constrained search paths. The algorithm also includes intensification and diversification that leads to higher quality solutions.

Concerning hybrid methods, Banos et al. [21] proposed a new Pareto-based hybrid metaheuristic, *multistart multiobjective evolutionary algorithm with simulated annealing* (MMOEASA). The algorithm combines SA and EA to provide the set of high-quality nondominated solutions. Those solutions might be later analyzed by the decision maker in terms of given criteria. The approach uses mutation and crossover operations typical for the evolutionary algorithm to improve the population of solutions and SA as the acceptance criterion. The solutions get accepted or rejected in accordance with a modified Pareto-dominance criterion, which includes the current temperature and the Metropolis function. The algorithm stores already found nondominated solutions and works in a multistart manner, that is, the process continues to improve already obtained solutions, and the approach itself aims to improve the efficiency of a Pareto-based multiobjective EA by using SA as the acceptance criterion. Gehring and Homberger [100], in turn, proposed a parallel two-phase procedural approach based on EA and

TS for solving the VRPTW. The first phase involves EA in order to minimize the number of vehicles, whereas the second one uses TS, and its aim is to minimize the total distance. Both metaheuristics create sequential, parallel hybrid procedures following the concept of cooperative autonomy, that is, various solution procedures cooperate with each other by exchanging the solutions. If the given acceptance conditions are met, then this cooperation leads to considerable changes in the solution space.

2.4.5 Pickup and Delivery Vehicle Routing Problem

The Pickup and Delivery Problem (PDP) is an extension of VRP that focuses on finding the set of routes for a fleet of vehicles that has to satisfy transportation requests. Each request is composed of the pickup location, the size of the loaded transport, and the delivery location [119]. If, additionally, each request has a specified time window for the pickup, and for the delivery, the problem becomes Pickup and Delivery Problem with Time Windows (PDPTW). The main objective of the problem is to minimize the size of the vehicle fleet and to optimize the travel distance.

2.4.5.1 Definition

The Pickup and Delivery Problem can be modeled mathematically using a directed graph $G = (V, E)$ [216]. In this setting the set of vertices can be defined as $V = V^* \cup M$, where V^* is the set of all pickup locations and all delivery destinations, whereas M is the set for all start and end locations for k vehicles. Each vehicle k has a capacity $Q_k > 0$. If Q_k is constant for all k, then PDP is said to model a homogeneous fleet of vehicles.

Let E be a set of edges that represents travel connections (roads) between the travel locations in V. Each edge (i, j), $i \neq j$, has a cost c_{ij}, travel distance d_{ij}, and travel time t_{ij}. It is assumed that $c_{ij} \geq 0, t_{ij} > 0$, and $d_{ij} > 0$ for all i, j.

Denote by N^+ the set of all origins and by N^- the set of all destinations. Let N be the total number of transportation requests. Similarly, let M^+ and M^- represent the starting and end locations, respectively. Finally, denote by R_k the route defined for a vehicle k that goes through $V_k \subset V$. R_k is said to be *well defined* if:

1. R_k starts in k^+ and finishes in k^-.
2. Vehicle k load never exceeds Q_k.
3. If there exists $i \in N$ such that $(N_i^+ \cup N_i^-) \cap V_k = (N_i^+)$ or $i \in N$ such that $(N_i^+ \cup N_i^-) \cap V_k = (N_i^-)$, then $(N_i^+ \cup N_i^-) \cap V_k = (N_i^+ \cup N_i^-)$, and the location N_i^+ is visited before N_i^-.

Condition 3 states that if vehicle k visits a pickup locations N_i^+, then it has to visit the corresponding delivery location N_i^-. The set $R = \{R_k : k \in K\}$ is a *solution* to the PDP if

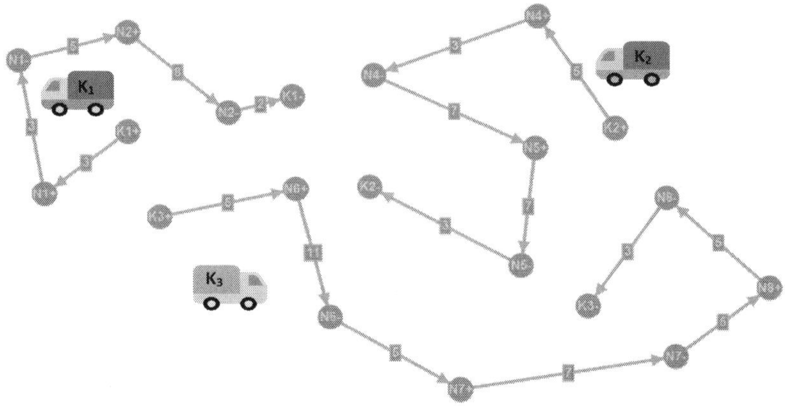

FIGURE 2.3 Example of a solution of PDP. The solution is composed of three routes for three different vehicles, denoted by K_1, K_2, and K_3. Nodes with the label $N(i)+$ symbolize the pickup location, and nodes with the label $N(i)-$ represent the corresponding delivery location ($\forall i \in \{1, \ldots, 8\}$). Nodes with the label $K(i)+$ represent the starting point of the route for the vehicle i. Nodes with the $K(i)-$ symbolize the end destination point for vehicle i. The number on each edge serves as the cost of given travel connection.

1. for all k, R_k is a well-defined route.
2. $\bigcup_{i=1}^{K} V_k = V$.

The main problem for the PDP is to minimize the size of M and the sum of travel distances for a given R [32], which can be expressed as

$$\min f(R) = \sum_{i=1}^{K_R} x_i, \tag{2.21}$$

where K_R is the size of a fleet, and x_i is the cost of route for vehicle i, which is in fact a function depending on cost c_{ij}, travel distance d_{ij}, and travel time t_{ij} [32]. Fig. 2.3 presents an example solution for PDP.

The definition of PDP can be extended to the definition of Pickup and Delivery Problem with Time Windows (PDPTW) by adding the following constraint to the definition of R_k:

3. For each $i \in V$, vehicle k must arrive at point i in a time window $[e_i, l_i]$.

This constraint states that each request $i \in V$ has to be realized in a specified time interval called time window $[e_i, l_i]$. It is assumed that $0 \le e_i \le l_i$ for all i. The vehicle can arrive at a given location i before the opening time e_i of time window, but it must wait to reach time e_i before the delivery could be performed. The vehicle must arrive at location i before time l_i. Fig. 2.4 presents an example solution for PDPTW.

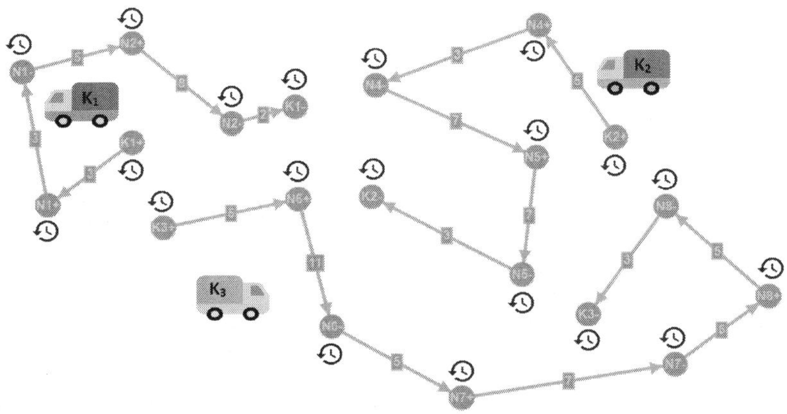

FIGURE 2.4 Example of a solution of PDPTW. Additionally to notation from Fig. 2.3, the clock icons symbolize the time windows constraint for each node.

2.4.5.2 *Heuristics for PDPTW*

Recall that exact methods usually are used for small problems (due to long computation time). Exact algorithms include dynamic programming techniques, column generation methods, branch-and-cut, branch-and-price solvers, and more [32]. Among approximate methods, the local search heuristics are commonly used to solve PDP or PDPTW.

The heuristics operators can be classified as follows [229]:

1. Intraroute operators, for example, 2-opt or rearrange operator.
2. Interroute operators, for example, an exchange or shift operator.

Ropke and Pisinger [210] used the heuristic called Adaptive Large Neighbor Search (ALNS) to solve PDPTW. It is an extended version of Large Neighbor Search (LNS), which removes and reinserts requests for a given solution to find a better solution. ALNS uses various removal and insertion heuristics. The removal heuristics use relatedness measure (the measure of similarity between two requests), worst removal measure (the cost of solution without the given request), or random selection. The insertion heuristic may use the regret heuristic, which improves the basic greedy algorithm by inserting the request at the position with the lowest cost.

2.4.5.3 *Metaheuristics for PDPTW*

The most common approach to solve PDP or PDPTW with approximate algorithms is a two-step process. First, the fleet size is minimized, and second, the cost function that depends on the total distance is optimized. This two-step approach permits to distinguish two various sets of algorithms for each step. Most common solutions implement guided ejection search for the first step and memetic algorithms for the second part.

2.4.5.3.1 Guided ejection search

The Guided Ejection Search (GES) focuses on minimizing the size of the fleet K. The baseline version of GES starts by randomly selecting a route and making unassigned all pickup and delivery pairs from it. For a fixed number of iterations, the algorithm tries to reinsert one of the unassigned pickup and delivery pairs into an existing route. If reinsertion is not possible, then the algorithm tries to eject some pairs from the fixed size set of pairs (k_{max}) of already assigned pairs. After reinsertion with the ejection, the solution is disarranged by randomly moving or swapping pairs between routes. This operation helps to reduce the risk of cycling and to diversify the search [55].

Parallel Guided Ejection Search (PGES) is an enhanced version of GES. PGES is formed of p components that are executed in parallel as the processes $P_1, P_2, \ldots, P_p$. The master process P_1 is responsible for stopping the iteration. The information about the termination is transferred to all others processes. Implementation presented by Curtois et al. [55] uses the ring and randomized ring cooperation. The ring topology guarantees that each P_i contains at least as good solution as during the previous cooperation. The cooperation process allows us to explore deeper the search space, which significantly improves the quality of final solution. PGES performs better for models with tight windows and smaller capacities than for problems with wide time windows and larger capacities.

Adaptive Guided Ejection Search is an improved version of GES, which uses preprocessing to determine the appropriate PDPTW problem instance class. The following classes of PDPTW are distinguished [179]:

1. Clustered customers with tight time windows and small vehicle capacities ($c1$).
2. Clustered customers with wide time windows and large vehicle capacities ($c2$).
3. Randomly scattered customers with tight time windows and small vehicle capacities ($r1$).
4. Randomly scatter customers with wide time windows and large vehicle capacities ($r2$).
5. Mixed customers with tight time windows and small vehicle capacities ($rc1$).
6. Mixed customers with wide time windows and large vehicle capacities ($rc2$).

A mixed class is composed of clustered and randomly scattered customers. Selection of appropriate GES affects the convergence efficiency of the algorithm. Each class requires different improvement version of GES. Nalepa and Blocho [179] identified the following possible improvements of GES for PDPTW:

1. S, squeezing of unfeasible solution.
2. R, randomizing the ejection pool.
3. M, imposing the maximum iteration limit on analysis of hard to inject requests.

Table 2.2 shows the best GES improvements for each problem instance class.

TABLE 2.2 Class and GES improvements.

Class	c1	c2	r1	r2	rc1	rc2
Minimize K	R	R	R	MR	R	M

Table shows best GES improvements for each class.

The pessimistic time complexity of the parallel algorithm for PDPTW which uses GES is $\mathcal{O}(n^{k_{\max}+2} + pn)$, where n is the data size input, p is the number of processors, and $k_{\max}$ is the number of ejected requests [31].

2.4.5.3.2 Memetic Algorithm

The Memetic Algorithm (MA) is used in the second step of finding the solution for PDP. It is composed of four procedures, which are repeated until the termination condition is not satisfied. Those procedures are selection, crossover, education, and mutation [178].

It is shown by Nalepa and Blocho [178] that the AB-selection [180] has high exploration capabilities. The AB-selection chooses two routes r_A and r_k, where $r_A \neq r_B$. Each route can be selected once for both of the roles. The time complexity of AB-selection is $\mathcal{O}(N)$, where N is the number of routes in the initial solution [178].

For the crossover part of the MA, there are various operators proposed, for example:

1. Edge Assembly Crossover Operator (EAX).
2. Selective Route Exchange Crossover (SREX).
3. Longest Common Subsequence based SREX (LCS-SREX).
4. Parallel LCSSREX (PLCSSREX).

The EAX operator was mainly used for VRPTW. It combines two solutions, R_A (with graph representation $G_A = (V_A, E_A)$) and R_B (with graph representation $G_B = (V_B, E_B)$), by creating a new solution R_{AB} ($G_{AB} = (V_{AB}, E_{AB})$). The graph representation of R_{AB} is composed of edges from the set

$$E_{AB} = (E_A \cup E_B) \setminus (E_A \cap E_B).$$

All edges from E_{AB} are divided into AB-cycles, where the edges from the graph G_A are traced in the forward direction, and the edges from the graph G_B are traced reversely. A new set $G_S = (V_S, E_S)$, composed of one or more AB-cycles, is created and used to construct intermediate solution that is constructed from edges from $(E_A \cap E_S) \cup (E_B \cap E_S)$. The intermediate solution can be composed of the subroutes that are merged with other routes by using 2-opt moves [178]. The EAX has time complexity $\mathcal{O}(M^2)$, where M is the total number of requests [178].

The SREX replaces some routes from parent R_A with the routes from parent R_B. It was used both for PDP and PDPTW, but its enhanced version (LCS-

SREX) performs better for the time-dependent problem [32]. In LCS-SREX the subset of routes from R_B is created by considering the longest common subsequence (LCS) value for the route represented as the arc sequence.

The repair, education, and mutation procedures use hill climbing techniques such as 2-opt (the time complexity of this operator is equal to $\mathcal{O}(1)$), out-exchange (the time complexity of this operator is equal to $\mathcal{O}(1)$), out-relocate (the time complexity of this operator is equal to $\mathcal{O}(1)$), in-exchange (the time complexity of this operator is equal to $\mathcal{O}(m)$), in-relocate (the time complexity of this operator is equal to $\mathcal{O}(1)$), or GENIUS-exchange (the time complexity of this operator is equal to $\mathcal{O}(1)$), where m is the route size [178].

2.5 Conclusions

Based on the conducted analysis, we conclude that all routing problems, ranging from ORC to VRP, are already well studied, and the amount of the existing literature is abundant. However, in some cases, there are differences in notations and even interpretation of the same concepts, so the metaresearch works aiming to standardize different approaches and objectively compare results are still needed. Also, in Section 2.3, we introduced algorithms that can be run on quantum computers, and we believe that once such machines become more reliable, they can enable developing better algorithms for combinatorial optimization problems such as TSP or VRP and their variants. Thus approaches based on quantum computers seem to be the natural and most disruptive research directions in the optimal fleet routing problems.

Acknowledgments

The presented research was carried out within the frame of the project "Green LAst-mile Delivery" (GLAD) realized at the University of Warsaw with the project partners: Colruyt Group, University of Cambridge and Technion. The project number: 19147. The project is partially supported by EIT Food, which is a Knowledge and Innovation Community (KIC) established by the European Institute for Innovation & Technology (EIT), an independent EU body set up in 2008 to drive innovation and entrepreneurship across Europe.

eit Food

EIT Food is supported by the EIT
a body of the European Union

References

[1] M.M. Abdulkader, Y. Gajpal, T.Y. ElMekkawy, Hybridized ant colony algorithm for the multi compartment vehicle routing problem, Applied Soft Computing 37 (2015) 196–203.

[2] Y. Agarwal, K. Mathur, H.M. Salkin, A set-partitioning-based exact algorithm for the vehicle routing problem, Networks 19 (7) (1989) 731–749.

[3] J. Albiach, J.M. Sanchis, D. Soler, An asymmetric TSP with time windows and with time-dependent travel times and costs: an exact solution through a graph transformation, European Journal of Operational Research 189 (3) (2008) 789–802.

[4] K. Altinkemer, B. Gavish, Parallel savings based heuristic for the delivery problem, Operations Research 39 (3) (1991) 456–469.

[5] J. Antes, U. Derigs, A New Parallel Tour Construction Algorithm for the Vehicle Routing Problem With Time Windows, Technical report, Lehrstuhl fur Wirtschaftsinformatik und Operations Research, Universitat zu Koln, 1995.

[6] D. Applegate, R. Bixby, V. Chvatal, W. Cook, Concorde TSP solver, Available at http://www.tsp.gatech.edu/concorde, 2006.

[7] C. Archetti, M. Speranza, Vehicle routing problems with split deliveries, International Transactions in Operational Research 19 (1–2) (2012) 3–22.

[8] Y. Arezki, D. Van Vliet, A full analytical implementation of the PARTAN/Frank–Wolfe algorithm for equilibrium assignment, Transportation Science 24 (1) (1990) 58–62.

[9] N. Ascheuer, M. Fischetti, M. Grotschel, Solving the asymmetric travelling salesman problem with time windows by branch-and-cut, Mathematical Programming 90 (3) (2001) 475–506.

[10] N. Ascheuer, M. Junger, G. Reinelt, A branch and cut algorithm for the asymmetric traveling salesman problem with precedence constraints, Computational Optimization and Applications 17 (1) (2000) 61–84.

[11] Y. Bai, W. Zhang, Z. Jin, An new self-organizing maps strategy for solving the traveling salesman problem, Chaos, Solitons and Fractals 28 (2006) 1082–1089.

[12] E. Baker, J. Schaffer, Solution improvement heuristics for the vehicle routing and scheduling problem with time windows constraints, American Journal of Mathematics and Management Sciences 6 (3) (1989) 261–300.

[13] E. Balas, N. Christofides, A restricted Lagrangean approach to the traveling salesman problem, Mathematical Programming 21 (1) (1981) 19–46.

[14] E. Balas, M. Fischetti, W.R. Pulleyblank, The precedence-constrained asymmetric traveling salesman polytope, Mathematical Programming 68 (1–3) (1995) 241–265.

[15] E. Balas, P. Toth, Branch and Bound Methods for the Traveling Salesman Problem, Technical Report MSRR-488, Carnegie-Mellon University, 1983.

[16] R. Baldacci, M. Battarra, D. Vigo, Routing a heterogeneous fleet of vehicles, in: B. Golden, S. Raghavan , E. Wasil (Eds.), The Vehicle Routing Problem: Latest Advances and New Challenges, Springer, New York, 2008, pp. 3–27.

[17] R. Baldacci, A. Mingozzi, R. Roberti, Recent exact algorithms for solving the vehicle routing problem under capacity and time window constraints, European Journal of Operational Research 218 (1) (2012) 1–6.

[18] R. Baldacci, D. Vigo, P. Toth, Exact solution of the capacitated vehicle routing problem, in: Wiley Encyclopedia of Operations Research and Management Science, American Cancer Society, 2011, pp. 1–12.

[19] J. Bang, J. Ryu, J. Lim, C. Lee, A quantum heuristic algorithm for the traveling salesman problem, Journal of the Korean Physical Society 61 (12) (2012) 1944–1949.

[20] M.J. Bannister, D. Eppstein, Randomized speedup of the Bellman–Ford algorithm, in: H.-K. Hwang, C. Martinez (Eds.), 2012 Proceedings of the Ninth Workshop on Analytic Algorithmics and Combinatorics, ANALCO, Society for Industrial and Applied Mathematics, 1 2012, pp. 41–47.

[21] R. Banos, J. Ortega, C. Gil, A.L. Marquez, F. De Toro, A hybrid meta-heuristic for multi-objective vehicle routing problems with time windows, Computers & Industrial Engineering 65 (2) (2013) 286–296.

[22] H. Bar-Gera, Origin-Based Algorithms for Transportation Network Modeling, Technical Report 103, National Institute of Statistical Sciences, 11 1999.

[23] J.E. Beasley, Route-first cluster-second methods for vehicle routing, Omega 11 (4) (1983) 403–408.

[24] M. Beckmann, C.B. McGuire, C.B. Winsten, Studies in the Economics of Transportation, Yale University Press, New Haven, Connecticut, USA, 1956.

[25] T. Bektas, G. Laporte, The pollution-routing problem, Transportation Research Part B 45 (2011) 1232–1250.

[26] R. Bellman, On a routing problem, Quarterly of Applied Mathematics 16 (1) (1958) 87–90.
[27] R. Bellman, Dynamic programming treatment of the travelling salesman problem, Journal of the ACM (JACM) 9 (1) (1962) 61–63.
[28] M. Ben-Akiva, M. Bierlaire, Discrete choice methods and their applications to short term travel decisions, in: R. Hall (Ed.), Handbook of Transportation Science, Springer, Boston, MA, 1999, pp. 5–33.
[29] J. Berger, M. Barkaoui, A hybrid genetic algorithm for the capacitated vehicle routing problem, in: Genetic and Evolutionary Computation, GECCO 2003, Springer, Berlin, Heidelberg, 2003, pp. 646–656.
[30] A. Björklund, T. Husfeldt, P. Kaski, M. Koivisto, The traveling salesman problem in bounded degree graphs, ACM Transactions on Algorithms (TALG) 8 (2) (2008) 198–209.
[31] M. Blocho, J. Nalepa, Complexity analysis of the parallel guided ejection search for the pickup and delivery problem with time windows, arXiv:1704.06724, 2017.
[32] M. Blocho, J. Nalepa, LCS-based selective route exchange crossover for the pickup and delivery problem with time windows, in: B. Hu, M. López-Ibáñez (Eds.), Evolutionary Computation in Combinatorial Optimization, 4 2017, pp. 124–140.
[33] C. Blum, A. Roli, Metaheuristics in combinatorial optimization: overview and conceptual comparison, ACM Computing Surveys (CSUR) 35 (3) (2001) 268–308.
[34] K. Braekers, K. Ramaekers, I. Van Nieuwenhuysec, The vehicle routing problem: state of the art classification and review, Computers & Industrial Engineering 99 (2015) 300–313.
[35] J. Bramel, D. Simchi-Levi, A location based heuristic for general routing problems, Operations Research 43 (4) (1995) 649–660.
[36] A. Brodnik, M. Grgurovic, Solving all-pairs shortest path by single-source computations: theory and practice, Discrete Applied Mathematics 231 (2017) 119–130.
[37] R. Burkard, M. Dell'Amico, S. Martello, Assignment Problems: Revised Reprint, SIAM, 2009.
[38] R.W. Calvo, A new heuristic for the traveling salesman problem with time windows, Transportation Science 34 (1) (2006) 111–120.
[39] G. Carpaneto, P. Toth, Primal-dual algorithms for the assignment problem, Discrete Applied Mathematics 18 (2) (1987) 137–153.
[40] A. Chen, D.H. Lee, R. Jayakrishnan, Computational study of state-of-the-art path-based traffic assignment algorithms, Mathematics and Computers in Simulation 59 (6) (2002) 509–518.
[41] H. Chen, X. Kong, B. Chong, G. Qin, X. Zhou, X. Peng, J. Du, Experimental demonstration of a quantum annealing algorithm for the traveling salesman problem in a nuclear-magnetic-resonance quantum simulator, Physical Review A 83 (3) (2011) 032314.
[42] H.K. Chen, C.W. Chang, M.S. Chang, Comparison of link-based versus route-based algorithms in the dynamic user-optimal route choice problem, Transportation Research Record 1667 (1) (1999) 114–120.
[43] C.B. Cheng, C.P. Mao, A modified ant colony system for solving the travelling salesman problem with time windows, Mathematical and Computer Modelling 46 (9–10) (2007) 1225–1235.
[44] W.-C. Chiang, R.A. Russell, Simulated annealing metaheuristics for the vehicle routing problem with time windows, Annals of Operations Research 63 (1) (1996) 3–27.
[45] W.-C. Chiang, R.A. Russell, A reactive tabu search metaheuristic for the vehicle routing problem with time windows, INFORMS Journal on Computing 9 (4) (1997) 417–430.
[46] Y.-C. Chiu, J. Bottom, M. Mahut, A. Paz, R. Balakrishna, T. Waller, J. Hicks, Dynamic Traffic Assignment: a Primer, Transportation research e-circular (e-c153), Transportation Research Board, 2011.
[47] N. Christofides, Worst-Case Analysis of a New Heuristic for the Travelling Salesman Problem, Technical Report RR-388, Graduate School of Industrial Administration, CMU, 1976.
[48] N. Christofides, A. Mingozzi, P. Toth, The vehicle routing problem, in: N. Christofides, A. Mingozzi, P. Toth (Eds.), Combinatorial Optimization, Wiley, Chichester, 1979, pp. 315–338.

[49] N. Christofides, A. Mingozzi, P. Toth, State-space relaxation procedures for the computation of bounds to routing problems, Networks 11 (2) (1981) 145–164.
[50] G. Clarke, J. Wright, Scheduling of vehicles from a central depot to a number of delivery points, Operations Research 12 (4) (1964) 568–581.
[51] T. Cook, R. Russell, A simulation and statistical analysis of stochastic vehicle routing with timing constraints, Decision Science 9 (1978) 673–687.
[52] J.-F. Cordeau, G. Desaulniers, J. Desrosiers, M.M. Solomon, F. Soumis, VRP with time windows, in: P. Toth, D. Vigo (Eds.), The Vehicle Routing Problem, Society for Industrial and Applied Mathematics, Philadelphia, PA, USA, 2001, pp. 157–193.
[53] J.-F. Cordeau, M. Gendreau, G. Laporte, J.-Y. Potvin, F. Semet, A guide to vehicle routing heuristics, Journal of the Operational Research Society 53 (5) (2002) 512–522.
[54] R. Cordone, R.W. Calvo, A Heuristic for Vehicle Routing Problem With Time Windows, Internal report 97.012, Tech. rep., Politecnico di Milano, Dipartimento di Elettronica e Informazione, 1997.
[55] T. Curtois, D. Landa-Silva, Y. Qu, W. Laesanklang, Large neighbourhood search with adaptive guided ejection search for the pickup and delivery problem with time windows, EURO Journal on Transportation and Logistics 7 (2) (2018) 151–192.
[56] H. Dai, Y. Yang, C. Li, J. Shi, S. Gao, Z. Tang, Quantum interference crossover-based clonal selection algorithm and its application to traveling salesman prob, IEICE Transactions on Information and Systems 92 (1) (2009) 78–85.
[57] G. Dantizg, J. Ramser, The truck dispatching problem, Management Science 6 (1) (1959) 80–91.
[58] G. Dantzig, R. Fulkerson, S. Johnson, Solution of a large-scale traveling-salesman problem, Operations Research 2 (4) (1954) 393–410.
[59] C. Demetrescu, G.F. Italiano, Experimental analysis of dynamic all pairs shortest path algorithms, ACM Transactions on Algorithms (TALG) 2 (4) (2006) 578–601.
[60] E. Demir, T. Bektas, G. Laporte, A review of recent research on green road freight transportation, European Journal of Operational Research 237 (2014) 775–793.
[61] M. Desrochers, T.W. Verhoog, A Matching Based Savings Algorithm for the Vehicle Routing Problem, Technical Report G-89-04, GERAD, 1989.
[62] R.B. Dial, Algorithm 360: shortest-path forest with topological ordering [H], Communications of the ACM 12 (11) (1969) 632–633.
[63] R.B. Dial, A path-based user-equilibrium traffic assignment algorithm that obviates path storage and enumeration, Transportation Research Part B: Methodological 40 (10) (2006) 917–936.
[64] E.W. Dijkstra, A note on two problems in connexion with graphs, Numerische Mathematik 1 (1) (1959) 269–271.
[65] A.V. Donati, R. Montemanni, N. Casagrande, A.E. Rizzoli, L.M. Gambardella, Time dependent vehicle routing problem with a multi ant colony system, European Journal of Operational Research 185 (3) (2008) 1774–1791.
[66] M. Dorigo, E. Bonabeau, G. Theraulaz, Ant algorithms and stigmergy, Future Generation Computer Systems 16 (8) (2000) 851–871.
[67] J. Dréo, A. Pétrowski, P. Siarry, E. Taillard, Metaheuristics for Hard Optimization: Methods and Case Studies, Springer, 2006.
[68] M. Drexl, Rich vehicle routing in theory and practice, Logistics Research 5 (1–2) (2012) 47–63.
[69] M. Dror, P. Trudeau, Savings by split deliveries, Naval Research Logistics 37 (3) (1989) 383–402.
[70] M. Dror, P. Trudeau, Split delivery routing, Naval Research Logistics 37 (3) (1990) 383–402.
[71] M. Duell, H. Grzybowska, D. Rey, S.T. Waller, Strategic dynamic traffic assignment incorporating travel demand uncertainty, Transportmetrica B: Transport Dynamics 7 (1) (2019) 950–966.

[72] E. Duman, I. Or, Precedence constrained TSP arising in printed circuit board assembly, International Journal of Production Research 42 (1) (2004) 67–78.
[73] W. Eastman, Linear Programming With Pattern Constraints, Ph.D. thesis, Department of Economics, Harvard University, 1958.
[74] R.W. Eglese, L.Y.O. Li, A tabu search based heuristic for arc routing with a capacity constraint and time deadline, in: I.H. Osman, J.P. Kelly (Eds.), Meta-Heuristics: Theory and Applications, Springer, 1996, pp. 633–649.
[75] B. Eksioglu, A.V. Vural, A. Reisman, The vehicle routing problem: a taxonomic review, Computers & Industrial Engineering 57 (4) (2009) 1472–1483.
[76] N. El-Sherbeny, Resolution of a Vehicle Routing Problem With Multi-Objective Simulated Annealing Method, Ph.D. thesis, Faculté Polytechnique de Mons, 2001.
[77] N.A. El-Sherbeny, Vehicle routing with time windows: an overview of exact, heuristic and metaheuristic methods, Journal of King Saud University-Science 22 (3) (2010) 123–131.
[78] D. Eppstein, The traveling salesman problem for cubic graphs, Journal of Graph Algorithms and Applications 11 (1) (2007) 61–81.
[79] L. Euler, Solutio problematis ad geometriam situs pertinentis, Commentarii academiae scientiarum Petropolitanae 8 (1741) 128–140.
[80] L. Euler, Solution d'une question curieuse que ne paroit soumise à aucune analyse, Mémoires de l' Academie des Sciences de Berlin 15 (1766) 26–56.
[81] J.-G. Fages, X. Lorca, Improving the asymmetric TSP by considering graph structure, arXiv:1206.3437, 2012.
[82] J. Fakcharoenphol, S. Rao, Planar graphs, negative weight edges, shortest paths, and near linear time, Journal of Computer and System Sciences 72 (5) (2006) 868–889.
[83] S.M. Ferrandez, T. Harbison, T. Weber, R. Sturges, R. Rich, Optimization of a truck-drone in tandem delivery network using k-means and genetic algorithm, Journal of Industrial Engineering and Management (JIEM) 9 (2) (2016) 374–388.
[84] M.A. Figliozzi, The time dependent vehicle routing problem with time windows: benchmark problems, an efficient solution algorithm, and solution characteristics, Transportation Research Part E 48 (2012) 616–636.
[85] M. Fischetti, P. Toth, A polyhedral approach to the asymmetric traveling salesman problem, Management Science 43 (11) (1997) 1520–1536.
[86] M.L. Fisher, R. Jaikumar, A generalized assignment heuristic for the vehicle routing problem, Networks 11 (2) (1981) 109–124.
[87] I. Fister Jr, X.-S. Yang, I. Fister, J. Brest, D. Fister, A brief review of nature-inspired algorithms for optimization, Elektrotehniski Vestnik/Electrotechnical Review 80 (3) (2013) 116–122.
[88] B. Fleischmann, Linear programming approaches to traveling salesman and vehicle scheduling problems, in: The XIth International Symposium on Mathematical Programming, Bonn, 1982.
[89] M. Florian, J. Guálat, H. Spiess, An efficient implementation of the "partan" variant of the linear approximation method for the network equilibrium problem, Networks 17 (3) (1987) 319–339.
[90] R.W. Floyd, Algorithm 97: shortest path, Communications of the ACM 5 (6) (1962) 345.
[91] L.R. Ford Jr, Network Flow Theory, Technical Report P-923, RAND Corp Santa Monica Ca., 1956.
[92] M. Frank, P. Wolfe, An algorithm for quadratic programming, Naval Research Logistics Quarterly 3 (1–2) (1956) 95–110.
[93] M.L. Fredman, R.E. Tarjan, Fibonacci heaps and their uses in improved network optimization algorithms, Journal of the ACM (JACM) 34 (3) (1987) 596–615.
[94] L. Fu, D. Sun, L.R. Rilett, Heuristic shortest path algorithms for transportation applications: state of the art, Computers & Operations Research 33 (11) (2006) 3324–3343.
[95] R. Fukasawa, H. Longo, J. Lysgaard, M.P. de Aragão, M. Reis, E. Uchoa, R.F. Werneck, Robust branch-and-cut-and-price for the capacitated vehicle routing problem, Mathematical Programming 106 (3) (2006) 491–511.

[96] M. Fukushima, A modified Frank-Wwolfe algorithm for solving the traffic assignment problem, Transportation Research Part B: Methodological 18 (2) (1984) 169–177.

[97] L.M. Gambardella, M. Dorigo, Solving symmetric and asymmetric TSPs by ant colonies, in: Proceedings of IEEE International Conference on Evolutionary Computation, 1996, pp. 622–627.

[98] L.M. Gambardella, E. Taillard, G. Agazzi, A multiple ant colony system for vehicle routing problem with time window, in: D. Corne, M. Dorigo, F. Glover (Eds.), New Ideas in Optimization, McGraw-Hill, London, UK, 1999, pp. 63–76.

[99] T.J. Gaskell, Bases for vehicle fleet scheduling, Operational Research Quarterly 18 (1967) 281–295.

[100] H. Gehring, J. Homberger, A Parallel Hybrid Evolutionary Metaheuristic for the Vehicle Routing Problem With Time Windows, Proceedings of EUROGEN99, vol. 2, Springer, Berlin, 1999, pp. 57–64.

[101] M. Gendreau, J.-Y. Potvin, O. Bräumlaysy, G. Hasle, A. Løkketangen, Metaheuristics for the vehicle routing problem and its extensions: a categorized bibliography, in: B. Golden, S. Raghavan, E. Wasil (Eds.), The Vehicle Routing Problem: Latest Advances and New Challenges, Springer, Boston, MA, 2008, pp. 143–169.

[102] S.O. Gharan, A. Saberi, The asymmetric traveling salesman problem on graphs with bounded genus, in: D. Randall (Ed.), Proceedings of the Twenty-Second Annual ACM-SIAM Symposium on Discrete Algorithms, Society for Industrial and Applied Mathematics, 1 2011, pp. 967–975.

[103] B.E. Gillett, L.R. Miller, A heuristic algorithm for the vehicle dispatch problem, Operations Research 22 (2) (1974) 340–349.

[104] F. Glover, G. Gutin, A. Yeo, A. Zverovich, Construction heuristics for the asymmetric TSP, European Journal of Operational Research 129 (3) (2001) 555–568.

[105] A.V. Goldberg, T. Radzik, A heuristic improvement of the Bellman-Ford algorithm, Applied Mathematics Letters 6 (3) (1993) 3–6.

[106] B.L. Golden, T.L. Magnanti, H.Q. Nguyen, Implementing vehicle routing algorithms, Networks 7 (2) (1977) 113–148.

[107] B. Goldengorin, G. Jäger, P. Molitor, Tolerance based contract-or-patch heuristic for the asymmetric TSP, in: Combinatorial and Algorithmic Aspects of Networking, vol. 4235, Springer, 2006, pp. 86–97.

[108] GoogleScholar, Search for vehicle routing problem articles, https://scholar.google.pl/scholar?q=allintitle:+%22vehicle+routing+problem%22&hl=pl&as_sdt=0,5&as_vis=1, 2019. (Accessed 13 May 2019).

[109] P. Gora, Simulation-based traffic management system for connected and autonomous vehicles, in: Road Vehicle Automation 4, Springer, 2017, pp. 257–266.

[110] D. Gulczynski, B. Golden, E. Wasil, Recent developments in modeling and solving the split delivery vehicle routing problem, in: Z.-L. Chen, S. Raghavan (Eds.), Tutorials in Operations Research: State-of-the-Art Decision-Making Tools in the Information-Intensive Age, INFORMS, 2008, pp. 170–180.

[111] P.E. Hart, N.J. Nilsson, B. Raphael, A formal basis for the heuristic determination of minimum cost paths, IEEE Transactions on Systems Science and Cybernetics 4 (2) (1968) 100–107.

[112] D.W. Hearn, S. Lawphongpanich, J.A. Venture, Restricted simplicial decomposition: computation and extensions, Mathematical Programming Study 31 (1987) 99–118.

[113] G.T. Heineman, G. Pollice, S. Selkow, Chapter 6: graph algorithms, in: G.T. Heineman, G. Pollice, S. Selkow (Eds.), Algorithms in a Nutshell, O'Reilly Media, 2008, pp. 142–177.

[114] M. Held, R.M. Karp, A dynamic programming approach to sequencing problems, Journal of the Society for Industrial and Applied Mathematics 10 (1) (1962) 196–210.

[115] M. Held, R.M. Karp, The traveling salesman problem and minimum spanning trees, Operations Research 18 (6) (1970) 1138–1162.

[116] K. Helsgaun, An effective implementation of the Lin–Kernighan traveling salesman heuristic, European Journal of Operational Research 126 (1) (2000) 106–130.

[117] S.C. Ho, D. Haugland, A tabu search heuristic for the vehicle routing problem with time windows and split deliveries, Computers & Operations Research 31 (2004) 1947–1961.

[118] J.J. Hopfield, D.W. Tank, Computing with neural circuits: a model, Science 233 (4764) (1986) 625–633.

[119] I.M. Hosny, C.L. Mumford, Constructing initial solutions for the multiple vehicle pickup and delivery problem with time windows, Journal of King Saud University – Computer and Information Sciences 24 (2012) 59–69.

[120] S. Ichoua, M. Gendreau, J.-Y. Potvin, Vehicle dispatching with time-dependent travel times, European Journal of Operational Research 144 (2003) 379–396.

[121] R. Jayakrishnan, W.T. Tsai, J.N. Prashker, S. Rajadhyaksha, A faster path-based algorithm for traffic assignment, Transportation Research Record 1443 (1994) 75–83.

[122] J. Jin, T.G. Crainic, A. Løkketangen, A parallel multi-neighborhood cooperative tabu search for capacitated vehicle routing problems, European Journal of Operational Research 222 (3) (2012) 441–451.

[123] D.S. Johnson, L.A. McGeoch, The traveling salesman problem: a case study in local optimization, in: E.H.L. Aarts, J.K. Lenstra (Eds.), Local Search in Combinatorial Optimization, John Wiley and Sons, 1997, pp. 215–310.

[124] D.S. Johnson, L.A. McGeogh, E.E. Rothberg, Asymptotic experimental analysis for the Held–Karp traveling salesman bound, in: Proceedings of the Seventh Annual ACM-SIAM Symposium on Discrete Algorithms, 2013, pp. 341–350.

[125] J. Jones, A. Adamatzky, Computation of the travelling salesman problem by a shrinking blob, Natural Computing 13 (1) (2014) 1–16.

[126] R. Jonker, T. Volgenant, Transforming asymmetric into symmetric traveling salesman problems, Operational Research Letters 2 (4) (1983) 161–163.

[127] M. Junger, G. Reinelt, G. Rinaldi, The traveling salesman problem, in: M.O. Ball, T.L. Magnanti, C.L. Monma, G.L. Nemhauser (Eds.), Handbooks in Operations Research and Management Science, vol. 7, Elsevier Science B.V., Boston, MA, 1995, pp. 225–330.

[128] D.R. Karger, D. Koller, S.J. Phillips, Finding the hidden path: time bounds for all-pairs shortest paths, SIAM Journal on Computing 22 (6) (1993) 1199–1217.

[129] R.M. Karp, Reducibility among combinatorial problems, in: R.E. Miller, J.W. Thatcher, J.D. Bohlinger (Eds.), Complexity of Computer Computations, in: The IBM Research Symposia Series, Springer, Boston, MA, 1972, pp. 85–103.

[130] R.M. Karp, A patching algorithm for the nonsymmetric traveling salesman problem, SIAM Journal on Computing 8 (4) (1979) 561–573.

[131] T. Kirkman, XVIII. On the representation of polyedra, Philosophical Transactions of the Royal Society of London 146 (1856) 413–418.

[132] C. Koc, T. Bektas, O. Jabali, G. Laporte, The fleet size and mix pollution-routing problem, Transportation Research Part B 70 (2014) 239–254.

[133] C. Koc, I. Karaoglan, The green vehicle routing problem: a heuristic based exact solution approach, Applied Soft Computing 39 (2016) 154–164.

[134] T. Kohonen, Self-organized formation of topologically correct feature maps, Biological Cybernetics 43 (1) (1982) 59–69.

[135] G. Kontoravdis, J.F. Bard, A grasp for the vehicle routing problem with time windows, ORSA Journal on Computing 7 (1995) 10–23.

[136] M. Kubo, H. Kasugai, The precedence constrained traveling salesman problem, Journal of the Operations Research Society of Japan 34 (2) (1991) 152–172.

[137] R.J. Kuo, F.E. Zulvia, K. Suryadi, Hybrid particle swarm optimization with genetic algorithm for solving capacitated vehicle routing problem with fuzzy demand – a case study on garbage collection system, Applied Mathematics and Computation 219 (5) (2012) 2574–2588.

[138] B. La Maire, V. Mladenov, Comparison of neural networks for solving the travelling salesman problem, in: 11th Symposium on Neural Network Applications in Electrical Engineering, NEUREL 2012 – Proceedings, 09 2012, pp. 21–24.

[139] R. Lahyani, L.C. Coelho, M. Khemakhem, G. Laporte, F. Semet, A multi-compartment vehicle routing problem arising in the collection of olive oil in Tunisia, Omega 51 (2015) 1–10.
[140] A.H. Land, A.G. Doig, An automatic method of solving discrete programming problems, Econometrica 28 (3) (1960) 497–520.
[141] G. Laporte, The traveling salesman problem: an overview of exact and approximate algorithms, European Journal of Operational Research 59 (2) (1992) 231–247.
[142] G. Laporte, Y. Nobert, Exact algorithms for the vehicle routing problem, in: S. Martello, G. Laporte, M. Minoux, C. Ribeiro (Eds.), Surveys in Combinatorial Optimization, in: North-Holland Mathematics Studies, vol. 132, North-Holland, 1987, pp. 147–184.
[143] G. Laporte, F. Semet, Classical heuristics for the capacitated VRP, in: P. Toth, D. Vigo (Eds.), The Vehicle Routing Problem, Society for Industrial and Applied Mathematics, Philadelphia, 2001, pp. 109–128.
[144] G. Laporte, P. Toth, D. Vigo, Vehicle routing: historical perspective and recent contributions, EURO Journal on Transportation and Logistics 1 (2) (2013) 1–4.
[145] T. Larsson, M. Patriksson, Simplicial decomposition with disaggregated representation for the traffic assignment problem, Transportation Science 26 (1) (1992) 4–17.
[146] T. Larsson, M. Patriksson, C. Rydergren, Applications of simplicial decomposition with non-linear column generation to nonlinear network flows, in: P.M. Pardalos, D.W. Hearn, W.W. Hager (Eds.), Network Optimization, Springer, Berlin, Heidelberg, 1998, pp. 346–373.
[147] L. LeBlanc, R. Helgason, D. Boyce, Improved efficiency of the Frank–Wolfe algorithm for convex network programs, Transportation Science 19 (4) (1985) 445–462.
[148] D.H. Lee, Y. Nie, A. Chen, Y.C. Leow, Link- and path-based traffic assignment algorithms: computational and statistical study, Transportation Research Record 1783 (1) (2002) 80–88.
[149] M. Lihoreau, L. Chittka, N.E. Raine, Travel optimization by foraging bumblebees through readjustments of traplines after discovery of new feeding locations, The American Naturalist 176 (6) (2010) 744–757.
[150] C. Lin, K.L. Choy, G.T.S. Ho, S.H. Chung, H.Y. Lam, Survey of green vehicle routing problem: past and future trends, Expert Systems with Applications 41 (4) (2013) 1118–1138.
[151] S. Lin, Computer solutions of the traveling salesman problem, Bell System Technical Journal 44 (10) (1965) 2245–2269.
[152] S. Lin, B.W. Kernighan, An effective heuristic algorithm for the traveling-salesman problem, Operations Research 21 (2) (1973) 498–516.
[153] Y. Lin, Z. Bian, X. Liu, Developing a dynamic neighborhood structure for an adaptive hybrid simulated annealing – tabu search algorithm to solve the symmetrical traveling salesman problem, Applied Soft Computing 49 (2016) 937–952.
[154] J.D.C. Little, K.G. Murty, D.W. Sweeney, C. Karel, An algorithm for the traveling salesman problem, Operations Research 11 (6) (1963) 972–989.
[155] Y. Liu, J. Bunker, L. Ferreira, Transit users' route-choice modelling in transit assignment: a review, Transport Reviews 30 (6) (2010) 753–769.
[156] M. Lupi, Convergence of the Frank–Wolfe algorithm in transportation networks, Civil Engineering Systems 3 (1) (1986) 7–15.
[157] G. Macrina, G. Laporte, F. Guerriero, L.D.P. Pugliese, An energy-efficient green-vehicle routing problem with mixed vehicle fleet, partial battery recharging and time windows, European Journal of Operational Research 276 (3) (2019) 971–982.
[158] K. Magzhan, H.M. Jani, A review and evaluations of shortest path algorithms, International Journal of Scientific & Technology Research 2 (6) (2013) 99–104.
[159] M. Maher, K. Stewart, A. Rosa, Stochastic social optimum traffic assignment, Transportation Research Part B: Methodological 39 (8) (2005) 753–767.
[160] S. Mancini, Time dependent travel speed vehicle routing and scheduling on a real road network: the case of Torino, Transportation Research Procedia 3 (2014) 433–441.
[161] S. Mandra, G.G. Guerreschi, A. Aspuru-Guzik, Faster than classical quantum algorithm for dense formulas of exact satisfiability and occupation problems, New Journal of Physics 18 (7) (2016) 073003.

[162] F. Margot, Quick updates for p-opt TSP heuristics, Operations Research Letters 11 (1) (1992) 45–46.

[163] R. Martonak, G.E. Santoro, E. Tosatti, Quantum annealing of the traveling salesman problem, Physical Review E 70 (5) (2004).

[164] R. Matai, S. Singh, M.L. Mittal, Traveling salesman problem: an overview of applications, formulations, and solution approaches, in: D. Davendra (Ed.), Traveling Salesman Problem, Theory and Applications, IntechOpen, 2010, pp. 1–24.

[165] J.E. Mendoza, B. Castanier, C. Guéret, A.L. Medaglia, N. Velasco, A memetic algorithm for the multi-compartment vehicle routing problem with stochastic demands, Computers & Operations Research 37 (11) (2010) 1886–1898.

[166] D. Merchant, G. Nemhauser, A model and an algorithm for the dynamic traffic assignment problems, Transportation Science 12 (3) (1978) 183–199.

[167] S. Miki, D. Yamamoto, H. Ebara, Applying deep learning and reinforcement learning to traveling salesman problem, in: 2018 International Conference on Computing, Electronics & Communications Engineering, iCCECE, 03 2019, pp. 65–70.

[168] P. Miliotis, Integer programming approaches to the travelling salesman problem, Mathematical Programming 10 (1976) 367–378.

[169] A. Mingozzi, L. Bianco, A. Ricciardelli, Dynamic programming strategies for the traveling salesman problem with time window and precedence constraints, Operations Research 45 (3) (1997) 19–32.

[170] S. Mitra, An algorithm for the generalized vehicle routing problem with backhauling, Asia-Pacific Journal of Operational Research 22 (2005) 153–169.

[171] R.H. Mole, S.R. Jameson, A sequential route-building algorithm employing a generalized savings criterion, Operational Research Quarterly 27 (2) (1976) 503–511.

[172] A. Montanaro, Quantum walk speedup of backtracking algorithms, Theory of Computing 14 (15) (2015) 1–24.

[173] D.J. Moylett, N. Linden, A. Montanaro, Quantum speedup of the travelling salesman problem for bounded-degree graphs, Physical Review A 95 (3) (2017) 032323.

[174] D. Naddef, Polyhedral theory and branch-and-cut algorithms for the symmetric tsp, in: G. Gutin, A.P. Punnen (Eds.), The Traveling Salesman Problem and Its Variations, in: Combinatorial Optimization, vol. 12, Springer, Boston, MA, 2007, pp. 29–116.

[175] D. Naddef, G. Rinaldi, The symmetric traveling salesman problem and its graphical relaxation, Mathematical Programming 51 (1–3) (1991) 359–400.

[176] H. Nagamochi, M. Xiao, An improved exact algorithm for TSP in graphs of maximum degree 4, Theory of Computing Systems 58 (2) (2016) 241–272.

[177] Y. Nagata, D. Soller, A new genetic algorithm for the asymmetric traveling salesman problem, Expert Systems with Applications 39 (2012) 8947–8953.

[178] J. Nalepa, M. Blocho, Co-operation in the parallel memetic algorithm, International Journal of Parallel Programming 43 (5) (2015) 812–839.

[179] J. Nalepa, M. Blocho, Adaptive guided ejection search for pickup and delivery with time windows, Journal of Intelligent & Fuzzy Systems 32 (2) (2017) 1547–1559.

[180] J. Nalepa, Z. Czech, New selection schemes in a memetic algorithm for the vehicle routing problem with time windows, in: International Conference on Adaptive and Natural Computing Algorithms, 04 2013, pp. 396–405.

[181] S. Nallaperuma, M. Wagner, F. Neumann, B. Bischl, O. Mersmann, H. Trautmann, A feature-based comparison of local search and the Christofides algorithm for the travelling salesperson problem, in: Proceedings of the Twelfth Workshop on Foundations of Genetic Algorithms XII, FOGA XII'13, 2013, pp. 147–160.

[182] D.N. Nawrocki, A brief history of downside risk measures, Journal of Investing 8 (1999) 9–25.

[183] M.D. Nelson, K.E. Nygard, J.H. Griffin, W.E. Shreve, Implementation techniques for the vehicle routing problem, Computers and Operations Research 12 (3) (1985) 273–283.

[184] S.U. Ngueveu, C. Prins, R.W. Calvo, An effective memetic algorithm for the cumulative capacitated vehicle routing problem, Computers & Operations Research 37 (11) (2010) 1877–1885.

[185] T.B.T. Nguyêñ, T. Bektas, T.J. Cherrett, F.N. McLeod, J. Allen, O. Bates, M. Piotrowska, M. Piecyk, A. Friday, S. Wise, Optimising parcel deliveries in London using dual-mode routing, Journal of the Operational Research Society (2018) 1–13.

[186] M.A. Nielsen, I.L. Chuang, Quantum Computation and Quantum Information, Cambridge University Press, 2010.

[187] B. Ombuki, B.J. Ross, F. Hanshar, Multi-objective genetic algorithms for vehicle routing problem with time windows, Applied Intelligence 24 (1) (2006) 17–30.

[188] I. Or, Traveling Salesman-Type Combinatorial Optimization Problems and Their Relation to the Logistics of Regional Blood Banking, Ph.D. thesis, Department of Industrial Engineering and Management Sciences, Northwestern University, 1976.

[189] E. Osaba, X.-S. Yang, F. Diaz, P. Lopez-Garcia, R. Carballedo, An improved discrete bat algorithm for symmetric and asymmetric traveling salesman problems, Engineering Applications of Artificial Intelligence 48 (2016) 59–71.

[190] I.H. Osman, J.P. Kelly, Meta-heuristics: an overview, in: Meta-Heuristics: Theory and Applications, Springer US, Boston, MA, 1996, pp. 1–21.

[191] M. Padberg, G. Rinaldi, A branch-and-cut algorithm for the resolution of large-scale symmetric traveling salesman problems, SIAM Review 33 (1) (1991) 60–100.

[192] M.W. Padberg, M.R. Rao, Odd minimum cut-sets and b-matchings, Mathematics of Operations Research 7 (1) (1982) 67–80.

[193] H. Paessens, The savings algorithm for the vehicle routing problem, European Journal of Operational Research 34 (3) (1988) 336–344.

[194] C.N. Papadimitrou, The Euclidean traveling salesman problem is NP-complete, Theoretical Computer Science 4 (3) (1977) 237–244.

[195] G. Passarelli, Adiabatic Quantum Annealing of Simple NP-Complete Problems, Master's thesis, Scuola Politecnica e delle Scienze di Base, Area Didattica di Scienze Matematiche Fisiche e Naturali, 2017.

[196] M. Patriksson, The Traffic Assignment Problem – Models and Methods, VSP, Utrecht, 1994.

[197] S. Peeta, H.S. Mahmassani, System optimal and user equilibrium time-dependent traffic assignment in congested networks, Annals of Operations Research 60 (1) (1995) 81–113.

[198] S. Peeta, A.K. Ziliaskopoulos, Foundations of dynamic traffic assignment: the past, the present and the future, Networks and Spatial Economics 1 (3–4) (2001) 233–265.

[199] O. Perederieieva, M. Ehrgott, J.Y. Wang, A. Raith, A computational study of traffic assignment algorithms, in: Australasian Transport Research Forum, Brisbane, Australia, 10 2013, pp. 1–18.

[200] G. Pesant, M. Gendreau, J.-Y. Potvin, J.M. Rousseau, An exact constraint logic programming algorithm for the traveling salesman problem with time windows, Transportation Science 32 (1) (1998) 3–85.

[201] A. Pessoa, M.P. de Aragão, E. Uchoa, Robust branch-cut-and-price algorithms for vehicle routing problems, in: B. Golden, S. Raghavan, E. Wasil (Eds.), The Vehicle Routing Problem: Latest Advances and New Challenges, Springer US, Boston, MA, 2008, pp. 297–325.

[202] J.-Y. Potvin, J.-M. Rousseau, A parallel route building algorithm for the vehicle routing and scheduling problem with time windows, European Journal of Operational Research 66 (3) (1993) 331–340.

[203] R.C. Prim, Shortest connection networks and some generalizations, Bell System Technical Journal 36 (6) (1957) 1389–1401.

[204] Z. Qian, H.M. Zhang, A hybrid route choice model for dynamic traffic assignment, Networks and Spatial Economics 13 (2013) 183–203.

[205] B. Ran, D. Boyce, Modeling Dynamic Transportation Networks: An Intelligent Transportation System Oriented Approach, Springer, 2012.

[206] S.S. Ray, S. Bandyopadhyay, S.K. Pal, Genetic operators for combinatorial optimization in TSP and microarray gene ordering, Applied Intelligence 26 (3) (2007) 183–195.
[207] M. Reed, A. Yiannkou, R. Evering, An ant colony algorithm for the multi-compartment vehicle routing problem, Applied Soft Computing 15 (2014) 169–176.
[208] C. Rego, C. Roucairol, A parallel tabu search algorithm using ejection chains for the vehicle routing problem, in: I.H. Osman, J.P. Kelly (Eds.), Meta-Heuristics, Springer, 1996, pp. 661–675.
[209] R. Roberti, P. Toth, Models and algorithms for the asymmetric traveling salesman problem: an experimental comparison, EURO Journal on Transportation and Logistics 1 (1–2) (2012) 113–133.
[210] S. Ropke, D. Pisinger, An adaptive large neighborhood search heuristic for the pickup and delivery problem with time windows, Transportation Science 40 (4) (2006) 455–472.
[211] J. Rudy, D. Zelazny, Multi-criteria fuel distribution: a case study, in: L. Rutkowski, M. Korytkowski, R. Scherer, R. Tadeusiewicz, L. Zadeh, J. Zurada (Eds.), Artificial Intelligence and Soft Computing. Part II, ICAISC 2015, in: Lecture Notes in Computer Science, Springer, Cham, 6 2015, pp. 272–281.
[212] R.A. Russell, Hybrid heuristics for the vehicle routing problem with time windows, Transportation Science 29 (1995) 156–166.
[213] T. Saradatta, P. Pongchairerks, A time-dependent ATSP with time window and precedence constraints in air travel, Journal of Telecommunication, Electronic and Computer Engineering 9 (2–3) (2017) 149–153.
[214] S.C. Sarin, H.D. Sherali, A. Bhootra, New tighter polynomial length formulations for the asymmetric traveling salesman problem with and without precedence constraints, Operations Research Letters 33 (1) (2005) 62–70.
[215] M.W.P. Savelsbergh, Local search in routing problems with time windows, Annals of Operations Research 4 (1985) 285–305.
[216] M.W.P. Savelsbergh, M. Sol, The general pickup and delivery problem, Transportation Science 29 (1) (1995) 17–29.
[217] K. Saw, B.K. Katti, G. Joshi, Literature review of traffic assignment: static and dynamic, International Journal of Transportation Engineering 2 (4) (2015) 339–347.
[218] J.P. Schmitt, F. Baldo, R.S. Parpinelli, A max-min ant system with short-term memory applied to the dynamic and asymmetric traveling salesman problem, in: 2018 7th Brazilian Conference on Intelligent Systems, BRACIS, 2018, pp. 1–6.
[219] A. Schrijver, On the history of combinatorial optimization (till 1960), in: K. Aardal, G.L. Nemhauser, R. Weismantel (Eds.), Handbooks in Operations Research and Management Science, vol. 12, Elsevier, 2005, pp. 1–68.
[220] D.B. Shmoys, D.P. Williamson, Analyzing the Held–Karp TSP bound: a monotonicity property with application, Information Processing Letters 35 (6) (1990) 281–285.
[221] M.M. Solomon, Algorithms for the vehicle routing and scheduling problems with time window constraints, Operations Research 35 (2) (1987) 254–265.
[222] M.M. Solomon, E. Baker, J. Schaffer, Vehicle routing and scheduling problems with time window constraints: efficient implementations of solution improvement procedures, Vehicle Routing: Methods and Studies (1988) 85–105.
[223] K. Srinivasan, S. Satyajit, B.K. Behera, P.K. Panigrahi, Efficient quantum algorithm for solving travelling salesman problem: an IBM quantum experience, arXiv:1805.10928v2, 2018.
[224] T. Stutzle, Adaptive Ant Colony Optimization for the Traveling Salesman Problem, Master's thesis, Technische Universitat Darmstadt, 2009.
[225] O. Svensson, J. Tarnawski, L.A. Vegh, A constant-factor approximation algorithm for the asymmetric traveling salesman problem, arXiv:1708.04215, 2017.
[226] W.Y. Szeto, S.C. Wong, Dynamic traffic assignment: model classifications and recent advances in travel choice principles, Central European Journal of Engineering 2 (1) (2012) 1–18.
[227] W.Y. Szeto, Y. Wu, S.C. Ho, An artificial bee colony algorithm for the capacitated vehicle routing problem, European Journal of Operational Research 215 (1) (2011) 126–135.

[228] H. Talbi, A. Draa, M. Batouche, A new quantum-inspired genetic algorithm for solving the travelling salesman problem, in: 2004 IEEE International Conference on Industrial Technology, vol. 3, 2004, IEEE ICIT'04, 12 2005, pp. 1192–1197.
[229] V. Tam, L.C. Tseng, Effective heuristics to solve pickup and delivery problems with time windows, in: 15th IEEE International Conference on Tools With Artificial Intelligence, 11 2003, pp. 184–188.
[230] M. Tatineni, H. Edwards, D. Boyce, Comparison of disaggregate simplicial decomposition and Frank–Wolfe algorithms for user-optimal route choice, Transportation Research Record 1617 (1) (1998) 157–162.
[231] R. Tavakkoli-Moghaddam, N. Safaei, M.M.O. Kah, M. Rabbani, A new capacitated vehicle routing problem with split service for minimizing fleet cost by simulated annealing, Journal of the Franklin Institute 344 (2007) 406–425.
[232] S.R. Thangiah, A. Fergany, S. Awan, Real-time split-delivery pickup and delivery time window problems with transfers, Central European Journal of Operational Research 15 (2007) 329–349.
[233] P.M. Thompson, H.N. Psaraftis, Cyclic transfer algorithms for multi-vehicle routing and scheduling problems, Operations Research 41 (5) (1993) 935–946.
[234] P. Toth, D. Vigo, An overview of vehicle routing problems, in: P. Toth, D. Vigo (Eds.), The Vehicle Routing Problem, Society for Industrial and Applied Mathematics, 2002, pp. 1–26.
[235] P. Toth, D. Vigo, The Vehicle Routing Problem, Society for Industrial and Applied Mathematics, 2002.
[236] C.L. Valenzuela, A.J. Jones, Estimating the Held–Karp lower bound for the geometric TSP, European Journal of Operational Research 102 (1) (1997) 157–175.
[237] A. Van Breedam, An Analysis of the Behavior of Heuristics for the Vehicle Routing Problem for a Selection of Problems With Vehicle-Related, Customer-Related, and Time-Related Constraints, Ph.D. thesis, University of Antwerp - RUCA, 1994.
[238] H.R.G. Van Landeghem, A bi-criteria heuristic for the vehicle routing problem with time windows, European Journal of Operational Research 36 (2) (1988) 217–226.
[239] M. Veenstra, L. Coelho, The Time-Dependent Shortest Path and Vehicle Routing Problem, Technical report, working paper, CIRRELT-2017-57, 2017.
[240] D. Venturelli, A. Kondratyev, Reverse quantum annealing approach to portfolio optimization problems, arXiv:1810.08584, 2018.
[241] T. Vidal, T.G. Cranic, M. Gendreau, C. Prins, Heuristics for multi-attribute vehicle routing problems: a survey and synthesis, European Journal of Operational Research 231 (1) (2013) 1–21.
[242] B. Von Hohenbalken, Simplicial decomposition in nonlinear programming algorithms, Mathematical Programming 13 (1) (1977) 49–68.
[243] Y. Wang, X.Y. Feng, Y.X. Huang, D.B. Pu, W.G. Zhou, Y.C. Liang, C.G. Zhou, A novel quantum swarm evolutionary algorithm and its applications, Neurocomputing 70 (4–6) (2007) 633–640.
[244] Y. Wang, W.Y. Szeto, K. Han, T.L. Friesz, Dynamic traffic assignment: a review of the methodological advances for environmentally sustainable road transportation applications, Transportation Research Part B: Methodological 111 (2018) 370–394.
[245] J. Wardrop, Some theoretical aspects of road traffic research, Proceedings of the Institute of Civil Engineers 1 (5) (1952) 767–768.
[246] P. Wark, J. Holt, A repeated matching heuristic for the vehicle routing problem, Journal of Operational Research Society 45 (10) (1994) 1156–1167.
[247] S. Warshall, A theorem on Boolean matrices, Journal of the ACM (JACM) 9 (1) (1962) 11–12.
[248] B.W. Wie, R.L. Tobin, D. Bernstein, T.L. Friesz, A comparison of system optimum and user equilibrium dynamic traffic assignments with schedule delays, Transportation Research Part C: Emerging Technologies 3 (6) (1995) 389–411.
[249] J.M. Wooldridge, Introductory Econometrics: A Modern Approach, 5th edition, Cengage Learning, 2012, pp. 583–585, Ch. 17.

[250] A. Wren, J.D. Carr, Computers in Transport Planning and Operation, Ian Allan, London, 1971.
[251] A. Wren, A. Holliday, Computer scheduling of vehicles from one or more depots to a number of delivery points, Operational Research Quarterly 23 (3) (1972) 333–344.
[252] X.-B. Wu, J. Liao, Z.-C. Wang, Water wave optimization for the traveling salesman problem, in: Intelligent Computing Theories and Methodologies, vol. 9225, 08 2015, pp. 137–146.
[253] M. Xiao, H. Nagamochi, An exact algorithm for TSP in degree-3 graphs via circuit procedure and amortization on connectivity structure, Algorithmica 74 (2) (2016) 713–741.
[254] P. Yellow, A computational modification to the savings method of vehicle scheduling, Operational Research Quarterly 21 (1970) 281–283.
[255] A. Yeo, Large Exponential Neighbourhoods for the TSP, Technical report, Department of Maths and CS, Odense University, Odense, Denmark, 1997.
[256] Y.J. Zheng, Water wave optimization: a new nature-inspired metaheuristic, Computers & Operations Research 55 (2015) 1–11.
[257] L. Zhu, S.J. Kim, M. Hara, M. Aono, Royal Society Open Science 5 (2018).

Chapter 3

Exact algorithms for solving rich vehicle routing problems

Miroslaw Blocho
ABB, Krakow, Poland

Exact approaches are mostly used when the number of customers in the problem is low and the optimal solution can be found in a reasonable time. This is due to NP-hardness of the routing problems. In the literature the most important approaches to solve these problems in optimal way are Integer Linear Programming (ILP) and Constraint Programming (CP). Most of the routing problems can be formulated as ILP where the variables are to be integers (e.g., the capacity of a vehicle, time window constraints, etc.) and the constraints are formulated in the way of linear equations or inequalities [17]. The strategies of branch-and-bound, branch-and-cut, branch-and-price, and branch-and-cut-and-price are among the mostly used strategies in ILP, and they are described in the next subsections. On the other hand, Constraint Programming is a process of identifying feasible solutions where the problem is modeled in terms of global constraints. CP is also used to solve the routing problems [27] and is described in Section 3.5. In Section 3.6, we briefly summarize the most widely used exact algorithms for solving rich vehicle routing problems.

3.1 Branch-and-bound methods

The branch-and-bound methods are based on partitioning a complete problem into subproblems and reducing the solution space by divide-and-conquer strategies [18]. In this method, the entire solution space is initially examined followed by bounding phases where a problem is relaxed by accepting infeasible solutions. A lower bound on value of optimal solution is obtained by solving such relaxation of the problem. In each step of the algorithm, one of the subproblems is selected and examined, and the attempt to solve the relaxed problem is carried out. By checking the value of solved relaxation, either subproblems are pruned or a new optimal solution is found, and the process can be continued by branching a given subproblem. The process is continued until all subproblems are examined or pruned, and the best solution achieved will be the final optimal solution. An illustration of the branch-and-bound method is shown in Fig. 3.1.

Smart Delivery Systems. https://doi.org/10.1016/B978-0-12-815715-2.00008-7

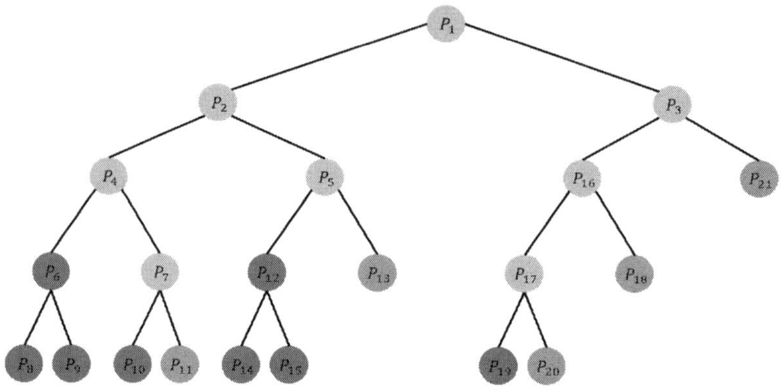

FIGURE 3.1 Illustration of the branch-and-bound method.

In 1983, Laporte and Nobert proposed the branch-and-bound algorithm for the Capacitated Vehicle Routing Problem (CVRP), where the problem was formulated as an integer program. The integrality was obtained by means of a branch-and-bound procedure. The capacity constraints of the CVRP were initially relaxed and then introduced back again only when they are found to be violated in the next phases. By applying the branch-and-bound method they were able to find exact solutions for problems ranging from 15 to 50 cities [20].

In 1994, Fisher proposed the K-tree method for solving the CVRP and used a novel branch-and-bound procedure in which problem is split by fixing the edge incidence of selected subsets of clustered clients [18]. The side constraints are dualized to obtain a Lagrangian relaxation, which is a minimum degree-constrained K-tree problem. By applying this technique Fisher succeeded in solving a problem with 71 clients [18].

3.2 Branch-and-cut methods

The combinatorial optimization problems formulated as linear integer programming problems can be solved by branch-and-cut algorithms. These algorithms can be described as exact algorithms being combination of the branch-and-bound algorithm and the cutting plane method [21]. The idea of branch-and-cut methods is to solve a sequence of linear programming relaxations of the integer programming problem. The cutting plane methods are used to improve the relaxation of the problem, and then branch-and-bound methods are used with divide-and-conquer strategies to finally solve the problem.

Cutting plane algorithms were first introduced by Gomory in 1963 and used as integer solutions to linear programs [13]. In the next decades this approach was evolving and together with branch-and-bound methods led to first branch-

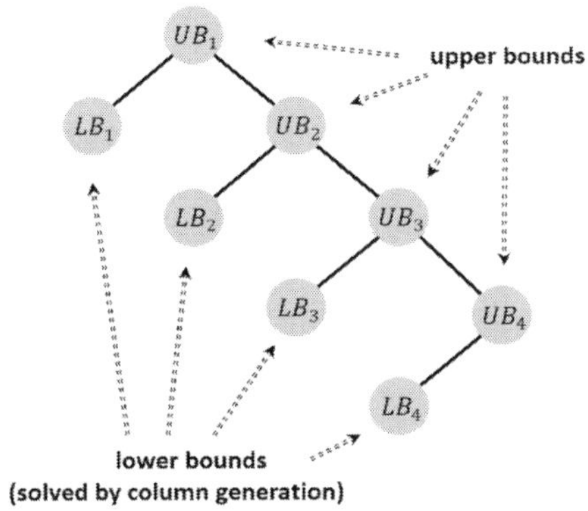

FIGURE 3.2 Illustration of the branch-and-price method.

and-cut algorithms for the traveling salesman problem. These algorithms were used to solve large-scale TSP problems, for example, by Padberg and Rinaldi [24] and Grotschel and Holland [14].

One important aspect of the branch-and-cut method is that it is used to provide bounds to the problem. This is very useful when minimizing but with no option to prove optimality. In such cases a lower bound on the optimal value can be deducted from the branch-and-cut algorithm, and it is used then to provide a guarantee on the distance from optimality. For this reason, branch-and-cut algorithms are often used in combination with metaheuristics to obtain near-optimal solution and to indicate how far such a solution is from optimality [21].

Considering routing problems, Contardo et al. [7] described in 2013 the branch-and-cut algorithm for capacitated routing problems providing optimal solutions for 8 depots and 88 customers.

3.3 Branch-and-price methods

The branch-and-price method is a generalization of the branch-and-bound procedure with linear programming relaxations, allowing column generation to be applied throughout a branch-and-bound tree [8]. The idea of branch-and-price algorithm is similar to the branch-and-cut procedure with such a difference that it focuses on column generation rather than on row generation. An illustration of the branch-and-price method is shown in Fig. 3.2.

Pricing and cutting are complementary methods in terms of the relaxation; however, branch-and-price is more suited for solving huge-scale integer programming problems [3]. Due to satisfactory results of the branch-and-price procedure, a number of various algorithms were proposed just recently for scheduling problems, including vehicle routing problems [23,22,1] and pickup and delivery problems [5,12,26,6].

3.4 Branch-and-cut-and-price methods

The branch-and-cut-and-price (BCP) method is the extension of branch-and-price method by adding cutting planes in order to strengthen the relaxation [9]. The cuts and variables can be dynamically generated over the whole search tree during linear programming branch and bound [19]. Combining two techniques of pricing and cutting requires much more sophisticated methods and high implementation effort, and thus number of algorithms in the literature is low.

The branch-and-cut-and-price algorithms for the routing problems are usually very robust and can solve to optimality many test instances as in case of robust BCP algorithm proposed by Fukasawa et al. to solve the CVRP where the problem with up to 135 vertices was solved [10]. In turn, Pecin et al. [25] proposed in 2017 the improved BCP algorithm for the CVRP problem introducing route enumeration and strong branching, which allowed them to solve to optimality all instances with up to 199 customers and some larger instances with up to 360 customers too, which were solvable before only by heuristic methods. The other BCP algorithms were proposed also recently for the vehicle routing problems by Gauvin et al. [11] and Bettinelli et al. [4]. In all the cases, computational results show that BCP methods are competitive with standard branch-and-cut and branch-and-price procedures in terms of time and power to find optimal solutions.

3.5 Constraint Programming

Constraint Programming (CP) is a method for representing and solving various optimization problems [27]. The optimization problems are expressed in terms of variables and constraints between the variables. The constraints do not specify the sequence of steps to be executed but the properties of the optimal solution. The technique of depth-first branch and bound is typically used in CP as the method of solving optimization problems [2]. The depth-first search is used for satisfaction, and branch and bound is used for optimization the problem. Constraint Programming is often used for many variants of the vehicle routing problems as these optimization problems can often be expressed in terms of capacity, time windows, service time, and precedence constraints, and CP is a natural problem representation. An example of the Constraint Programming method is shown in Fig. 3.3.

Recently, constraint programming technique is often combined with metaheuristics approach giving substantial improvements in the area of vehicle routing problems. Guimarans et al. presented the approach of Constraint Programming and Lagrangean Relaxation (LR) methods to improve the efficiency of algorithms. The complete problem was decomposed into two separated subproblems, to which the mentioned techniques were applied to obtain a final solution. With such a decomposition, the methodology provides a quick initial feasible solution, which is quickly improved by metaheuristics. CP and LR were also embedded within such a structure to ensure constraints satisfaction and reduc-

$$\textit{Objective function:}\ f(x,y,z) = 2x + 3y - 4\log(z^2 + x^y)$$

$$\textit{Constraints:}$$
$$x - y \leq 10$$
$$\frac{2x}{z^3} - y \geq -24$$
$$\sqrt{x} - 3z^2 = 10$$

FIGURE 3.3 An example of the Constraint Programming.

ing calculation time. Such combination allowed for new optimal solutions on 200-customer test instances [15]. In 2018, Hojabri et al. proposed a combination of large neighborhood search with constraint programming to solve the VRP with synchronization constraints. The search abilities of the large neighborhood search and the constraint propagation abilities of constraint programming are combined to determine the feasibility of searched solutions [16].

3.6 Summary

In this chapter, we described the most widely used exact algorithms for solving rich vehicle routing problems. The branch-and-bound methods are based on partitioning the complete problem into subproblems and reducing the solution space by divide-and-conquer strategies. The branch-and-cut method can be described as a combination of the branch-and-bound algorithm and cutting plane methods. In turn, the branch-and-price method is a generalization of the branch-and-bound procedure with linear programming relaxations combined with column generations, and the branch-and-cut-and-price method is an extension of the branch-and-price method by adding cutting planes to strengthen the relaxation. In constraint programming the optimization problems are expressed in terms of variables and constraints between the variables. Due to very large time complexity of the exact approaches, they are mostly used only when number of customers in particular problems is relatively low and thus the optimal solution can be found in a reasonable time. The recent trend in solving rich vehicle routing problems is hybrid algorithms combining approximation algorithms to explore the solution space with exact algorithms used to exploit solution regions of highest probability to obtain very good quality solutions.

References

[1] Z. Akca, R.T. Berger, T.K. Ralphs, A branch-and-price algorithm for combined location and routing problems under capacity restrictions, in: Proceedings of the Eleventh INFORMS Computing Society Meeting, 2009, pp. 309–330.

[2] Bruno De Backer, Vincent Furnon, Paul Shaw, Philip Kilby, Patrick Prosser, Solving vehicle routing problems using constraint programming and metaheuristics, Journal of Heuristics 6 (4) (Sep 2000) 501–523.

[3] Cynthia Barnhart, Ellis L. Johnson, George L. Nemhauser, Martin W.P. Savelsbergh, Pamela H. Vance, Branch-and-price: column generation for solving huge integer programs, Operations Research 46 (1998) 316–329.
[4] Andrea Bettinelli, Alberto Ceselli, Giovanni Righini, A branch-and-cut-and-price algorithm for the multi-depot heterogeneous vehicle routing problem with time windows, Transportation Research Part C: Emerging Technologies 19 (5) (2011) 723–740, Freight Transportation and Logistics (selected papers from ODYSSEUS 2009 – the 4th International Workshop on Freight Transportation and Logistics).
[5] Andrea Bettinelli, Alberto Ceselli, Giovanni Righini, A branch-and-price algorithm for the multi-depot heterogeneous-fleet pickup and delivery problem with soft time windows, Mathematical Programming Computation 6 (2) (Jun 2014) 171–197.
[6] Marilène Cherkesly, Guy Desaulniers, Gilbert Laporte, Branch-price-and-cut algorithms for the pickup and delivery problem with time windows and last-in-first-out loading, Transportation Science 49 (4) (November 2015) 752–766.
[7] Claudio Contardo, Jean-Francois Cordeau, Bernard Gendron, A computational comparison of flow formulations for the capacitated location-routing problem, Discrete Optimization 10 (4) (2013) 263–295.
[8] Jacques Desrosiers, Marco E. Lübbecke, A Primer in Column Generation, Springer US, Boston, MA, 2005, pp. 1–32.
[9] Jacques Desrosiers, Marco E. Lübbecke, Branch-Price-and-Cut Algorithms, American Cancer Society, 2011.
[10] Ricardo Fukasawa, Humberto Longo, Jens Lysgaard, Marcus Poggi de Aragão, Marcelo Reis, Eduardo Uchoa, Renato F. Werneck, Robust branch-and-cut-and-price for the capacitated vehicle routing problem, Mathematical Programming 106 (3) (May 2006) 491–511.
[11] Charles Gauvin, Guy Desaulniers, Michel Gendreau, A branch-cut-and-price algorithm for the vehicle routing problem with stochastic demands, Computers & Operations Research 50 (10 2014).
[12] Veaceslav Ghilas, Emrah Demir, Tom Van Woensel, The pickup and delivery problem with time windows and scheduled lines, INFOR: Information Systems and Operational Research 54 (2) (2016) 147–167.
[13] Ralph E. Gomory, Outline of an Algorithm for Integer Solutions to Linear Programs and an Algorithm for the Mixed Integer Problem, Springer, Berlin, Heidelberg, 2010, pp. 77–103.
[14] Martin Grötschel, Olaf Holland, Solution of large-scale symmetric travelling salesman problems, Mathematical Programming 51 (1) (Jul 1991) 141–202.
[15] Daniel Guimarans, Rosa Herrero, Juan Ramos, Silvia Padrón, Solving vehicle routing problems using constraint programming and Lagrangean relaxation in a metaheuristics framework, International Journal of Information Systems and Supply Chain Management 4 (04 2011) 61–81.
[16] Hossein Hojabri, Michel Gendreau, Jean-Yves Potvin, Louis-Martin Rousseau, Large neighborhood search with constraint programming for a vehicle routing problem with synchronization constraints, Computers & Operations Research 92 (2018) 87–97.
[17] Victor Klee, in: George B. Dantzig (Ed.), Linear Programming and Extensions, Princeton University Press, Princeton, N.J., 1963, xviii + 625 pp. illus., Science 146 (3651) (1964) 1572.
[18] Marshall L. Fisher, Optimal solution of vehicle routing problems using minimum k-trees, Operations Research 42 (08 1994) 626–642.
[19] Laszlo Ladányi, Ted K. Ralphs, Leslie E. Trotter, Branch, Cut, and Price: Sequential and Parallel, Springer, Berlin, Heidelberg, 2001, pp. 223–260.
[20] G. Laporte, Y. Nobert, A branch and bound algorithm for the capacitated vehicle routing problem, Operations-Research-Spektrum 5 (2) (Jun 1983) 77–85.
[21] John Mitchell, Branch-and-cut algorithms for combinatorial optimization problems, in: Handbook of Applied Optimization, 01 2002.
[22] Ibrahim Muter, Jean-Francois Cordeau, Gilbert Laporte, A branch-and-price algorithm for the multidepot vehicle routing problem with interdepot routes, Transportation Science 48 (08 2014) 425–441.

[23] Gizem Ozbaygin, Oya Ekin Karasan, Martin Savelsbergh, Hande Yaman, A branch-and-price algorithm for the vehicle routing problem with roaming delivery locations, Transportation Research Part B: Methodological 100 (C) (2017) 115–137.

[24] Manfred Padberg, Giovanni Rinaldi, A branch-and-cut algorithm for the resolution of large-scale symmetric traveling salesman problems, SIAM Review 33 (1) (February 1991) 60–100.

[25] Diego Pecin, Artur Pessoa, Marcus Poggi, Eduardo Uchoa, Improved branch-cut-and-price for capacitated vehicle routing, Mathematical Programming Computation 9 (1) (Mar 2017) 61–100.

[26] Stefan Ropke, Jean-François Cordeau, Branch and cut and price for the pickup and delivery problem with time windows, Transportation Science 43 (3) (August 2009) 267–286.

[27] Paul Shaw, Using constraint programming and local search methods to solve vehicle routing problems, in: Michael Maher, Jean-Francois Puget (Eds.), Principles and Practice of Constraint Programming — CP98, Springer, Berlin, Heidelberg, 1998, pp. 417–431.

Chapter 4

Heuristics, metaheuristics, and hyperheuristics for rich vehicle routing problems

Miroslaw Blocho
ABB, Krakow, Poland

4.1 Heuristics for rich vehicle routing problems

The heuristics is an approach to solve a given problem that does not guarantee obtaining the optimal solution. However, they allow us to elaborate high-quality feasible solutions that meet the problem objectives. The main families of heuristics are classical heuristics developed mostly between 1960 and 1990 and metaheuristics with rapid development lasting until now [103].

The classical heuristics can be grouped into two main groups, construction and improvement techniques. Both methods perform a limited exploration of the solution space and usually produce good quality solutions within reasonable time. Most of them can be relatively easily extended for various constraints encountered in real life, and for this reason, they are still used in many commercial applications [103].

The main reason to develop and use heuristic approaches is finding good-quality solutions of various problems in short time. Other reasons are the nonexistence of an exact method to solve the problem and flexibility of a heuristic algorithm to handle additional side constraints. Summing up, a heuristics is considered "good" when a solution can be computed with reasonable computational effort, the resulting solution is near-optimal, and the low probability of obtaining solution that is far from optimal [111].

4.1.1 Construction heuristics

Construction heuristics build feasible solutions while trying to minimize its cost but often do not consider any improvement phases. Route construction heuristics usually start from an empty solution and build iteratively subsequent routes by inserting customers one by one until all customers are served. They are considered the simplest and also fastest methods to solve the problem; however, resulting solutions are usually far from optimal solutions. Due to that fact, they

Smart Delivery Systems. https://doi.org/10.1016/B978-0-12-815715-2.00009-9

are commonly used only as the baseline to find an initial solution to the problem, which is later improved in the next phases of the algorithm [103]. The main methods used in construction heuristics for the routing problems are nearest-neighbor, insertion, and savings techniques.

In the nearest-neighbor method, the solution is constructed by a greedy algorithm of selecting the closest unserved customer each time. This method was proposed in 1956 by Flood [52] and is considered the simplest way of creating initial solutions. Although the method was proposed many years ago, it still attracts interest. Kizilateş and Nuriyeva [94] provided in 2013 a modification of the nearest-neighbor algorithm for the traveling salesman problem. In 2015, Joshi and Kaur proposed the nearest-neighbor insertion algorithm for solving Capacitated Vehicle Routing Problem (CVRP), where they investigated a practical scenario in which different college buses take students from different bus stops.

Insertion heuristics generate feasible solutions by repeatedly inserting unserved customers into partial feasible routes. Various variants of the insertion heuristics were created due to two key decisions made at every iteration: selection of customer to be inserted and choosing a place of insertion [31]. Insertion heuristics proved to be very fast and able to produce relatively good initial solutions. On the other hand, they are usually easy to be implemented and can be easily extended to handle more complex constraints. Most of the insertion methods were created several decades ago for the family of the vehicle routing problems by Solomon [172], Vigo [192], Liu and Shen [106], and Salhi and Nagy [164]. In 2006, Joubert and Classen [87] proposed a new sequential insertion heuristic for the initial solution to a constrained VRP introducing a new concept of time window compatibility. More recently, in 2015, Pinto et al. [147] proposed an insertion heuristics for the CVRP with loading constraints and mixed linehauls and backhauls. Their approach was based on standard insertion heuristics extended to tackle the explicit consideration of loading constraints.

The savings algorithm was originally developed for the VRP with one central repository and variable number of vehicles by Clarke and Wright [35] in 1964. Its main idea is to compute savings for combining two customers into the same route. The main advantage of that algorithm is that it can be applied for both directed and undirected problems. The savings method was proposed more than fifty years ago, but it still attracts some interest. Gajpal and Abad [56] proposed in 2010 saving-based algorithms for the vehicle routing problem with simultaneous pickup and delivery introducing cumulative net-pickup approach for checking the feasibility when two existing routes are merged. In 2012, Pichpibul and Kawtummachai [146] proposed a new enhancement for Clarke–Wright savings algorithm to optimize the CVRP. They suggested two-phase probabilistic mechanism and the route postimprovement, which gave good results for benchmarks involving from 16 to 135 customers.

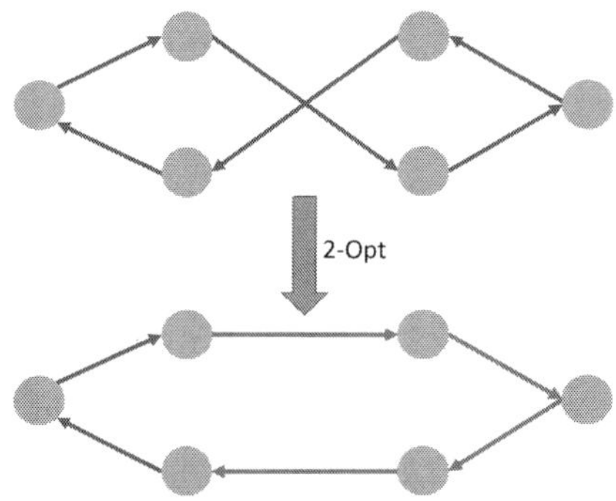

FIGURE 4.1 2-Opt local search move.

4.1.2 Improvement heuristics

The improvement heuristics are usually used for already generated solutions by other heuristics or exact algorithms. Local search methods are typically applied for simple local modifications such as customer or arc exchanges to generate neighboring solutions of possibly better quality. In case better solution is found, it replaces the current one, and the process is continued until local minimum has been found [103]. The main idea driving local search is that by repeatedly improving the quality of solution by making small changes (called moves) it is often possible to find very good solutions. Different heuristics are mostly characterized by different types of local search moves and by obtained neighborhoods. Considering created neighborhoods, there are usually two strategies considered, first-accept (FA) and best-accept (BA). In the FA strategy the first neighbor satisfying acceptance criteria is selected, whereas in the BA strategy, all neighbors have to be examined, and the best one is chosen [27].

The most famous local search moves are 2-Opt and 3-Opt related to edge-exchange neighborhoods for a single route. The 2-Opt method was proposed by Flood [52] in 1956 and Croes [39] in 1958 and is based on deleting two edges from the same route followed by reconnecting the subtours in the other way. A sample 2-Opt local search move is shown in Fig. 4.1. The 3-Opt neighborhood was proposed by Bock [20] in 1958 and is similar to 2-Opt except that three edges are deleted and rejoined with three new links. A sample 3-Opt local search move is shown in Fig. 4.2. In turn, k-Opt neighborhood is a generalization of the 2-Opt and 3-Opt neighborhoods to k-edge exchanges in one local search move. It is important to note that checking k-Opt neighborhoods requires $O(n^k)$ computational time. The first works for application of k-Opt improvement heuristics for the vehicle routing problems were conducted by Russel [161] in 1977. They were further improved mostly in the area of checking the infeasible neighboring solutions by Savelsbergh [165], Solomon and Desrosiers [173], and Baker and Schaffer [10].

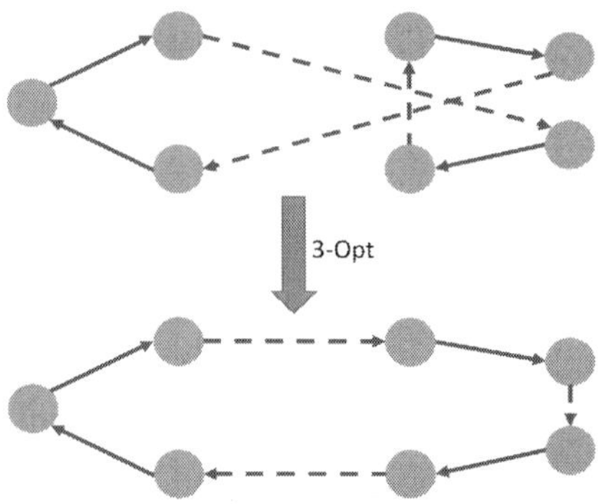

FIGURE 4.2 3-Opt local search move.

The OR-Opt technique proposed in 1976 by Or [134] is a modification of the 3-Opt by taking into account only such 3-Opt exchanges that result in one, two, or three adjacent customers inserted between two other customers. Sample OR-Opt local search move is shown in Fig. 4.4. Replacing up to three edges in the original tour by three new edges does not modify the orientation of the route, which is very crucial in many vehicle routing problems. The cross-exchange neighborhood is constructed by exchanging two subroutes of length at most k, and thus it may be considered as a specific subset of the k-Opt neighborhood [178].

In 1992, Savelsbergh [165] proposed interroute operators for relocation and exchange. The relocate operator is used to move one customer from one route to another, whereas the exchange operator swaps two customers in two different routes. The 2-Opt* neighborhood is a specific variant of 2-Opt, which is a set of solutions obtained by removing two edges from two different routes, followed by adding two different edges to reconnect broken subroutes [153]. A sample 2-Opt* local search move is shown in Fig. 4.5.

This type of neighborhood was introduced in 1995 by Potvin and Rousseau [153], who were doing extensive studies on 2-Opt, 3-Opt, and OR-Opt techniques. The key point in 2-Opt* moves is that the order of customers in subroutes is maintained. Potvin and Rousseau proposed also a hybrid approach comprising both OR-Opt and 2-Opt* methods together, which proved to be competitive to these methods applied separately. In this hybrid approach the operator (OR-Opt, 2-Opt*) changes each time local minimum is reached.

The more specialized neighborhoods are GENI-exchange by Gendreau et al. [62], λ-interchange by Osman [140], CROSS-exchange by Taillard et al. [178], and ejection chains by Glover [66]. The GENI operator is an extension of the standard relocate operator by Savelsbergh with such a difference that an out-relocated customer may be inserted between two customers in the destination route even violating the order of the customers in that route. A sample GENI exchange is shown in Fig. 4.3. The λ-interchange defines the neighbor-

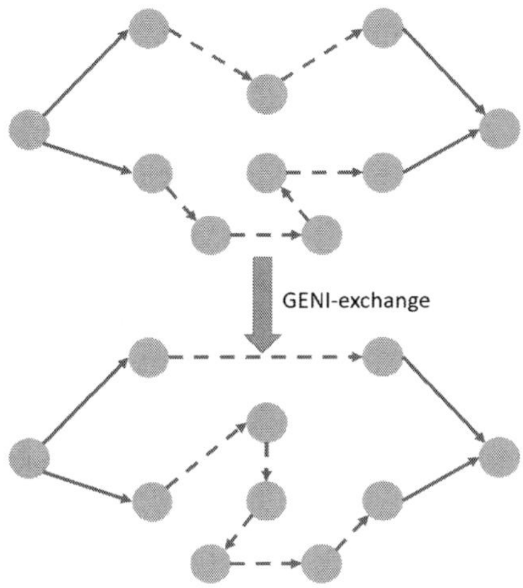

FIGURE 4.3 GENI local search move.

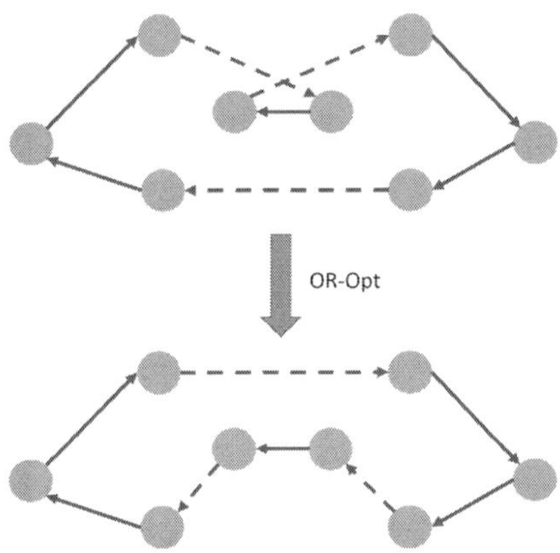

FIGURE 4.4 OR-Opt local search move.

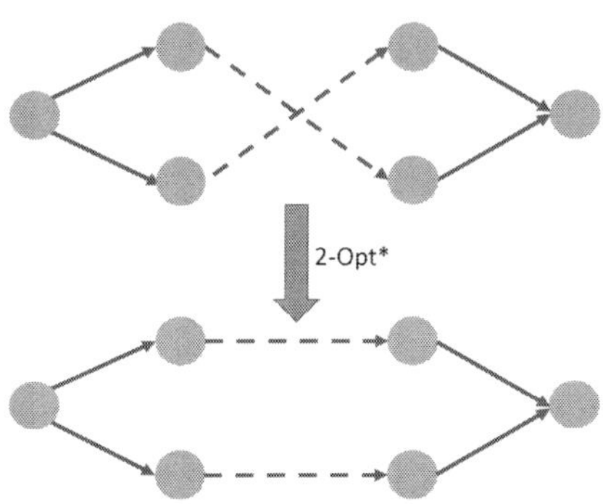

FIGURE 4.5 2-Opt* local search move.

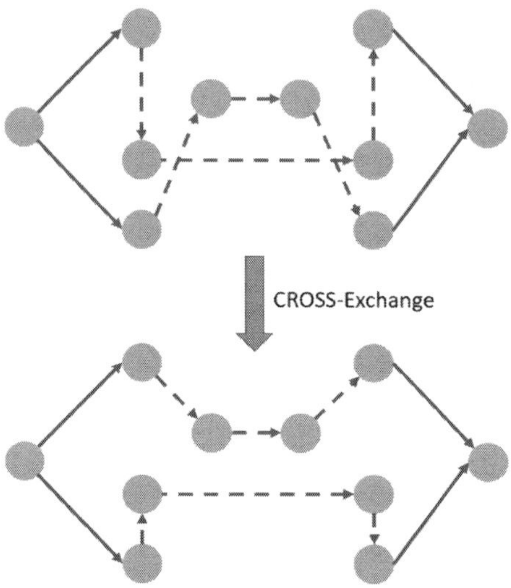

FIGURE 4.6 CROSS-exchange local search move.

hood with λ customers shifted from one route to another or exchanged between two routes with a specific order on each pair of routes. The basic idea of the CROSS-exchange is to exchange two consecutive customers from one route with two consecutive customers from the second route preserving the original order. A sample CROSS-exchange is shown in Fig. 4.6. In turn, ejection chains are based on a sequence of removal and insertion moves repeated until an unrouted customer can be inserted into the destination route without the need to eject any customer.

4.2 Metaheuristics for rich vehicle routing problems

Metaheuristics is an enhancement of classical heuristics with emphasis on deep exploration of the solution space. They usually combine sophisticated neighborhood search rules and recombination of solutions [103]. The quality of the solutions produced by metaheuristics is typically much higher than that obtained by classical heuristics techniques but with a price of increased computing time. The metaheuristic techniques are context dependent and usually require finely tuned parameters, which unfortunately make their extensions to other problems difficult. Many various metaheuristics have been proposed for the vehicle routing problems, and they can be widely divided into local search, population search, and learning mechanism groups; however, best metaheuristics merge ideas from different approaches [38]. These different approaches will be described in the following subsections.

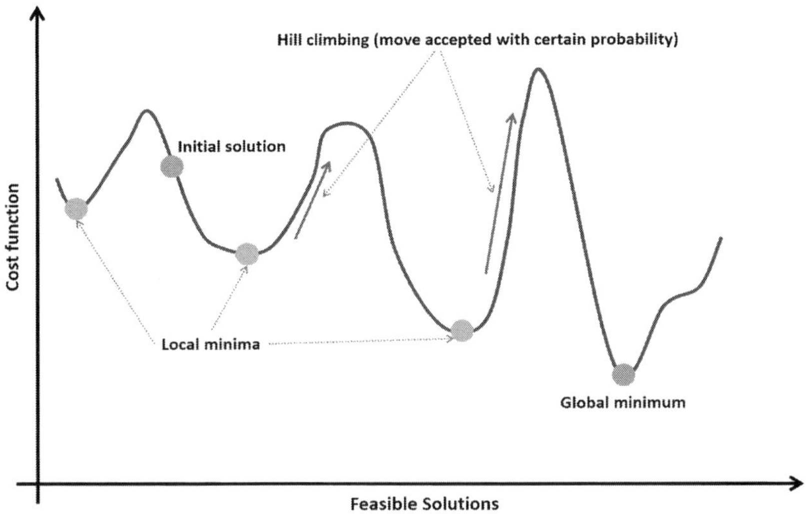

FIGURE 4.7 Simulated Annealing procedure (blue (mid gray in print version) dot – initial solution, orange (light gray in print version) dot – local minimum, green (mid gray in print version) dot – global minimum, red (dark gray in print version) arrow – hill climbing (move accepted with certain probability)).

4.2.1 Simulated Annealing

Simulated Annealing (SA) is a probabilistic technique proposed in 1983 by Kirkpatrick et al. [93] and in 1985 by Černý [213]. Its main idea is to find a global minimum of a specific objective function attempting to escape local minima in the search process. Due to low complexity, it can be used in many various optimization problems, not only related to the VRPs. This method takes its name from the annealing in metallurgy, involving heating a solid material to a certain temperature at which it becomes liquid, followed by slow and controlled cooling down until the solid state is reached again and the metal particles are rearranged in a minimal energy molecular structure. An example of simulated annealing procedure is shown in Fig. 4.7.

The SA is a stochastic algorithm involving asymptotic convergence and allowing random movements in the searched neighborhood in order to escape local minima [1]. Although proposed more than 40 years ago, it still attracts some attention and is broadly used in many existing solutions for different variants of the vehicle routing problems.

Woch and Łebkowski [212] presented in 2009 a standard sequential simulated annealing algorithm applied to the VRPTW, which gave two new world's best solutions to Solomon benchmark set. In 2013, Afifi et al. [4] developed a simulated annealing algorithm for the VRPTW with synchronization constraints incorporating several local search techniques to deal with this problem. By implementing it the researchers were able to produce high-quality solutions in very short computational time detecting some new world's best solutions too.

Yu et al. [207] suggested in 2016 a simulated annealing heuristic for the hybrid vehicle routing problem (HVRP), an extension of the Green Vehicle Routing Problem. They focused on vehicles that use a hybrid power source, and the model considered the utilization of electric and fuel power depending on the availability of either electric charging or fuel stations. By developing SA with restart strategy combined with Boltzmann and Cauchy functions for determining the acceptance probability of a worse solution, the researchers were able to effectively solve the HVRP test instances.

In 2017, Wei et al. [198] proposed a simulated annealing algorithm for the CVRP with two-dimensional loading constraints combining mechanism of repeatedly cooling and rising the temperature with a new open space-based packing method. The computational results proved that such combination of approaches allows for reaching or improving best-known solutions for most instances.

Just recently in 2018, Linan-Garcia et al. [50] proposed another multi-phases metaheuristic algorithm based on Simulated Annealing to solve the CVRP with stochastic demands. Their algorithm comprises very custom phases of annealing including Fast Quenching Phase, the Annealing Boltzmann Phase, the Bose–Einstein Annealing Phase, and the Dynamical Equilibrium Phase applied in different ranges of temperature in the Simulated Annealing.

The SA was also used to solve pickup and delivery problems. The SA algorithm developed by Wang et al. [194] in 2013 for the VRPTW with simultaneous pickup-delivery, which also proved to be an effective metaheuristic due to results of computational tests on Wang and Chen benchmark set. There exists also a combination of the SA with the other heuristics such as Two-Stage Hybrid Local Search algorithm for the VRPTW proposed in 2001 by Bent and Hentenryck [15]. Their algorithm used the simulated annealing method to minimize the number of vehicles in the first phase of the VRPTW, whereas the large neighborhood search (LNS, see Section 4.2.5 for more details) was used to minimize total travel cost.

4.2.2 Tabu Search

Tabu Search (TS) was introduced and formalized by Glover [65] in 1959 as a metaheuristic search technique comprising local search methods and memory structures called tabu-list. Its main idea is to avoid cycling by inserting recently checked solution on the tabu-list, so during the search process, the solutions marked with tabu label are not taken into consideration. Such approach helps in getting out from the local minima and increases chances to find global, optimal solution. An example of tabu search method is shown in Fig. 4.8.

The initial implementations of the tabu search method to the routing problems were conducted in 1990s by Taillard [179] and Gendreau et al. [63] for the

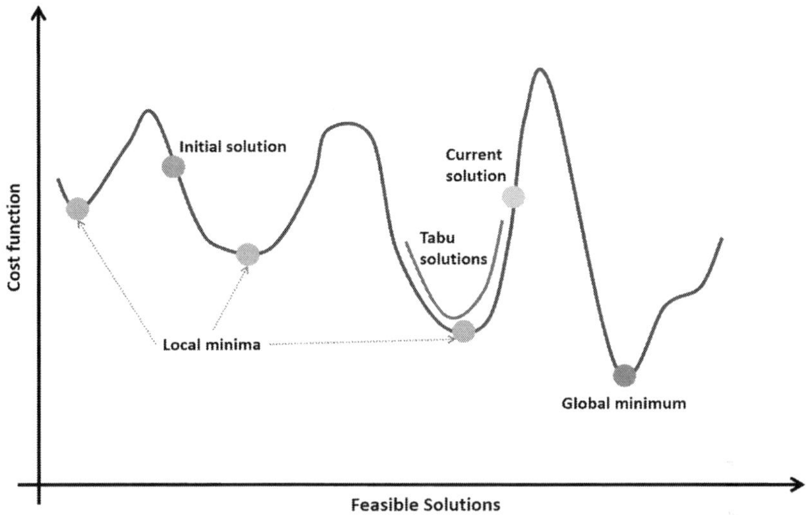

FIGURE 4.8 Tabu Search method (blue (mid gray in print version) dot – initial solution, orange (light gray in print version) dot – current solution, orange (light gray in print version) dot – local minimum, green (dark gray in print version) dot – global minimum, red (dark gray in print version) curve – tabu solutions).

CVRP and by Semet and Taillard [169] and Potvin et al. [152] for the VRPTW. The TS algorithm by Taillard [179] for the CVRP is still considered one of the best methods to solve it, and its main ideas comprise random tabu durations combined with continuous diversification mechanism penalizing frequently performed moves, making broader exploration of the search space. In turn, the tabu search algorithm by Gendreau et al. [63] is based on moving in each iteration a customer from one route to another where at least one neighbor is present. The consecutive insertions are carried out simultaneously with a local route improvement based on the GENI-exchanges (see Section 4.1.2 for more details). Similarly to Taillard's algorithm, infeasible solutions are also penalized, and random tabu durations are combined with continuous diversification mechanism.

Based on the initial tabu search algorithms, many other researchers conducted extensive studies and enhancements producing many different versions of the TS method applied to probably all variants of the vehicle routing problems. A good review on various tabu search heuristics for the VRPTW was given by Bräysy and Gendreau [26] in 2002. In 2004, Ho and Haugland proposed tabu search heuristics for the VRPTW with Split Deliveries, a variant where customer demands exceed vehicle capacity. Cordeau and Laporte [37] gave in 2005 a very good review on the most important heuristics for the whole family of the VRPs focusing on short/long memory structures, neighborhood structures, intensification, and comparison of computational results of various algorithms. A combination of the Tabu Search with guided local search principles were proposed in 2009 by Zachariadis et al. [208] for the VRP with two-dimensional loading constraints, and its implementation led to several new best known solutions. Moccia et al. [117] described in 2010 the incremental tabu search algorithm for the Generalized VRPTW introducing an incremental procedure to compute successive neighborhoods and demonstrating through extensive computational experiments that such general algorithm is competitive with other specialized heuristics for the VRPTWs. Brandao [23] reported in 2011 four new world's best solutions for the heterogeneous fixed fleet VRP by implementing a tabu search algorithm for this variant of the VRP.

More recently in 2012, Tarantilis et al. [184] proposed a tabu search solution framework for the Consistent VRP adopting a two-level master–slave decomposition scheme. The tabu search heuristic approach was employed on a dual mode basis to modify both the template routes and current daily schedules proving the competitiveness of such an approach by computational experiments on benchmark datasets. In 2013, Nguyen et al. [129] described a novel tabu search heuristic for the Time-dependent Multizone Multitrip VRPTW, where two types of neighborhoods related to the two sets of decisions of the problem together with an approach controlling the selection of the neighborhood type for particular phases of the search are able to combine exploration and exploitation capabilities for the search. Moreover, to drive the search to potentially unexplored good regions, the diversification strategy, guided by an elite solution set, is used. Another, very competitive iterated tabu search heuristic for the multicompartment VRP was introduced in 2017 by Silvestrin and Ritt [171], which dominated all existing approaches in nearly all cases for this variant of the VRP.

An attempt to combine tabu search heuristics with simulated annealing was suggested by Wang et al. [196] in 2017 for a vehicle routing problem with cross docks and split deliveries by introducing a constructive heuristics with two layers. The first layer was used to optimize the allocation of trucks to cross docks, whereas the second layer was used to optimize the visitation order to suppliers and retailers for trucks assigned to each cross dock. Another interesting combination of tabu search with ejection chains for the multidepot Open VRP was proposed also in 2017 by Soto et al. [174]. They hybridized this approach also with multiple neighborhood search being able to generate neighborhoods from path moves and ejection chains, and by numerical and statistical tests they proved that such a combination outperforms state-of-the-art methods.

4.2.3 Adaptive Memory Procedures

The Adaptive Memory Procedure (AMP) was introduced in 1995 by Rochat and Taillard [159] and is based on the idea of the so-called *adaptive memory* being pool of good-quality solutions. The solutions in *adaptive memory* are always replaced with the new better-quality solutions coming from recombination of existing ones. The AMP by Rochat and Taillard was initially developed for the VRP. In this context, they proposed the process of extracting nonoverlapping

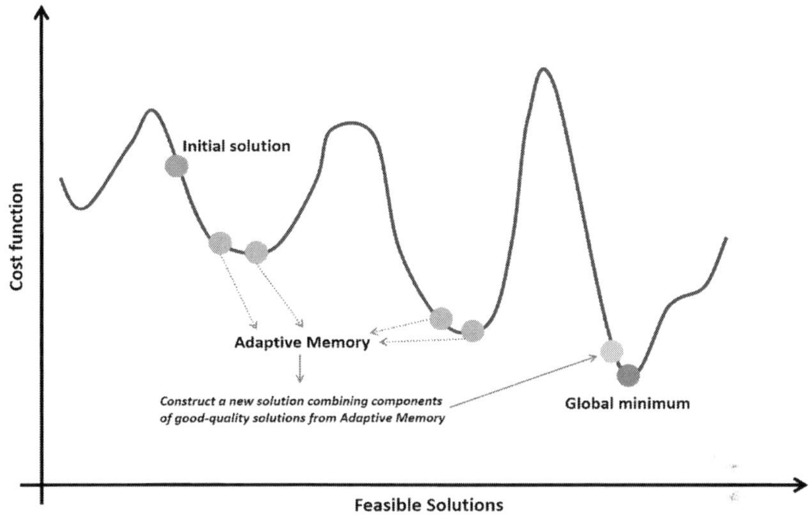

FIGURE 4.9 Adaptive Memory Procedure (blue (mid gray in print version) dot - initial solution, orange (light gray in print version) dots - good quality solutions, orange (light gray in print version) dot - new solution combining components of good-quality solutions from Adaptive Memory, green (dark gray in print version) dot - global minimum).

routes from the solutions followed by insertions of unrouted customers to create new feasible offsprings. Each time new offspring solution has a better quality than the worst one in the adaptive memory, and then such a better offspring solution replaces the worst solution in the adaptive memory. The AMP is illustrated in Fig. 4.9.

Bozkaya et al. [22] developed in 2003 an interesting combination of the adaptive memory procedure with tabu search algorithm for political districting optimization problem, in which the objective is to partition a territory into electoral constituencies subject to contiguity, population equality, and compactness side constraints. The authors performed the computational experiments on the real example of Edmonton city.

Tarantilis and Kiranoudis [183] modified in 2002 the original AMP by changing initial solution generation with the Clarke–Wright savings method combined with 2-Opt local moves, customer swaps between routes, and reinsertions of customers. Instead of extracting full routes from the adaptive memory as in original AMP technique by Rochat and Taillard, they suggested extracting particular route segments (so-called bone-routes) to generate new offspring solutions. The bone-routes proved to be very efficient by executing computational experiments over benchmark sets of the Capacitated VRP. This method was further improved in 2005 by Tarantilis [182] by introducing Solutions Elite Parts Search (SEPAS), an iterative method generating initial solutions by a systematic diversification technique and storing them in the adaptive memory. Generation of the new solutions was carried out using the tabu search heuristics, and such combination allowed the author to find several new best known solutions for the CVRP. The AMP was also applied to the other variants of the VRP such as VRP with multiple trips by Olivera and Viera [132] in 2007 or for the heterogeneous fixed fleet VRP by Li et al. [105] in 2010. In the latter one, Li et al. proposed a multistart adaptive memory programming (MAMP) and path relinking to solve this variant of the VRP. At each iteration, the MAMP was used to construct multiple temporary solutions, further improved by modified tabu search heuristics integrated with path relinking methods.

For the Fleet Size and Mix VRPTW variant, Repoussis and Tarantilis [157] proposed in 2010 an interesting AMP solution utilizing the basic concept of AMP framework combined with probabilistic semiparallel construction heuristic, a novel solution reconstruction mechanism, an innovative Iterated Tabu Search algorithm tuned for intensification local search and frequency-based long-term memory structures. The authors proved to improve the best reported cumulative and mean results for almost all problem instances in reasonable computational time.

More recently, in 2016, Gounaris et al. [72] presented the AMP method to solve the Robust CVRP under demand uncertainty. By focusing on verifying the robust feasibility of candidate routes and uncertainty sets, the authors were able to obtain new best known solution to 123 benchmark instances.

4.2.4 Variable Neighborhood Search

The Variable Neighborhood Search (VNS) was proposed in 1997 by Mladenovic and Hansen [115], which can be described as a framework for building heuristics exploiting the neighborhoods for both finding local optima and getting out of them by perturbation moves. Contrary to classical local search methods, VNS is not based on following the trajectory but rather on exploring increasingly distant neighborhoods of a given solution and moving to the new one only in case of improvement. Such a method usually leads to maintaining best characteristics of current solution and helps obtain neighboring solutions of better quality. After VNS was proposed, a lot of other researchers used this idea to combine VNS with other techniques, usually to create hybrids of VNS with tabu search and other improvement heuristics.

In 2003, Braysy [24] presented a novel Reactive VNS method for the VRPTW based on the modification of the original procedure combined with a new four-phase approach. In the first phase an initial solution is created using specialized route-construction heuristic, and a route-elimination method is applied in the second phase to further decrease the number of routes. The total traveled distance is minimized in the third phase by four new local-search procedures, and finally in the fourth phase the best solution is enhanced by modifying the objective function to escape from a local minimum.

The first attempt to solve the Multi-Depot VRPTW by a Variable Neighborhood Search method was undertaken by Polacek et al. [148] in 2004, and it was proven to be more effective than the baseline Tabu Search algorithm for this variant of the VRPTW.

Paraskevopoulos et al. [144] described in 2008 a very interesting solution methodology for the heterogeneous fleet VRPTW. Their approach was based on the idea of two-phase solution framework built upon a hybridized Tabu Search within a new Reactive Variable Neighborhood Search metaheuristic algorithm. Computational experiments proved the effectiveness of the approach and its applicability to realistic routing problems.

The VNS method was also applied to very large-scale vehicle routing problems, in particular, by Kytojoki et al. [101] in 2007, who applied it to the CVRP. The VNS procedure was used to guide standard improvement heuristics, and a strategy reminiscent of the guided local search metaheuristic was applied to help escape local minima. The authors proved that such a method is very robust, and by performing computational experiments they were able to find high-quality solutions for the problem instances with up to twenty thousands customers, which is considered very large scale in the CVRP.

More recently, in 2015, Wei et al. [199] presented a VNS for the CVRP with two-dimensional loading constraints, in which customer demand is a set of two-dimensional rectangular weighted items. Apart from the VNS to address the routing aspect, they used also a skyline heuristic to examine the loading constraints. Moreover, they utilized Trie data structure to speed up the search process to record the loading feasibility information of routes and to control the computational effort of the skyline spending on the same route. Based on the results of the extensive computational experiments, the authors proved that the proposed method outperformed all existing methods at that time for this variant of the CVRP.

4.2.5 Large Neighborhood Search

The Large Neighborhood Search (LNS) was introduced in 1998 by Shaw [170], initially for the VRPTW, and can be described as an iterative way of destroying and repairing the solution in the neighborhood. Destroying methods have some randomness such that different parts of the solution are destroyed and broader parts of the search tree are visited, and thus the searched neighborhood is larger than in classical local search methods. The removed customers from routes are usually reinserted back using Constraint Programming; however, other methods can be used too. The Constraint Programming strategies are used to check feasibility of the moves, to handle side constraints, and to allow iterative improvements in the search procedures.

Ergun et al. [49] have extended this concept further by proposing Very Large Neighborhood Search (VLNS) for the VRP by introducing operation on multiple routes simultaneously, which allowed even broader search. This advantage comes however with much higher effort required to perform extensive repairing moves in each iteration.

In 2006, Ropke and Pisinger [160] extended the idea of the LNS to the Adaptive Large Neighborhood Search (ALNS) for the Pickup and Delivery Problem with Time Windows. In their concept, multiple destroy and repair methods were allowed within the same search iteration. Another adaptive LNS heuristic but for Two-Echelon VRP (2E-VRP) was proposed in 2012 by Hemmelmayr et al. [75], who proposed new neighborhood search operators by exploiting the structure of the problem, and these operators were used in a hierarchical scheme reflecting the multilevel nature of the problem. Ribeiro and Laporte [158] in 2012 also developed an adaptive LNS for the Cumulative Capacitated VRP, which is a variation of the classical CVRP where the objective is to minimize the sum of arrival times at customers, instead of the total routing cost. The VRP with multiple routes was solved by an adaptive LNS by Azi et al. [8] in 2014, who developed various destruction and reconstruction operators working either at the customer, route, or workday level.

More recently, in 2016, Grangier et al. [73] developed an adaptive LNS for the Two-Echelon Multi-Trip VRP with satellite synchronization, addressing constraints arising in city logistics such as time window constraints, synchronization constraints, and multiple trips. They suggested adaptive LNS combined with custom destruction and repair heuristics and efficient feasibility checks for moves.

4.2.6 Greedy Randomized Adaptive Search Procedure

The Greedy Randomized Adaptive Search Procedure (GRASP) was introduced by Feo and Resende [51] in 1995 and is based on iterative two-phase search algorithm comprising construction and local search phases applied to various combinatorial optimization problems. In each iteration in the construction phase, a feasible solution is constructed by a randomized greedy function. The solution is then iteratively improved in the second phase by local search movements. An illustration of the GRASP is shown in Fig. 4.10.

The GRASP method was applied to the VRPTW by Kontoravdis and Bart [95] in 1995, and by developing additionally three lower bounding heuristics, they found optimum solutions for almost all test instances up to 100 customers. GRASP is also very often combined with evolutionary local search (ELS) methods such as VRP heuristic by Prins [154] from 2009, who proposed solutions encoded as giant tours and a robust local search based on a sequential decomposition of moves. Based on the experimental studies, their approach outperformed all previous methods at that time except Active Guided Evolution Strategy (AGES) algorithm by Mester and Bräysy (see Section 4.2.16 for more details). In 2010, Duhamel et al. [47] addressed the GRASP-ELS approach to the Capacitated Location Routing Problem (CLRP), which is defined by a set of depot locations with opening costs and limited capacities, a homogeneous fleet of vehicles, and a set of customers with known demands. The objective of the CLRP is to assign customers to opened depots and to design vehicle routes in order

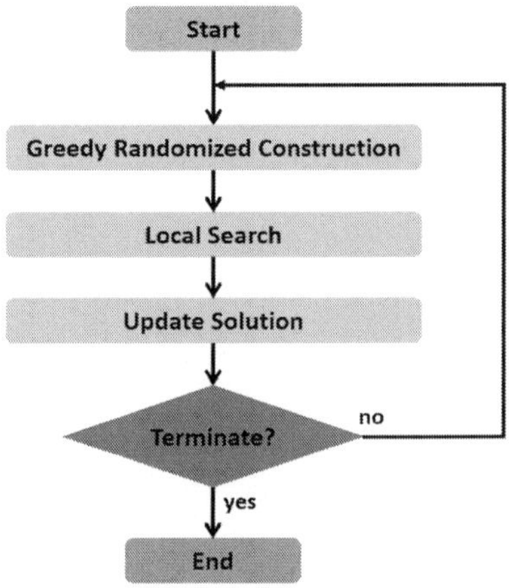

FIGURE 4.10 Illustration of the Greedy Randomized Adaptive Search Procedure (GRASP).

to minimize both the cost of open depots and the total cost of the routes. Their method comprises GRASP with ELS combined with searching within solutions identified by giant tours without trip delimiters. Giant tours were evaluated by a splitting procedure minimizing the total cost due to vehicle capacity, fleet size, and depot capacities.

The combination of GRASP with Variable Neighborhood Search (VNS) (see Section 4.2.4 for more details) and path relinking methods was proposed by Villegas et al. [193] in 2011 to solve the Truck and Trailer Routing Problem (TTRP). The TTRP comprises a heterogeneous fleet composed of trucks and trailers serving a set of customers, some only accessible by truck and others accessible with a truck pulling a trailer.

In 2012, Marinakis [109] developed a modified version of GRASP, called Multiple Phase Neighborhood Search-GRASP (MPNS-GRASP) for solving the Capacitated VRP. He utilized a stopping criterion based on Lagrangean Relaxation and Subgradient Optimization in addition to the Circle Restricted Local Search Moves strategy being a novel way for expanding the neighborhood search.

4.2.7 Particle Swarm Optimization

The Particle Swarm Optimization (PSO) introduced in 1995 by Kennedy and Eberhart [92] aimed at producing computational intelligence by exploiting analogues of social interaction rather than individual cognitive abilities [149]. It is an optimization method based on the idea of iterative improvements of a solution with regard to certain quality measures. PSO is using a population of particles representing solutions to move them around in the search space due to some mathematical rules over particles velocity and position. All particles movements are guided toward best known positions in the search space, thus moving the whole swarm toward the global minimum [149]. Similarly to flock of birds collectively searching for food, the swarm is likely to move close to an optimum of the fitness function. An illustration of the Particle Swarm Optimization is shown in Fig. 4.11.

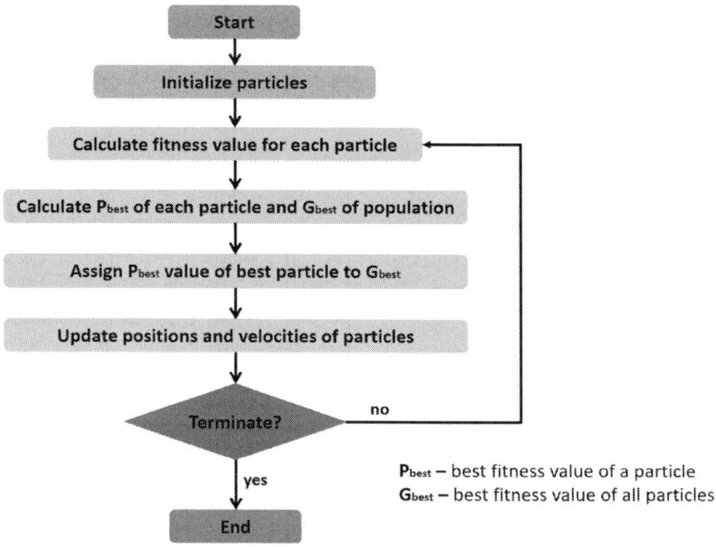

FIGURE 4.11 Illustration of the Particle Swarm Optimization.

The PSO was originally designed for solving continuous optimization problems, but there was created also a mechanism to tackle discrete optimization problems, the so-called discrete PSO [200]. This discrete PSO method was taken into account by many researchers for different variants of the vehicle routing problems. In each PSO algorithm for the VRP variants, most important questions are how to represent a solution by a particle and how to build an efficient decoding method to decode particle back into a solution.

The discrete PSO was used initially in 2004 by Yang et al. [200] to solve the CVRP where particles were represented as an array of dimensions and positions related to customers served by particular vehicles.

Ai and Kachitvichyanukul [5] developed in 2009 a PSO algorithm with multiple social structures to solve the VRP with simultaneous pickup and delivery. They proposed a random key-based representation of solution and corresponding decoding method. The decoding method was used to transfer the particle to a priority list of customers and to a priority matrix of vehicles to serve all customers. The routes were then constructed based on the customer priority list and vehicle priority matrix. The authors proved the competitiveness of such approach and were able to find some new best known solutions.

In 2010, Marinakis et al. [110] developed an interesting combination of the PSO algorithm with the multiple phase neighborhood search-greedy randomized adaptive search procedure (MPNS-GRASP), expanding the neighborhood search strategy and path relinking to solve the classical VRP. It proved to be suitable and effective for solving very large-scale problems within short computational time and was ranked in the fifth place among the 39 most known and effective algorithms in the literature and in the first place among all nature inspired methods for the VRP.

In 2011, MirHassani and Abolghasemi [114] developed a PSO algorithm for the Open VRP, where vehicles do not return to depot after serving last customer in the route. They proposed a particular decoding method in which the customer position vector is generated in descending order and all customers are assigned to a route always taking into account feasibility constraints. The extra one-point moves were also applied on constructed routes, which seems promising in achieving better solutions.

A combination of the PSO algorithm with a genetic algorithm to solve the Capacitated VRP with fuzzy demands was developed in 2012 by Kuo et al. [100]. This algorithm used the idea of a particle best solution and the best global solution combined with crossover and mutation of Genetic Algorithm. The modification of particle coding was also performed to ensure that particle always generates a new feasible solution.

In 2013, Goksal et al. [68] developed a very interesting combination of the PSO with Variable Neighborhood Search to solve the VRP with simultaneous pickup and delivery. Moreover, they introduced a solution represented by a giant tour without trip delimiters, and an annealing-like strategy was applied to preserve the swarm diversity. Their algorithm was able to improve 104 best known solutions for this variant of the VRP.

An interesting solution for particle coding and decoding in the PSO algorithm was also proposed by Ho et al. [81] in their hybrid chaos-particle PSO algorithm for the VRPTW. The chaos algorithm was employed to reinitialize the particle swarm, and an insertion heuristic algorithm was incorporated to build the feasible vehicle routes in the particle decoding process. Moreover, a particle swarm premature convergence judgment mechanism was combined with Gaussian mutation into the hybrid chaos algorithm and used when the particle swarm falls into the local convergence.

More recently, in 2016, Yao et al. [205] proposed an improved PSO algorithm for carton heterogeneous VRP with a collection depot utilizing a self-adaptive inertia weight and a local search strategy. The computational results showed that developed model is feasible with savings of about 28% in total delivery cost.

4.2.8 Ant Colony Algorithms

The Ant Colony Algorithm (ACO) was first introduced in 1992 by Dorigo in his PhD thesis [44] and was based on the behavior of ants seeking paths between their colony and sources of food, laying some pheromone on trails. The fact that

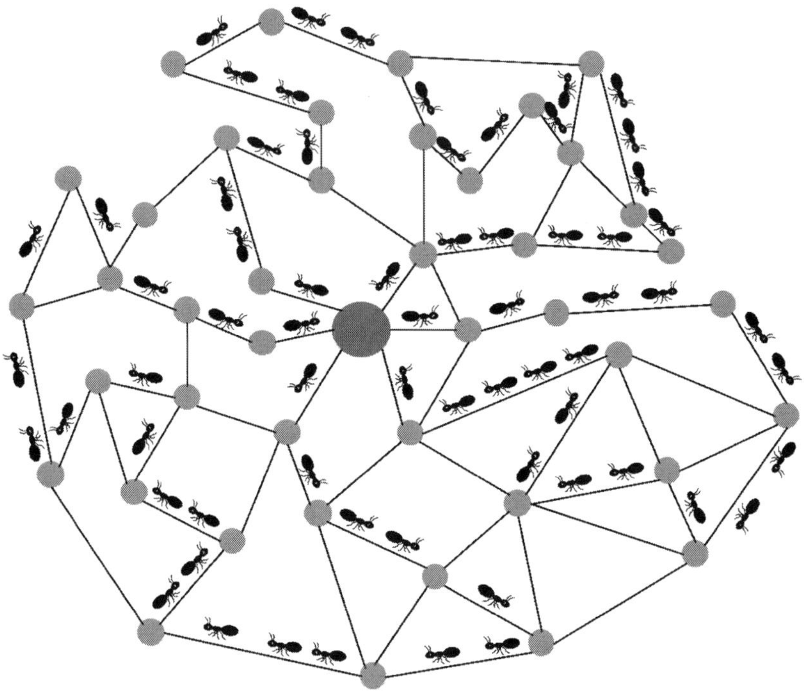

FIGURE 4.12 Illustration of the Ant Colony algorithm (ants seeking paths between their colony (blue (dark gray in print version) dot) and sources of food (green (mid gray in print version) dots), laying some pheromone on trails with information of quantity and quality of food; ants always follow the same path, which is the shortest path).

ants always follow the same path, which is indeed the shortest path, was the main motivation to take advantage of such real natural behavior of ants. The walking ants mark trails by laying down pheromones with information of quantity and quality of food. Such idea could be translated to the vehicle routing problems as searching in the neighborhood for good-quality solutions. An illustration of the Ant Colony Algorithm is shown in Fig. 4.12.

It was initially used to solve TSP problem instances by Colorni et al. [36] and then designed for the VRP and CVRP in 1999 by Bullnheimer et al. [28]. The Ant Colony method was evolving and the so-called *reinforcement learning* mechanism emerged there meaning automatic adjustments of heuristic components as the search process evolves [188]. This idea incorporates learning mechanism how to make proper decisions based on behavior and introducing reinforcement methods to improve the quality of solutions.

In the next years, the Ant Colony algorithms became very popular methods for the other variants of the VRP. In 1999, Gambardella et al. [58] proposed an ant colony optimization algorithm for the VRPTW with an idea of association the attractiveness measure with the arcs. They treated artificial ants as parallel processes, and their role was to create feasible solutions. Two ant colonies were used, the first for minimizing the number of routes and the second for minimizing the total distance traveled. Similarly to natural behavior of ants, also such artificial ants were cooperating by exchanging information about solutions through pheromone updates. Each time a new solution with smaller number of routes or smaller total distance was found, such an information was spread out to other ants through pheromones. Baran and Schaerer [12] further improved this algorithm in 2003 by developing only one colony to get a set of Pareto optimal solutions considering three objectives at the same time, the number of vehicles, the total traveling time, and the total delivery time. The computational experiments proved that the new technique outperforms the original approach. Modification to the original ACO was tackled also by Bell and McMullen [14] in 2004, who focused on the size of the candidate lists used within the algorithm, which appeared a significant factor in finding improved solutions.

In 2008, Donati et al. [43] developed multi-ant colony systems for the Time-Dependent VRP by combining the original ACO approach with discretizing the time space in a suitable number of subspaces. Moreover, they introduced new time dependent local search procedures and conditions guaranteeing that feasible moves are checked in constant time. Such a model was also integrated with a robust shortest path algorithm to compute time-dependent paths between each customer pairs.

The ACO approaches were further investigated and enhanced by Fuellerer et al. [55] in 2009 for the two-dimensional loading VRP, excellent behavior of which was proven through extensive computational results. Their algorithm was flexible enough in handling different loading constraints, including items rotation and rear loading so that it allowed for qualitative conclusions of practical interest in transportation, such as evaluating the potential savings by more flexible loading configurations. Moreover, it was one of the first ACO methods successfully combining two totally different heuristic measures of loading and routing within one pheromone matrix. Based on the computational results, their algorithm was able to reach optimal solutions on small-size instances, and for larger-size instances, it clearly outperformed all other heuristics at that time.

In 2009, Gajpal and Abad [57] took advantages of the ACO for solving VRP with simultaneous pickup and delivery, getting very competitive results with the other approaches. They used construction rules and two multiroute local search schemes to solve this variant of the VRP. It is remarkable to note that their version was able to solve also the other VRP with backhaul and mixed load.

More recently, in 2014, Reed et al. [156] developed the ACO algorithm for the multicompartment VRP associated for example with a collection of recycling waste from households, treated as nodes in a spatial network. They introduced preprocessing by k-means clustering, greatly reducing the computation time and producing improved routings for networks where the nodes are concentrated in separate clusters. The algorithm was also extended to model the use of multicompartment vehicles with kerbside sorting of waste into separate compartments for each category. Abdulkader et al. [2] also tackled the multicompartment VRP in 2015 creating a hybridized version of the ant colony algorithm. They combined the local search method with existing ACO algorithm and proved by computational experiments that the new approach gives better results.

Kalayci and Kaya [90] empowered in 2016 Variable Neighborhood Search (see Section 4.2.4 for more details) method with the ACO resulting in a very competitive algorithm to solve the VRP with simultaneous pickup and delivery. The application of the VNS provided intensive local search, and its weakness of lack of memory structure was minimized by utilizing long-term memory structure of Ant Colony System, and the overall performance of the algorithm was boosted. It is remarkable to note that in the proposed algorithm, instead of ants, the pheromones were released on edges. The ants provide a perturbation mechanism for the integrated algorithm using the pheromone information to explore search space broader and jump out from local minima. Numerical experiments proved that such a combination of the VNS with ACO is very robust and efficient in terms of both quality and CPU time, and the authors reported some new best known solutions too.

In 2018, Goel and Maini [67] developed a very interesting hybridization of the Ant Colony System with firefly algorithms (HAFA) for solving the vehicle routing problems. The ACS provided the basic framework to proposed algorithm, and Firefly Algorithm (FA) was used to search for the unexplored solution space. Moreover, a novel pheromone shaking technique was incorporated in ACS to escape local minima due to avoiding pheromone stagnation on the exploited regions. The computational experiments proved that such combination outperforms other FA-based approaches in terms of convergence time.

4.2.9 Artificial Bee Colony Algorithms

The Artificial Bee Colony (ABC) algorithm was developed in 2005 by Karaboga [91]. Similarly to other swarm optimization algorithms, it is also based on two fundamental concepts, namely self-organization and division of labor, which allow problem solving systems to self-organize and adapt to the environment. The ABC model related to collective intelligence of honey bees comprises employed and unemployed bees and food sources [91]. It defines also recruitment to a nectar source and abandonment of such source as two leading methods of the honey bees behavior. The value of the food source for honey bees usually depends on its richness, proximity to the nest, and the ease of forage extracting. The employed bees are associated with a certain food source they are exploiting, and they share the information about such source with other bees. In turn, unemployed bees are constantly searching the neighborhood of the nest for new sources of forage and finally establishing new directions by gathering all information together shared by all the bees [91]. The information exchange among bees is happening in the so-called dancing area, and dance is called a waggle

FIGURE 4.13 Illustration of the Artificial Bee Colony algorithm (the value of the food source (flowers) depends on its richness, proximity to the nest, and the ease of forage extracting and can be associated with good solutions; the information exchange among bees is happening in the so-called waggle dance (illustrated as two dancing bees in a circle).

dance. Based on that, the profitability of new food sources is judged and chosen. An illustration of the Artificial Bee Colony algorithm is shown in Fig. 4.13.

The ABC algorithm by Karaboga simulates both behavior of honey bees and foraging in order to solve potential optimization problems. A possible solution is represented by a food source, whereas the amount of food in the source is related to the solution quality. The search process is associated with selecting various sources of food and abandoning them if the solution represented by the food source is not improved for a certain number of trials [91].

In the field of swarm intelligence the ABC algorithms and Bee Colony Optimization (BCO) methods have much less attention than the other methods such as Ant Colony Optimization mostly because of more complex implementation; however, the obtained results are still competitive and worth further investigations. In 2012, Bhagade et al. [18] developed the ABC algorithm for the classical TSP enhancing it additionally by using the nearest-neighbor method. The effectiveness of the bee paths was evaluated with tour length and bee travel time. Based on the computational results, the authors proved that the ABC algorithm can be efficiently used for solving the TSP and moreover is highly flexible and can be extended to other optimization problems by considering relatively few control parameters.

The ABC algorithm was also taken into account for solving the Capacitated VRP by Szeto et al. [176] in 2011 and for solving the Periodic VRP by Yao et al. [203] in 2013. To improve the performance of the ABC algorithm, Yao et al. combined it with multidimensional heuristic information and a local optimization based on a scanning strategy. By running numeric experiments they proved that the proposed modification of the ABC algorithm is a powerful method for solving the PVRP.

Zhang et al. [209] developed in 2014 a hybrid artificial bee colony algorithm with hybrid operators for the Environmental VRP. The performance of this approach was evaluated through computational tests on CVRP instances, and the results showed that a hybrid approach outperforms the original ABC algorithm by 5% on average. An interesting combination of the adaptive memory with the ABC algorithm for the Green VRP with cross-docking was proposed in 2016 by Yin et al. [206]. Their hybrid metaheuristic was able to reach higher fuel efficiency than the tabu search algorithm by managing the loading along the route and yielding less total cost. Such a method proved to be robust against the problem size, and convergence of the objective value was guaranteed with high confidence.

The application of the ABC algorithm for the VRPTW was undertaken by Alzaqebach et al. [6] in 2016, who proposed a modified version of the ABC tackling the issue that high exploration ability of the ABC usually slows down its convergence speed. In their approach a list of abandoned solutions is used by the scout bees to memorize the abandoned solutions. The scout bees select then a random solution from the list and replace by a new solution with random routes chosen from the best solution. Yao et al. [204] also used in 2017 a modification of the ABC algorithm to solve the VRPTW. They proposed the enhancements of a local optimization based on a crossover operation and a scanning strategy.

Just recently, Sedighizadeh and Mazaheripour [168] developed in 2017 and optimization technique of multiobjective VRP using a hybrid algorithm based on Particle Swarm Optimization combined with ABC algorithm considering precedence constraints. In particular, they proposed a solver algorithm in which the idea is to consider different constraints of the problem, use penalty method, and additional segmentation constraint methods to obtain the best vehicle routes.

Ng et al. [128] also developed (very recently in 2017) a multiple-colony ABC algorithm for the CVRP and rerouting strategies under time-dependent traffic congestion. They took into account a rerouting strategy to solve the problem of inefficient vehicle routing caused by traffic congestion and proposed a flexible delivery rerouting strategy aiming at reducing the risk of late delivery. They introduced a novel scheme using a Multiple Colonies Artificial Bee Colony algorithm to solve the problem of solutions trapped in local minima. Such a design of the outstanding bee selection for colony communication proved to be superior in exploitation.

4.2.10 Bat Algorithms

The Bat Algorithm (BA) is a relatively new bioinspired metaheuristic introduced in 2010 by Xin-She Yang [202] and is based on echolocation behavior of microbats when searching for their prey in nature [177]. An illustration of the Bat algorithm is shown in Fig. 4.14.

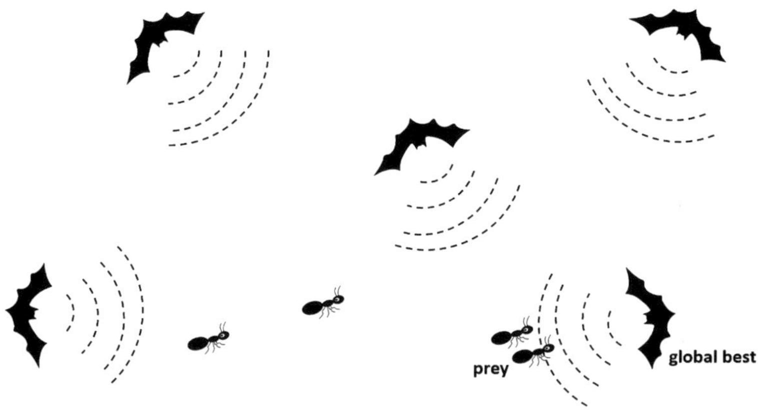

FIGURE 4.14 Illustration of the Bat algorithm (microbats use echolocation when searching for their prey (insects); good-quality solutions are associated with bigger quantity and quality of prey).

Taha et al. [177] developed in 2010 a hybrid algorithm executing a discrete version of the bat algorithm combined with the LNS (see Section 4.2.5 for more details) to solve the VRPTW. Their algorithm aimed at improving the performance of the BA by the destroy and repair paradigm of the LNS, allowing bats to explore a large part of the solution space.

Many more Bat Algorithms for various variants of the routing problems were developed just in a few years. In 2016, Zhou et al. [211] proposed a hybrid BA with Path Relinking for the Capacitated VRP, which is based on the framework of the continuous bat algorithm, GRASP, and path relinking method built into the BA. Additionally, the random subsequences and single-point local search are operated with certain loudness to further improve the performance of the developed algorithm. Osaba et al. [138] proposed in 2017 the BA with random reinsertion operators to solve the VRPTW. This algorithm comprises both improved movement strategy and diverse heuristic operators to deal with VRPTW. They unified the search process and the minimization phases by using selective node extractions and subsequent reinsertions. In 2018, Wang et al. [195] presented a self-adaptive BA algorithm applied for the first time for the truck and trailer routing problem, a generalization of the VRP involving a group of geographically scattered customers served by the vehicle fleet including trucks and trailers. They developed the BA combined with a local search procedure performed by five different neighborhood structures. Additionally, a self-adaptive tuning strategy was implemented to preserve the swarm diversity.

4.2.11 Cuckoo search

The Cuckoo Search is a metaheuristic optimization algorithm introduced by Yang and Deb [7] in 2009 inspired by the cuckoo birds being the "brood parasites" birds. They never build they own nests and instead lay their eggs in nests

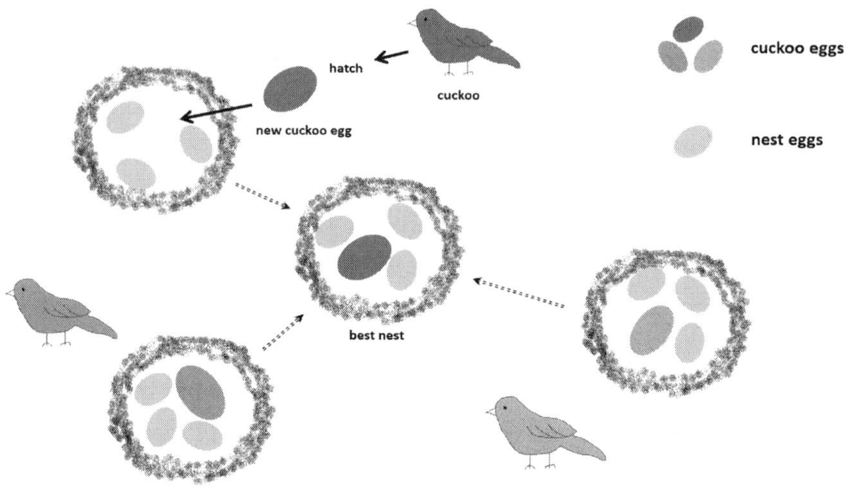

FIGURE 4.15 Illustration of the Cuckoo Search (orange (light gray in print version) – nest eggs, red/blue/green (dark/mid/light gray in print version) – cuckoo eggs; each egg represents a solution, and each cuckoo egg represents a potentially better solution; the Cuckoo Search algorithm is based on the idea of creating subsequent generations of nests containing best eggs and thus highest quality solutions).

of other host bird nests [86]. Moreover, if the host bird identifies eggs that are not theirs, then it throws them away from nest or leaves its nest and simply builds a new nest somewhere else. In terms of optimization algorithms, each egg represents a solution, and each cuckoo egg represents a new potentially better solution. The Cuckoo Search algorithm is based on the idea of creating subsequent generations of nests containing best eggs and thus highest quality solutions. In each iteration, each cuckoo lays one egg at a time and leaves it in a randomly selected nest, followed by selection of nests with highest quality eggs. Moreover, the number of host nests is fixed, and the cuckoo eggs are identified by host birds with a certain probability. In case cuckoo eggs are identified, the host bird either throws such solutions away or leaves its nest and builds a new nest. An illustration of the Cuckoo Search is shown in Fig. 4.15.

Initially, there were attempts to take advantage of the Cuckoo Search algorithm to the Traveling Salesman Problem. In 2012, Jati et al. [82] developed a discrete cuckoo search for TSP considering discrete step sizes and the cuckoo's updating scheme; however, the results showed that it was trapped into local minima for some TSP instances. Ouyang et al. [142] proposed in 2013 a novel discrete cuckoo search algorithm for Spherical TSP based on the Levy flight and brood parasitic behavior. They applied study, A, 3-Opt, and search-new-nest operators to speed up the convergence. An interesting combination of the Genetic Algorithm with Cuckoo Search and 2-Opt movements was presented in 2013 by Abu-Srhan and Al Daoud [3] to avoid the local minima problem and take advantage of the powerfulness of the GA, all applied to the TSP. More recently, in 2015 Ouaarab et al. [141] developed the random-key cuckoo search (RKCS) algorithm for solving the TSP by introducing a simplified random-key encoding scheme to pass from a continuous space (real numbers) to a combinatorial space. They also applied displacement of a solution in both spaces using Levy flights.

Zheng et al. [210] proposed an interesting combination of GRASP with Cuckoo Search applied to the Vehicle Routing Problem. In their version, called CS-GRASP (more information on GRASP can be found in Section 4.2.6), they utilized path relinking, swap, and inversion strategy. In 2016, Chen and Wang [33] developed a hybrid cuckoo search for the VRP based on the combination of Particle Swarm Optimization (PSO), Optical Optimization, and Cuckoo Search method. The Optical Optimization was used to initialize population with high-quality solutions optimized further by PSO. Cuckoo Search was then iteratively used to optimize the rest of the individual solutions. Teymourian et al. [185] developed in 2016 hybrid methods of Cuckoo Search combined with Local Search and Improved Intelligent Water Drops algorithm to control the balance between diversification and intensification of the search process.

4.2.12 Firefly Algorithms

The Firefly Algorithm (FA) was introduced in 2008 by Yang [201] inspired by flashing behavior of fireflies. The main purpose why fireflies flash is to signal and attract other flies. Yang introduced the FA assuming that all fireflies are unisexual, attractiveness is proportional to brightness, and that fireflies move randomly when there are no brighter fireflies in the neighborhood.

In 2011, Yati and Suyanto [83] developed an evolutionary discrete FA algorithm to solve the TSP by examining discrete distance between two fireflies and movement schemes. Their algorithm however without combination with the other methods was trapped in local minima in case of some TSP test instances. This idea was further explored by Kumbharana and Pandey [99] in 2013, who proposed the FA for the TSP comprising constructing a suitable conversion of the continuous functions of attractiveness, distance, and movement into new discrete functions. In 2017, Jie et al. [84] presented an improved version of the FA algorithm for the TSP by redefining the distance of FA by introducing a swap operator and a swap sequence to avoid falling into local minima and to accelerate convergence speed. They also adopted dynamic mechanism based on neighborhood search algorithm, proving by experiments that such approach is very competitive.

The initial research on the application of the FA for the routing problems was started just a few years ago by Pan et al. [143] in 2013, who developed the FA for the VRPTW. They introduced the coding and design of disturbance mechanism of elicit fireflies in their FA. In 2016, Osaba et al. [139] developed the first FA algorithm to solve the Rich VRP by introducing the Hamming Distance function to measure the distance between two fireflies of the swarm. Osaba et al.

[137] designed also an evolutionary discrete FA to the VRPTW by introducing some novel route optimization operators. These route optimization operators were used to incorporate minimizing the number of routes for a solution in the search process and to analyze all routes of the solution. They allowed them to increase the diversification capacity of the search process contrary to classical node and arc exchange-based operators.

4.2.13 Golden Ball Algorithms

The Golden Ball (GB) is a multipopulation metaheuristic based on soccer concept introduced very recently in 2014 by Osaba et al. [135]. In the initialization phase of the GB, the whole population of players (indicated as solutions) is created, followed by random division of players into fixed number of subpopulations called teams. Each team has its own couch related to training method, which is also randomly assigned to each time. The training method can be also associated with the way how each player evolves individually in the team, and in terms of the vehicle routing problems, it might be associated, for example, with 2-Opt moves. Osaba et al. introduced also the so-called Custom Training, which is a special training applied to the player trapped in local minimum and is based on a training with cooperation with the best player in the team, usually by Crossover operation. The second phase is called the competition phase and is divided into seasons and weeks. In each week, all teams train independently, and they face each other by creating a soccer league. Moreover, there are transfers at the end of each seasons, where both coaches and players can switch teams. An illustration of the Golden Ball Algorithm is shown in Fig. 4.16.

Initially, the GB was introduced for the TSP in 2014 by Osaba et al. [136], and it was proved to be competitive with the Evolutionary Simulated Annealing and Tabu Search approaches. This algorithm was later improved by Ruttanateerawichien et al. [163] to handle Capacitated VRP, and based on experiments, it was able to find new best known solutions for three benchmark instances. The random keys representation to encode solutions in the GB algorithm was proposed in 2015 by Sayoti and Riffi [166], who applied this technique for the TSP.

More recently, in 2016, Ruttanateerawichien et al. [162] introduced a new technique, where a team represents the CVRP solution, and the players represent routes in the team, whereas in the previous work the CVRP solution was modeled by a player. Additionally, the solution quality of players and teams was improved by intraroute and interroute improvement algorithms. In 2018, Guezouli et al. [74] designed efficient GB algorithm based on clustering to solve the Multi-Depot VRP with Time Windows. They proposed a different solution representation from the original one and embedding a clustering algorithm to solve the problem more efficiently.

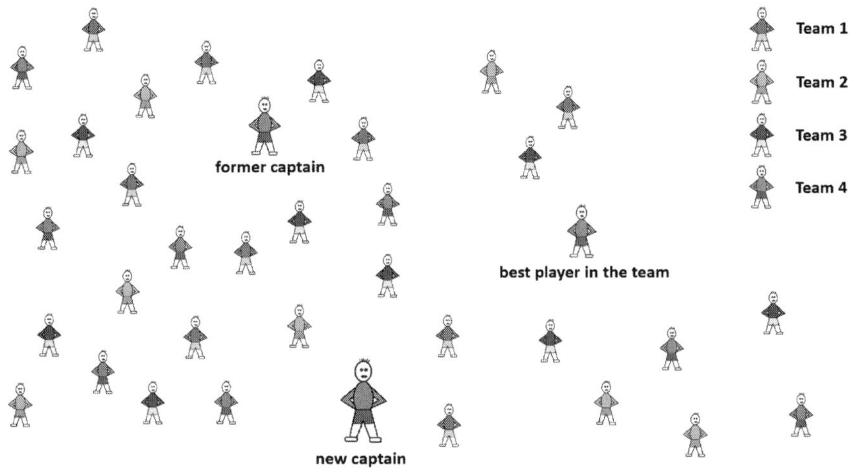

FIGURE 4.16 Illustration of the Golden Ball algorithm (each soccer represents a solution; soccers are divided into teams (populations); each team has its own captain (related to training method) and each player evolves individually in the team and may become the best player in the team (best solution in population).

4.2.14 Gravitational Search Algorithm

The Gravitational Search Algorithm (GSA) was introduced in 2009 by Rashedi et al. [155] based on the gravity law and mass interactions. The GSA was inspired by gravitation being the tendency of masses to accelerate toward each other, which is also one of the four fundamental interactions occurred in nature next to the electromagnetic force and the weak and strong nuclear forces [167]. Based on the fundamental Newton law of gravity, each particle attracts other particles with a gravitational force, which is directly proportional to the product of their masses and inversely proportional to the square of the distance between them [155]. In the GSA, agents are considered as objects, and their masses are related to their performance. All the agents attract each other by the gravity force communicating each other through gravitational force. The heavy objects—related to high-quality solutions—move slower than lighter objects, and this guarantees the exploitation step of the GSA algorithm [155]. Moreover, each agent can be characterized by position, inertial mass, and passive and active gravitational mass. The position of the agent is related with a solution, whereas the inertial and gravitational masses are determined by a fitness function so that the algorithm is navigated by properly adjusting these masses. It is expected in the GSA that during execution all the objects will be attracted by the heaviest one corresponding to a global minimum in the search space. An illustration of the Gravitational Search algorithm is shown in Fig. 4.17.

Chen et al. [32] proposed in 2011 a hybrid GSA algorithm for the TSP based on random-key encoding scheme of the solutions combined with simulated annealing. They incorporated into it also a multitype local improvement scheme

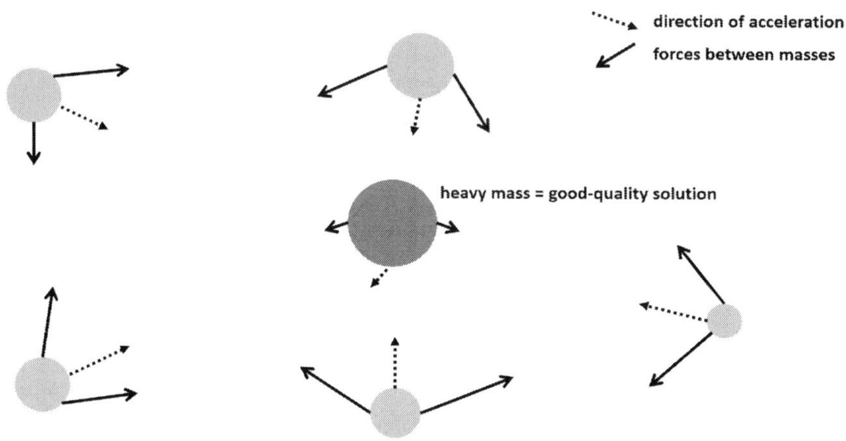

FIGURE 4.17 Illustration of the Gravitational Search Algorithm (bigger dots represent heavier masses, thus better solutions; solid arrows indicate forces between mases; dotted arrows indicate an acceleration direction).

used as a local search operator. The simulated Annealing method was utilized into proposed algorithm too, in order to manipulate the iteration progress algorithmically. Hosseinabadi et al. [80] presented in 2012 the GSA for the TSP using velocity and gravitational force in physics based on random search concepts. In 2014, Dowlatshahi et al. [45] developed a Discrete GSA for the TSP with path relinking strategy instead of original agent movements. Their algorithm was ranked ninth compared with 54 different algorithms for the TSP.

The GSA approach was also used to solve the Open VRP by Hosseinabadi et al. [79] in 2015. Their algorithm was based on random search concepts utilizing speed and gravity parameters, and the researcher agents were connected each other based on Newton's gravity and motion laws.

4.2.15 Bacterial Foraging Optimization Algorithm

The Bacterial Foraging Optimization Algorithm (BFOA) was introduced in 2002 by Passino [145] to mimic biological principles shown in the foraging behavior of E. coli bacteria for distributed optimization and control. E. coli bacteria live in intestines of most animals on the Earth, and they have a control system directing its behaviors in food foraging. The foraging process of E. coli bacteria comprises searching for food, deciding whether to enter possible food region or not, careful searching if to enter into a new area, and deciding whether to stay in actual region or move on to a new better area after consumption some food in the current region [76]. In the BFOA a swarm of bacteria is used as a searching agent for a solution to a specific optimization problem. A bacteria position represents a solution of the problem by a simple sequence of customer nodes. The bacteria direction vector is used to represent its ability to search for solution and is driving bacteria movement. In each step of the BFOA, all bacterium moves from one position to another are based on these directions, and by moving various solutions of the problem are evaluated.

The solution of the TSP using BFOA was carried out by Verma et al. [189], where the key step used in the algorithm was computation of chemotaxis, where a bacterium takes steps over the foraging landscape to reach the areas with high-nutrient food. The movement of a bacterium was led by computing direction and distance matrices, and additionally in case the bacterium does not reach a city within the minimum distance constraint in its first step, it searches for other cities in the next steps.

Much more BFOA algorithms were proposed for variants of the VRP family. In 2011, Hezer and Kara [76] applied for the first time the BFOA technique to solve the VRP with Simultaneous Delivery and Pick-up. Niu et al. [130] developed in 2012 the BFOA with adaptive chemotaxis step to solve the VRPTW, in which they applied additionally a nonlinearly decreasing exponential modulation model to further improve the efficiency of the algorithm.

In 2014, Tan et al. [180] proposed another BFOA for the VRPTW comprising two versions with linear and nonlinear decreasing chemotaxis steps to obtain the best solutions of a given VRPTW problem. Tan et al. [181] further improved their solution by developing adaptive comprehensive learning BFOA for the VRPTW to keep a good balance between the exploration and exploitation. They proposed also the comprehensive learning mechanism maintaining the diversity of the bacterial population and thus alleviating the premature convergence.

The solution of the Heterogeneous Fixed Fleet VRP was proposed in 2015 by Gan et al. [59], who developed a new method based on structure-redesign-based bacterial foraging optimization (SRBFO) combined with time decreasing chemotaxis step size mechanism. Additionally, the position of bacteria was encoded by $2N$ dimensions, where the first N dimensional vectors indicated the corresponding vehicle, and the next N dimensional vectors were related to the execution order of the corresponding vehicle routing.

In 2017, Li et al. [104] proposed the BFOA for the VRP with Pickup and Delivery. They established the mathematical model aiming at minimizing the total travel time and the total, which was combined with dynamic variable step factor and propagation threshold and death threshold to copy the best individuals and eliminate the inferior individuals.

4.2.16 Genetic and Evolutionary Algorithms

The Genetic Algorithms (GAs) are optimization search methods based on the evolution process in nature, and they imitate the biological process of natural selection where stronger populations among different species survive [69]. The main concepts of GAs were developed by Holland [78] in the 1960s and 1970s, and practicality of using GAs for solving complex problems was elaborated later in 1975 by De Jong [42] and in 1989 by Golberg [69]. The GAs

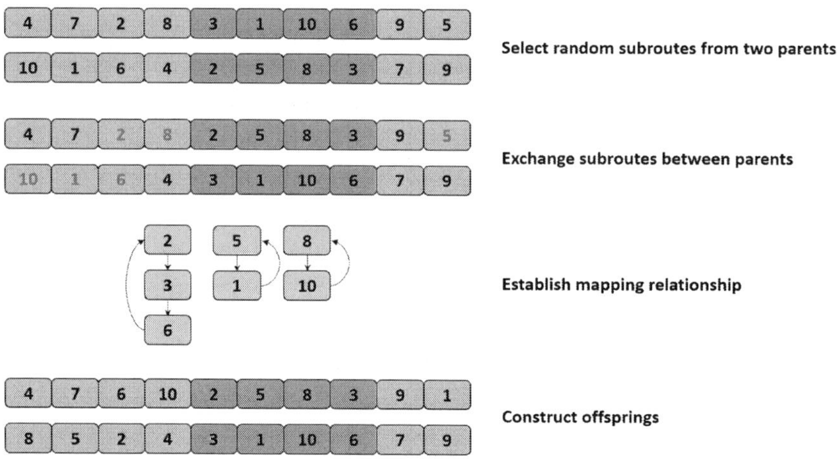

FIGURE 4.18 Illustration of the PMX (Partially Mapped Crossover).

belong to the larger class of Evolutionary Algorithms (EAs), which are generic population-based metaheuristic optimization algorithms. All EAs and thus also GAs use biological evolution mechanisms such as representation, selection, recombination, and mutation. In terms of TSP and VRP the representation of the solution comprises encoding most important features of the solution as genes in a chromosome, which identifies the so-called individual in the population [25]. The recombination is carried out on two selected parent solutions by combining genes of parent chromosomes to create offspring solutions with potentially better quality. In turn, mutation is performed on the offspring solutions by random modification of genes to further explore the solution space and ensure genetic diversity. The GA is based on creating subsequent generations, and each new generation is constructed by selection, recombination, and mutation of all the solutions in the population. A properly designed GA should maintain a right balance between solution quality and diversity within the population to support efficient search. A common problem in the GAs for the routing problems is the feasibility of the solution created after recombination and mutation phases. An important issue is also the way how the offspring solutions are created, which led to designs of different crossover operators. Due to high efficiency of the GAs for solving various optimization problems, they are probably the most explored group of metaheuristics. Some of the most important crossover operators designed for the rich vehicle routing problems are shown in Table 4.1 [150]. Various genetic and evolutionary strategies and methods are described in the remaining part of this section. An illustration of the PMX crossover is shown in Fig. 4.18.

The edge assembly crossover operator (EAX) originally designed in 1997 for the TSP by Nagata and Kobayashi [125] was later extended to handle the VRP in 2007 by Nagata [122]. This powerful operator is based on specific se-

TABLE 4.1 Most important crossover operators.

Shortcut	Crossover name
EAX	edge assembly crossover
NX	natural crossover
ER	edge recombination crossover
MBX	matrix-based crossover
OX1	order crossover
OX2	order-based crossover
SMX	sorted match crossover
PMX	partially mapped crossover
MPX	maximal preservative crossover
MX1, MX2	merge crossover operators
CX	cycle crossover
POS	position-based crossover
AP	alternating position crossover
RC	route crossover
IB_X	insertion-based crossover
CPX	cluster preserving crossover
CEPX	common edges preserving crossover
BCRC	best cost route crossover

quence of steps combining edges from two parent solutions and in case of the VRP is described in the following way. In the first stage a graph is constructed containing edges from two parent solutions, followed by constructing cycles generated by alternately selecting a new edge from the first and second parents. Next, the subset of cycles is selected, and an intermediate solution is constructed by taking one parent and removing all edges available in the subset of cycles and by adding all edges in the subset of cycles from the second parent. Such an intermediate solution is then a combination of set of routes connected to the depot and subtours not connected to the depot. Next, a complete solution is created by repeatedly merging a subtour to a route with least cost strategy combined with a greedy heuristic. In the last stage the constraint violations are eliminated by applying penalty function, 2-Opt exchanges, and customer exchanges until solution becomes feasible [122]. An illustration of the EAX crossover is shown in Fig. 4.19.

The natural crossover was originally developed for the TSP in 2000 by Jung and Moon [88] and further extended to the VRPTW in 2002 [89]. The main idea of this crossover is to partition the set of customers from two parent solutions into two classes by drawing curves or geometric figures on graphical representation of the problem. As a result, the customers are located either on one side or the other of a curve, or the customers are either outside or inside the

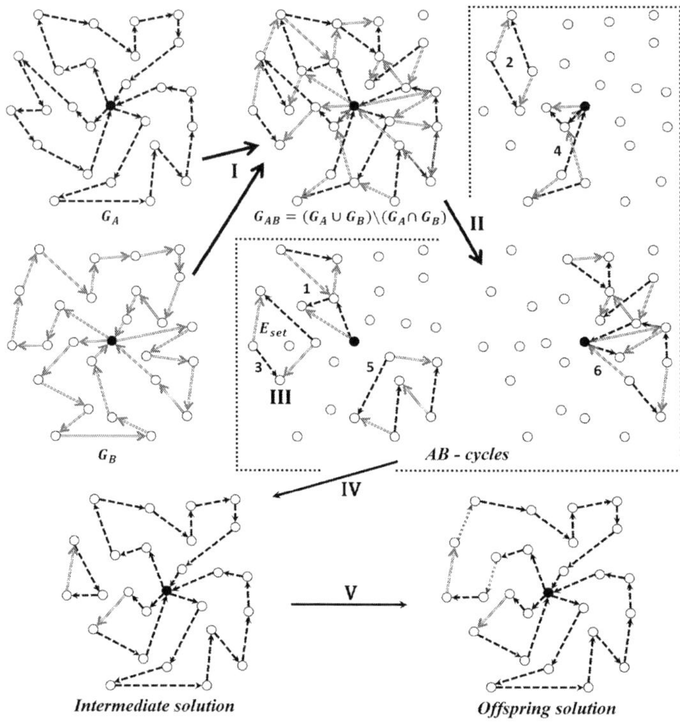

FIGURE 4.19 Illustration of the EAX (Edge Assembly Crossover).

geometric figures, such as ellipses, circles, or squares. The initial offspring solutions are created by transferring arcs from the first/second solution with both endings in the first/second class. They are further repaired by iterative attempts to connect disconnected segments from both parent solutions into the offsprings using the least cost strategy. In case a feasible solution is obtained, it is further improved by local optimization procedures based on 2-Opt, interroute, and intraroute search moves.

The Edge Recombination Crossover (ER) was originally developed for the TSP, and its main idea is to progressively extend tours by adding edges either from the first or second parent. Initially, all the edges from both parents are kept in a special edge table, in which all parental edges neighboring to a customer are grouped together. The ER was extended to the VRP in 1995 by Krajcar et al. [96] by using depot as a starting point of each route, by including the depot in the edge table, by considering only edges linked to the closest customers, and by considering capacity constraints to decide when to finish current route and start a new one.

The matrix-based crossover was originally developed for the TSP in 1991 by Fox and McMahon [53] and further extended in 1999 for the VRPTW by

Chin et al. [34]. In this crossover a matrix of size of customer numbers is used to indicate larger values in case customers *a* and *b* are in the same routes in two parent solutions and moreover they have the same precedence relationship. The values in the matrix for a given customer pair are smaller in case customers are not in the same route and intermediate when customers are in same routes but do not have the same precedence relationship. In the matrix-based crossover approach the offspring solutions are generated by iteratively inserting unserved customers into the routes with least insertion cost modified by the values in the matrix by checking all feasible insertion places.

The Order Crossover (OX1) was proposed in 1985 by Davis [40] and is used to build offspring solutions by selecting a subsequence of route from one parent and preserving the precedence order of customers from the other parent. The OX1 is used to copy the subtour between the crossover points directly the offspring solution, placing customers in same absolute position.

The Order-Based Crossover (OX2) was proposed in 1991 by Syswerda [175] and works by randomly selecting several customer positions in one parent route, and the corresponding order of customers is imposed on the second parent to create offspring solutions.

The Sorted Match Crossover (SMX) developed in 1985 by Brady [107] works by identifying subtours in both parent routes having same length, starting and ending at same customers, and containing the same customers. In case such subtours can be identified, the offspring solutions are created by replacing such a subtour with one having lower cost.

The Partially Mapped Crossover (PMX) was introduced in 1985 by Goldberg and Lingle [70] for the TSP, and it is based on selecting two random cut points on two parent solutions followed by creating offspring solutions by exchanging these subsequences between two parents. It is important to note that the order of remaining customers is tried to be preserved in as many customer cases as possible.

The Maximal Preservative Crossover (MPX) introduced in 1988 by Muhlenbein et al. [121] works similarly to the PMX by imposing additionally random subtour length to contain at least ten customers and to be smaller than or equal to the problem size divided by two. Having a subtour selected, all customers from it are removed from the second parent, and the offspring is generated by completing the random subtour with unserved customers in the same order as is in the second parent solution.

The Merge Crossover operators MX1 and MX2 for the VRPTW were proposed in 1993 by Blanton and Wainwright [19], and they were used to exploit a global precedence relationship between customers. Their main advantage especially for the VRPTW or PDPTW is that they rely on the assumption that is beneficiary for customer *a* to appear before customer *b* in a route in case time window of customer *a* starts earlier than time window of customer *b*. In MX1 and MX2 operators, two parent routes are checked position by position, selecting at each position customer with earlier start time window, who is later

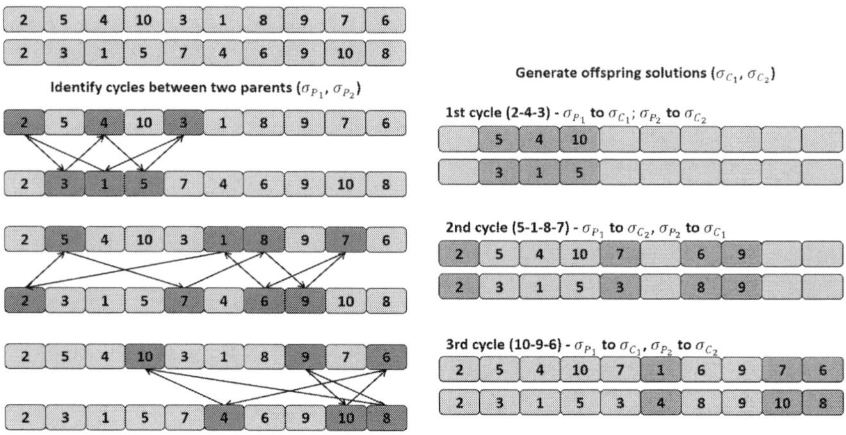

FIGURE 4.20 Illustration of the CX (Cycle Crossover).

transferred to the constructed offspring solution. By such a technique the generated offspring solution is biased toward solution containing customers sorted according to ascending order of their start time windows.

The Cycle Crossover (CX) was originally introduced in 1987 by Oliver et al. [131] for the TSP, and it is based on identifying cycles between two parent solutions using the connections between customer numbers and their positions in the routes. The customers that are included in the cycles will remain, whereas the other customers will be swapped to create offspring solutions. An illustration of the Cycle crossover is shown in Fig. 4.20.

The Position Based Crossover (POS) was developed in 1991 by Syswerda [175] and works similarly to the OX2 operator by randomly selecting customer positions in parent routes, but it imposes the position of selected customers on the corresponding customers in the second parent solution.

The Alternating Position Crossover (AP) was introduced in 1996 by Larranaga, and it creates offspring solutions by alternately selecting customers from the first and select parents, omitting customers already available in constructed offspring.

The Route Crossover (RC) was introduced in 1999 by Maeda et al. [108] based on the idea of bit masks corresponding to the number of routes in the first parent solution. The offspring solution is initially generated by selecting routes from first parent depending on bit mask values. All unserved customers are inserted into a temporary list, which is further sorted according to their order in the second part and then transferred one by one into generated offspring solution.

The Insertion-Based Crossover (IB_X) was proposed in 1998 by Berger et al. [17] based on iterative building and ruining solution taking into account large distances to successive customers and large waiting times. Initially, the first route is randomly selected from the first parent solution preferring routes containing customers with largest waiting times. Next, some customers are removed from such a route taking into account their large travel distance to successive customers or large waiting times to be served. In the next step a subset of routes from the second parent is selected that are close to this first route, and all such customers are further considered to be inserted into best positions in the first route of the first parent. Such a method is repeated for all routes from the first parent, resulting in an offspring solution. The second offspring solution is created by interchanging role of the parent from the first to the second one.

The Cluster-Preserving Crossover (CPX) and the Common Edges-Preserving Crossover (CEPX) operators were proposed in 2004 by Kubiak [97]. The CPX is used to preserve clusters or groups of customers that are common to both parent solutions. The routes having the largest number of customers are intersecting next to form clusters in the offspring solutions. Such clusters are used further to create offspring solutions by sequencing customers from clusters. The CEPX works similarly to CPX, but instead of clusters, it is used to preserve the common edges. Moreover, the longest subtours that are common to both parent solutions are used to generate routes in the offspring solutions.

The Best Cost Route Crossover (BCRC) was developed in 2006 by Ombuki et al. [133] for the VRP with Time Windows. In the first stage, two routes are randomly chosen from two parent solutions, and the customers from first route are removed from the second route, and in a similar way, customers from the second route are removed from the first route. The removed customers are reinserted back to original routes using the least-cost insertion methods.

Most of the genetic algorithms use one or more crossover operators combined with other methods such as guided local search, tabu search, simulated annealing, and other metaheuristics to provide best results for a given problem. The first genetic algorithms for the Vehicle Routing Problems were proposed in the 1990s, and since then, we may observe rapid development of the GAs until nowadays.

In 1991, Thangiah et al. [186] developed a genetic algorithm heuristic to solve the VRPTW, called GIDEON. This algorithm is a cluster-first route-second technique assigning customers to vehicles by the so-called Genetic Sectoring improved further by local optimization methods. The crossover used in GIDEON divides chromosomes at random points and exchanges subtours with other divided chromosomes.

Thangiah et al. [187] extended further this approach in 1994 by developing a hybrid genetic algorithm for the VRPTW. The initial solutions were created by Genetic Sectoring techniques and improved further by simulated annealing and tabu search methods.

In 1996, Potvin et al. [151] developed a new search technique for the VRPTW called the GENEtic ROUting System (GENEROUS) based on the paradigm of natural evolution. In this approach a population of solutions was evolving from one generation to another by merging two solutions into a single one, which is likely to be feasible with respect to the time window and capacity constraints.

Berger et al. [17] proposed in 1998 a new hybrid genetic algorithm for the VRPTW taking into account impact of explicit domain knowledge about and a priori characteristics of expected solutions used during the recombination and mutation phases of this algorithm. Their conceptually simple and easy-to-implement algorithm was designed to support time-constrained tasks and allowing for fast computation of near-optimal solutions, and thus it was used further in many other hybrid algorithms as the base model.

In 1999, Jih and Hsu [85] proposed a novel approach to solve single-vehicle Pickup and Delivery Problem with Time Windows by hybrid genetic algorithm. They incorporated usage of four crossover operators in the recombination phase of the GA: order crossover (OX1), order-based crossover (OX2), and merge crossover operators MX1 and MX2. Additionally, they included probability of mutation of a solution by three mutation schemes. The first mutation was based on randomly selecting two customers followed by interchanging their positions. The idea of second mutation was to randomly choose two cut customers and inverting the order of selected subroute. The third mutation was in turn based on checking if a vehicle arriving at the customer violated any of the constraints and then the order of customers in this subroute was randomly disturbed.

In 2003, Berger et al. [16] proposed a genetic algorithm for the VRPTW based on concurrently evolving two distinct populations of solutions, where the first population aimed the total travel distance, and the objective of the second population was to minimize the violations of the time window constraints. In the same year, Baker and Ayechew [9] compared a pure GA for the VRP with a hybrid GA combined with neighborhood search methods, proving that the hybrid method outperforms pure GA as well as tabu search and simulated annealing techniques.

Ho et al. [77] developed in 2008 two hybrid genetic algorithms for the Multi-Depot VRP, including in the first HGA random generating initial solutions, whereas the Clarke–Wright saving methods and the nearest neighbor heuristics were incorporated into the second HGA for the initialization phase. The results proved that simple generating random solutions in the first HGA was superior to the second considered technique.

In 2010, Ghoseiri and Ghannadpour [64] proposed a new solution for the Multi-Objective VRP with Time Windows by an interesting combination of goal programming and genetic algorithm in which the decision maker specifies optimistic aspiration levels to the objectives and deviations. In their GA, various heuristics incorporated local exploitation in the evolutionary search and the concept of Pareto optimality.

More recently, in 2015, Wang et al. [197] proposed a fitness-scaling adaptive GA with local search in which the fitness-scaling technique converts the raw fitness value to a new value suitable for further selection. Additionally, adaptive rates strategy was used to change the crossover and mutation probabilities depending on the fitness value, and a local search mechanism was applied to further exploit the problem space.

In 2017, Belhaiza et al. [13] developed a novel hybrid genetic variable neighborhood search algorithm for VRP with Multiple Time Windows. This algorithm encompasses combination of genetic crossover operators applied on list of best parent solutions with new implementations of local search operators.

Kumar and Panneerselvam [98] presented in 2017 a study of crossover operators for genetic algorithms to solve different variants of the VRP, and they also introduced a new Sinusoidal Motion Crossover (SMC) operator. This operator works analogously to sinusoidal motion of waves by alternately selecting consecutive customers from the first and second parents producing at the end two offspring solutions.

Baniamerian et al. [11] just recently in 2018 developed a hybrid metaheuristic combining the genetic algorithm hybridized with a modified variable neighborhood search to solve the vehicle routing problems with cross-docking. They suggested usage of some new four shaking and two neighborhood structures in a modified version of the VNS to solve the problem more efficiently in their combination with the GA.

4.2.17 Memetic Algorithms

The Memetic Algorithms (MAs) are population-based hybrid genetic algorithms hybridized with local search refinement procedures. They were introduced in 1989 by Moscato [118], and they are also commonly known as Genetic Local Search or Hybrid Evolutionary Algorithms because from optimization point of view, the evolutionary algorithm is used to perform exploration, whereas the local search methods are used to perform exploitation of the solution space. The MAs are inspired by models of adaptation in natural systems combining evolutionary adaptation of populations of individuals with individual learning within a lifetime [123]. The word *memetic* has its origin in word *meme* introduced in 1976 by Dawkins [41] to indicate the unit of imitation in cultural transmission and thus encompassing other forms of population-based techniques for optimization coming from a source different from genetic algorithms [120]. An illustration of the Memetic Algorithm is shown in Fig. 4.21.

The MAs are relatively a new group of specialized metaheuristics to solve various optimization problems, and initially they were considered to solve the TSP and simplest variants of the Vehicle Routing Problems [119]. In 1992, Moscato and Tinetti [120] developed a tree-structured memetic algorithm for the TSP, in which the optimizing population is divided into subpopulations and agents optimizing their current tours. Each agent is handling two tours, called *the pocket tour* and *current tour*, indicating best tour found so far and other tour being actually optimized by the heuristic assigned to the agent, respectively. All agents optimize their current tours with periods of local search, followed by recombination with other tours in case local minima is reached. The propagation of representative *memes* between subpopulations is taking place to spread out information about low-cost tours among all the agents.

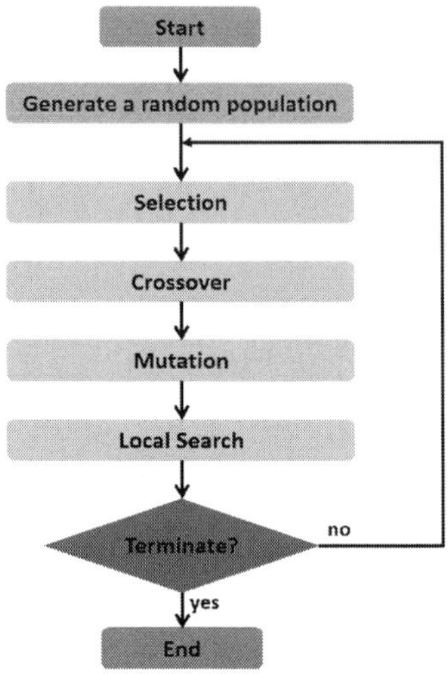

FIGURE 4.21 Illustration of the Memetic Algorithm.

In 1997, Nagata [125] proposed the Edge Assembly Crossover (EAX) operator for the TSP, which proved to be very powerful and was further extended and successfully used to the variants of the VRPs. This operator uses the edges from two parents to construct initial disjoint subtours, followed by connecting subtours in a greedy fashion using a construction similar to a minimal spanning tree to create offspring tours.

Merz and Freisleben [113] prepared in 2001 a good review on memetic algorithms for the TSP comparing the Maximum Preservative Crossover (MPX) by Gorges-Schleuter [71] and the Distance Preserving Crossover (DPX) by Freisleben and Merz [54]. The MPX generates offspring tours by copying a subtour between two randomly selected crossover points from the first parent and extending such partial tour by copying edges from the second parent. The DPX operator is very specific for the MAs as it is useful only if combined with local search [113]. It tries to generate offsprings by copying all items from the first parent and removing all edges not common between parents followed by reconnecting broken tour by local search methods. In 2002, Merz [112] further extended research on the MAs for the TSP by introducing the Generic Recombination Operator (GX), always preserving all common edges in offsprings. GX comprises four phases controlled by parameters reflecting the most important properties of recombination operators, and it proved to be superior to both MPX and DPX.

Labadi et al. [102] developed in 2008 a very efficient memetic algorithm for the VRP with Time Windows, which was able to minimize the total distance traveled also during the first phase of route minimization and improved 20 world's best-known solutions. The MA for the multicompartment VRP was proposed in 2008 by El Fallahi et al. [48], who combined memetic approach with a postoptimization phase based on path relinking and tabu search methods.

In 2009, Nagata and Braysy [123] developed a very powerful edge-assembly-based memetic algorithm for the Capacitated VRP. This algorithm combined the EAX crossover with well-known local search methods and additionally allowing for infeasible solutions due to capacity and route duration constraints after crossover operations. Their algorithm was able to find 20 new world's best-known solutions out of 47 standard benchmarks within very reasonable computational time. Nagata et al. [124] further extended the EAX crossover to be applied for the VRP with Time Windows in their penalty-based edge assembly memetic algorithm. They introduced adjustments of the EAX operator to the VRPTW and a penalty function to eliminate violations of both time window and capacity constraints from offspring solutions generated by the EAX operator. Their algorithm proved to be extremely powerful by finding 184 world's best-known solutions out of 356 benchmark instances.

Nalepa and Blocho [127] developed in 2016 an extension of the above method by introducing the adaptive version of the memetic algorithm for the VRPTW solving problem of automatic tuning of numerous hyperparameters. In their version the parameters of the algorithm, including the selection scheme, population size, and the number of child solutions, generated for each pair of parents, were adjusted dynamically during the search. Moreover, they introduced a new adaptive selection scheme to balance the exploration and exploitation of the solution space, which proved to be competitive and confirmed efficacy and convergence capabilities of the proposed approach.

A different approach was proposed in 2012 by Vidal et al. [190], who developed a very competitive hybrid genetic algorithm for solving the Capacitated VRP. They combined the exploration breadth of population-based evolutionary search with aggressive-improvement capabilities of neighborhood-based metaheuristics and advanced population-diversity management schemes, which allowed them to create competitive GA in terms of both computational efficiency and solution quality. They further extended these technique by introducing in 2013 [191] the Hybrid Genetic Search with Advanced Diversity Control for the VRPTW. This algorithm comprised new move evaluation techniques accounting for penalized infeasible solutions with respect to time window and duration constraints and allowing evaluation of moves from classical neighborhoods based on arc or node exchanges in amortized constant time. Moreover, they developed also geometric and structural problem decompositions to address large problems of the VRPTW efficiently.

The memetic algorithms were also taken into account while developing methods for the Pickup and Delivery Problems. In 2010, Nagata and Kobayashi [126] developed the MA for the PDP with Time Windows using Selective Route Exchange Crossover (SREX). Their algorithm allowed them to improve 146 best known solutions out of 298 instances. More recently, in 2017, Blocho and Nalepa [21] proposed a modification to the SREX based on the Longest

Common Subsequence (LCS) method combined with a new technique for representation of a solution to handle the crossover efficiently. Their approach was applied to the memetic algorithm for the PDPTW, proving ability to obtain high-quality feasible solutions.

4.3 Hyperheuristics for rich vehicle routing problems

The hyperheuristic is a new optimization paradigm, commonly described as heuristic to choose heuristics. It comprises search methods, learning techniques for generating or selecting heuristics to solve optimization problems [30]. The main difference between metaheuristics and hyperheuristics is that metaheuristics directly search a space of a problem, whereas hyperheuristics search a space of heuristics [116]. Heuristic selection and heuristic generation are two main classes of hyperheuristics approaches. The heuristic selection approach selecting and applying the most suitable heuristics from a set of problem-specific low-level heuristics (LLHs) at each problem solving state. On the other hand, heuristics generation techniques are used to automatically construct new heuristics from components of already existing heuristics, adjusted for the specific problem [29].

The heuristics selection techniques can be further divided into constructive and local-search hyperheuristics depending on the type of used LLHs. Similarly to construction heuristics (described in Section 4.1.1), the constructive hyperheuristics start with an empty solution building gradually a complete solution by subsequently choosing a proper low-level construction heuristic (selected from a set of LLHs) and by using it to enhance the quality of built solution. In turn, local-search hyperheuristics, similarly to improvement heuristics (described in Section 4.1.2), start from already generated solution and then try to gradually improve the quality of the solution by applying local searches and proper neighborhood structure.

The hyperheuristics usually can provide much better solutions than classical heuristics or metaheuristics applied to the specific problem individually. This is due that hyperheuristics can discover a good combination of best characteristics of individual low-level heuristics [116]. The crucial idea in hyperheuristics is solving optimization problems using various simple and flexible LLHs and developing a framework used to control the selection ad application of LLHs [60]. Typically, high-level heuristics and metaheuristics (variable neighborhood search, simulated annealing, tabu search, evolutionary algorithms, and many others described in Section 4.2) are used as search methods across the search space of heuristics.

In 2009, Garrido and Castro [60] proposed the Hill-climbing-based hyperheuristic for the CVRP problem. This technique works with ordered sequence of the so-called structures, and each structure defines one constructive and one improvement heuristic components. Initially, a random sequence of structures is generated with equal probability of each constructive and improvement heuristic. This sequence is next iteratively enhanced in a hill-climbing way trying to find a new neighbor sequence having better fitness than actual one. They also proposed some perturbing moves to add, delete, and replace structures and reallocating moves for exchanging customers between two adjacent structures.

Another hyperheuristic for the CVRP based on evolutionary algorithm was proposed by Garrido et al. [61] in 2009. This technique works with a population of sequences of structures, but contrary to the hill-climbing method described previously, structures may contain only one heuristic component and only of the same type, either constructive or improvement heuristics. This evolutionary hyperheuristic uses one recombination operator and eight mutation operators to generate new individuals. The population of individuals is evolving using a steady-state evolutionary model, where in each generation, one or two offsprings are constructed, which are further inserted into the main population.

In 2013, Mlejnek and Kubalík [116] proposed an evolutionary-based constructive heuristic selection hyperheuristic for the CVRP (HyperPOEMS), based on iterative local search algorithm. The HyperPOEMS operates on an ordered sequence of units containing selected constructive improvement heuristics. The design of that method allows for autonomous searching a structured space of various low-level heuristics to find proper combinations producing good solutions to the problem. The LLHs used in HyperPOEMS are various constructive (saving concept, sweep mechanism, cluster-first route-second, route-first cluster-second methods), improvement (2-opt, 3-opt, Or-opt, Van Breedam's moves) and order heuristics (increasing/decreasing demand, increasing/decreasing distance to the depot, radial sweeps). HyperPOEMS technique outperformed both hyperheuristics proposed by Garrido et al. [61].

Addressing limited availability of the constructive heuristics used by both Garrido et al. and Mlejnek and Kubalík, Drake et al. [46] proposed in 2013 a hyperheuristics methodology using Grammatical Evolution to simultaneously evolve constructive and perturbing heuristics. In this method a single initial solution is created by the construction heuristics, which is next iteratively enhanced by perturbation heuristics.

In 2016, Sim and Hart proposed a novel method for generating new constructive heuristics in a combined generative and selective hyperheuristic for the VRP. They tried to address the lack of available constructive methods and the potentially limiting quality of creating a weak candidate solution. This new construction heuristics may be used with any hyperheuristic methods requiring the creation of candidate solutions. Another important point is that they proposed also a new multipoint hyperheuristic method, called GP-MHH, comprising genetic programming in the first phase to evolve a population of construction heuristics, and a population of candidate solutions is constructed in the second phase by the evolved heuristics. The Memetic Hyperheuristic (MHH) is eventually applied on the population, which is a sophisticated perturbative-selection hyperheuristic. Based on extensive experimental studies, this method outperformed the grammatical evolution approach proposed by Drake et al. [46].

4.4 Summary

In this chapter, we described the most important heuristic methods for solving different variants of the Vehicle Routing Problems and Pickup and Delivery Problems. A heuristic approach to solve an optimization problem does not guarantee obtaining the optimal solution but enables to elaborate a feasible routing schedule in a very efficient way and in short time. Due to NP-hardness of the rich vehicle routing problems, solving problem instances to optimality is possible only in case of small-size tests. The main challenge in most of these problems is finding balance between available CPU time, size of the problem, and the quality of the approximated solution or need to find exact solution. Rapid development of various heuristic, metaheuristic, and hyperheuristic approaches helps in achieving both reduced computation time and better results.

In the first part of this chapter, we described the construction and improvement heuristics as two main groups of the classical heuristics. The construction heuristics build feasible solutions while trying to minimize its cost but often do not consider any improvement phases. The improvement heuristics are usually used for already generated solutions by other heuristics or exact algorithms. Local search methods are typically applied for simple local modifications, such as customer or arc exchanges, to generate neighboring solutions of possibly better quality.

The metaheuristics can be described as an enhancement of classical heuristics with emphasis on deep exploration of the solution space, and they usually combine sophisticated neighborhood search rules and recombination of solutions. The quality of the solutions produced by metaheuristics is typically much higher than that obtained by classical heuristics techniques but with a price of increased computing time. It is worth noting that the metaheuristic techniques are context dependent and usually require finely tuned parameters, which unfortunately make their extensions to other problems difficult.

Many various metaheuristics have been proposed for the vehicle routing problems, and they can be widely divided into local search, population search and learning mechanism groups; however, best metaheuristics merge ideas from different approaches. In the second part of this chapter, we described a wide range of the most popular metaheuristics algorithms.

The Simulated Annealing is a stochastic algorithm involving asymptotic convergence and allowing random movements in the searched neighborhood to escape local minima. Due to low-complexity, it can be used in many various optimization problems, not only related to the rich vehicle routing problems.

The Tabu Search is a search technique comprising local search methods and memory structures called tabu-list. Its main idea is to avoid cycling by inserting recently checked solution on the tabu-list, so during the search process, the solutions marked with tabu label are not taken into consideration. Such approach helps in getting out from the local minima and increases chances to find the global optimal solution.

The Adaptive Memory Procedure is based on the idea of the so-called adaptive memory being pool of good-quality solutions. The solutions in adaptive memory are always replaced with the new better-quality solutions coming from recombination of existing ones.

The Variable Neighborhood Search can be described as a framework for building heuristics, exploiting the neighborhoods both finding local optima and getting out of them by perturbation moves. In contrary to classical local search methods, the Variable Neighborhood Search is not based on following the trajectory but rather on exploring increasingly distant neighborhoods of a given solution and moving to the new one only in case of improvement. Such a method usually leads to maintaining best characteristics of current solution and helps obtain neighboring solutions of better quality.

The Large Neighborhood Search can be described as an iterative way of destroying and repairing the solution in the neighborhood. Destroying methods have some randomness such that different parts of the solution are destroyed and broader parts of the search tree are visited, and thus the searched neighborhood is larger than in classical local search methods.

The Greedy Randomized Adaptive Search Procedure is based on iterative two-phase search algorithm comprising construction and local search phases applied to various combinatorial optimization problems. In each iteration in the construction phase a feasible solution is constructed by a randomized greedy function. The solution is then iteratively improved in the second phase by local search movements.

The Particle Swarm Optimization aimed at producing computational intelligence by exploiting analogues of social interaction rather than individual cognitive abilities. It is an optimization method based on the idea of iterative improvements of a solution with regard to certain quality measures. The Particle Swarm Optimization is using a population of particles representing solutions to move them around in the search space due to some mathematical rules over particles velocity and position. Similarly to flock of birds collectively searching for food, the swarm is likely to move close to an optimum of the fitness function.

The Ant Colony Algorithm is based on the behavior of ants seeking paths between their colony and sources of food and laying some pheromone on trails. The fact that ants always follow the same path, which is indeed the shortest path, was the main motivation to take advantage of such real natural behavior of ants. The walking ants mark trails by laying down pheromones with information of quantity and quality of food. Such idea could be translated to the vehicle routing problems as searching in the neighborhood for good-quality solutions.

The Artificial Bee Colony algorithm is based on two fundamental concepts, namely self-organization and division of labor, which allow problem-solving systems to self-organize and adapt to the environment. The Artificial Bee Colony model related to collective intelligence of honey bees comprises employed and unemployed bees and food sources. It defines also recruitment to a nectar source and abandonment of such source as two leading methods of the honey bees behavior. The value of the food source for honey bees usually depends on its richness, proximity to the nest, and the ease of forage extracting, and they can be identified with good-quality solutions in the Artificial Bee Colony algorithm.

The Bat Algorithm is based on echolocation behavior of microbats when searching for their prey in nature.

The Cuckoo Search is a metaheuristic optimization algorithm inspired by the cuckoo birds being the "brood parasites" birds as they never build they own nests and instead lay their eggs in nests of other host bird nests. In terms of optimization algorithms, each egg represents a solution, and each cuckoo egg represents a new potentially better solution. The Cuckoo Search algorithm is based on the idea of creating subsequent generations of nests containing best eggs and thus highest quality solutions.

The Firefly Algorithm is inspired by flashing behavior of fireflies. The main purpose why fireflies flash is to signal and attract other flies.

The Golden Ball is a multipopulation metaheuristic based on soccer concept. The whole population of players (indicated as solutions) is created followed by random division of players into a fixed number of subpopulations called teams. Each team has its own couch related to training method, which is also randomly assigned to each time. The training method can be also associated with the way how each player evolves individually in the team, and in terms of the vehicle routing problems, it may be associated with local search moves.

The Gravitational Search Algorithm is based on the gravity law and mass interactions and was inspired by gravitation being the tendency of masses to accelerate toward each other. In the Gravitational Search Algorithm, agents are considered as objects, and their masses are related to their performance. All the agents attract each other by the gravity force communicating each other through gravitational force. The heavy objects, related to high-quality solutions, move slower than lighter objects, and this guarantees the exploitation step of this algorithm.

The idea of the Bacterial Foraging Optimization Algorithm is to mimic biological principles shown in the foraging behavior of E. coli bacteria for distributed optimization and control. In the Bacterial Foraging Optimization Algorithm a swarm of bacteria is used as searching agent for a solution to a specific optimization problem. A bacteria position represents a solution of the problem by a simple sequence of customer nodes. The bacteria direction vector is used to represent its ability to search for solution and is driving bacteria movement. In each step of the Bacterial Foraging Optimization Algorithm, all bacterium moves from one position to another are based on these directions, and by moving there are evaluated various solutions of the problem.

The Genetic Algorithms are optimization search methods based on the evolution process in nature, and they imitate the biological process of natural selection where stronger populations among different species survive. The Genetic Algorithms belong to the larger class of Evolutionary Algorithms, which are generic population-based metaheuristic optimization algorithms. All Evolutionary Algorithms and thus also Genetic Algorithms use biological evolution mechanisms such as representation, selection, recombination, and mutation. In terms of the rich vehicle routing problems, the representation of the solution comprises encoding most important features of the solution as genes in a chromosome, which identifies the so-called individual in the population. The recombination is carried out on two selected parent solutions by combining genes of parent chromosomes to create offspring solutions with potentially better quality. In turn, mutation is performed on the offspring solutions by random modification of genes to further explore the solution space and ensure genetic diversity. The Genetic Algorithm is based on creating subsequent generations, and each new generation is constructed by selection, recombination, and mutation of all the solutions in the population.

The Memetic Algorithms are population-based hybrid genetic algorithms hybridized with local search refinement procedures. From optimization point of view, the evolutionary algorithm is used to perform exploration, whereas the local search methods are used to perform exploitation. The MAs are inspired by models of adaptation in natural systems combining evolutionary adaptation of populations of individuals with individual learning within a lifetime.

The hyperheuristics is a new optimization paradigm, commonly described as heuristics to choose heuristics, comprising search methods, and learning techniques for generating or selecting heuristics to solve optimization problems. The main difference between metaheuristics and hyperheuristics is that metaheuristics directly search the solution space of a problem, whereas hyperheuristics search the space of heuristics.

References

[1] Emile Aarts, Jan Korst, Wil Michiels, Simulated Annealing, Springer US, Boston, MA, 2005, pp. 187–210.

[2] Mohamed M.S. Abdulkader, Yuvraj Gajpal, Tarek Y. ElMekkawy, Hybridized ant colony algorithm for the multi compartment vehicle routing problem, Applied Soft Computing 37 (2015) 196–203.

[3] Ala'a Abu-Srhan, E.A. Daoud, A hybrid algorithm using a genetic algorithm and cuckoo search algorithm to solve the traveling salesman problem and its application to multiple sequence alignment, 2014.

[4] Sohaib Afifi, Duc-Cuong Dang, Aziz Moukrim, A simulated annealing algorithm for the vehicle routing problem with time windows and synchronization constraints, in: Giuseppe Nicosia, Panos Pardalos (Eds.), Learning and Intelligent Optimization, Springer, Berlin, Heidelberg, 2013, pp. 259–265.

[5] The Jin Ai, Voratas Kachitvichyanukul, A particle swarm optimization for the vehicle routing problem with simultaneous pickup and delivery, Computers & Operations Research 36 (5) (2009) 1693–1702, Selected papers presented at the Tenth International Symposium on Locational Decisions (ISOLDE X).

[6] Malek Alzaqebah, Salwani Abdullah, Sana Jawarneh, Modified artificial bee colony for the vehicle routing problems with time windows, SpringerPlus 5 (1) (Aug 2016) 1298.

[7] X. Yang, S. Deb, Cuckoo search via Lévy flights, in: 2009 World Congress on Nature Biologically Inspired Computing, NaBIC, Dec 2009, pp. 210–214.

[8] Nabila Azi, Michel Gendreau, Jean-Yves Potvin, An adaptive large neighborhood search for a vehicle routing problem with multiple routes, Computers & Operations Research 41 (2014) 167–173.
[9] Barrie M. Baker, M.A. Ayechew, A genetic algorithm for the vehicle routing problem, Computers & Operations Research 30 (5) (2003) 787–800.
[10] Edward K. Baker, Joanne R. Schaffer, Solution improvement heuristics for the vehicle routing and scheduling problem with time window constraints, American Journal of Mathematical and Management Sciences 6 (3–4) (1986) 261–300.
[11] Ali Baniamerian, Mahdi Bashiri, Fahime Zabihi, A modified variable neighborhood search hybridized with genetic algorithm for vehicle routing problems with cross-docking, Electronic Notes in Discrete Mathematics 66 (2018) 143–150, 5th International Conference on Variable Neighborhood Search.
[12] Benjamin Baran, Matilde Schaerer, A multiobjective ant colony system for vehicle routing problem with time windows 21 (01 2003) 97–102.
[13] S. Belhaiza, R. M'Hallah, G.B. Brahim, A new hybrid genetic variable neighborhood search heuristic for the vehicle routing problem with multiple time windows, in: 2017 IEEE Congress on Evolutionary Computation, CEC, June 2017, pp. 1319–1326.
[14] John E. Bell, Patrick R. McMullen, Ant colony optimization techniques for the vehicle routing problem, Advanced Engineering Informatics 18 (1) (2004) 41–48.
[15] Russell Bent, Pascal Van Hentenryck, A two-stage hybrid local search for the vehicle routing problem with time windows, Transportation Science 38 (4) (November 2004) 515–530.
[16] Jean Berger, Mohamed Barkaoui, A route-directed hybrid genetic approach for the vehicle routing problem with time windows, INFOR 41 (2003) 179–194.
[17] Jean Berger, Martin Salois, Regent Begin, A hybrid genetic algorithm for the vehicle routing problem with time windows, in: Robert E. Mercer, Eric Neufeld (Eds.), Advances in Artificial Intelligence, Springer, Berlin, Heidelberg, 1998, pp. 114–127.
[18] Ashita S. Bhagade, Vinayak Puranik, Artificial bee colony (ABC) algorithm for vehicle routing optimization problem.
[19] Joe L. Blanton Jr., Roger L. Wainwright, Multiple vehicle routing with time and capacity constraints using genetic algorithms, in: Proceedings of the 5th International Conference on Genetic Algorithms, Morgan Kaufmann Publishers Inc., San Francisco, CA, USA, 1993, pp. 452–459.
[20] Andrius Blazinskas, Combining 2-opt, 3-opt and 4-opt with k-swap-kick perturbations for the traveling salesman problem, 2011.
[21] Miroslaw Blocho, Jakub Nalepa, LCS-based selective route exchange crossover for the pickup and delivery problem with time windows, in: Bin Hu, Manuel López-Ibáñez (Eds.), EvoCOP, in: Lecture Notes in Computer Science, vol. 10197, 2017, pp. 124–140.
[22] Burcin Bozkaya, Erhan Erkut, Gilbert Laporte, A tabu search heuristic and adaptive memory procedure for political districting, European Journal of Operational Research 144 (2003) 12–26.
[23] José Brandão, A tabu search algorithm for the heterogeneous fixed fleet vehicle routing problem, Computers & Operations Research 38 (2011) 140–151.
[24] Bräysy Olli, A reactive variable neighborhood search for the vehicle-routing problem with time windows, INFORMS Journal on Computing 15 (4) (2003) 347–368.
[25] Olli Bräysy, Wout Dullaert, Michel Gendreau, Evolutionary algorithms for the vehicle routing problem with time windows, Journal of Heuristics 10 (6) (2004) 587–611.
[26] Bräysy Olli, Michel Gendreau, Tabu search heuristics for the vehicle routing problem with time windows, Top 10 (2) (Dec 2002) 211–237.
[27] Olli Bräysy, Michel Gendreau, Vehicle routing problem with time windows, part I: route construction and local search algorithms, Transportation Science 39 (1) (February 2005) 104–118.
[28] Bernd Bullnheimer, Richard F. Hartl, Christine Strauss, Applying the ANT System to the Vehicle Routing Problem, Springer US, Boston, MA, 1999, pp. 285–296.

[29] Edmund K. Burke, Mathew R. Hyde, Graham Kendall, Gabriela Ochoa, Ender Ozcan, John R. Woodward, Exploring Hyper-Heuristic Methodologies With Genetic Programming, Springer, Berlin, Heidelberg, 2009, pp. 177–201.

[30] Edmund K. Burke, Matthew Hyde, Graham Kendall, Gabriela Ochoa, Ender Özcan, John R. Woodward, A Classification of Hyper-Heuristic Approaches, Springer US, Boston, MA, 2010, pp. 449–468.

[31] Ann Melissa Campbell, Martin Savelsbergh, Efficient insertion heuristics for vehicle routing and scheduling problems, Transportation Science 38 (3) (August 2004) 369–378.

[32] Huiqin Chen, Sheng Li, Zheng Tang, Hybrid gravitational search algorithm with random-key encoding scheme combined with simulated annealing, International Journal of Computer Science and Network Security 11 (01 2011).

[33] Xichun Chen, Junli Wang, A novel hybrid cuckoo search algorithm for optimizing vehicle routing problem in logistics distribution system, Journal of Computational and Theoretical Nanoscience 13 (1) (2016).

[34] A. Chin, H. Kit, A. Lim, A new GA approach for the vehicle routing problem, in: Proceedings 11th International Conference on Tools With Artificial Intelligence, Nov 1999, pp. 307–310.

[35] G. Clarke, J.W. Wright, Scheduling of vehicles from a central depot to a number of delivery points, Operations Research 12 (4) (August 1964) 568–581.

[36] Alberto Colorni, Marco Dorigo, Vittorio Maniezzo, Distributed optimization by ant colonies, 1991.

[37] Jean-Francois Cordeau, Gilbert Laporte, Tabu Search Heuristics for the Vehicle Routing Problem, Springer US, Boston, MA, 2005, pp. 145–163.

[38] Jean-Francois Cordeau, Martin W.P. Savelsbergh, Chapter 6: Vehicle routing, 2006.

[39] A. Croes, A method for solving traveling salesman problems, Operations Research 5 (1958) 791–812.

[40] Lawrence Davis, Applying adaptive algorithms to epistatic domains, in: Proceedings of the 9th International Joint Conference on Artificial Intelligence – vol. 1, IJCAI'85, Morgan Kaufmann Publishers Inc., San Francisco, CA, USA, 1985, pp. 162–164.

[41] R. Dawkins, The Selfish Gene, Oxford University Press, Oxford, UK, 1976.

[42] Kenneth Alan De Jong, An Analysis of the Behavior of a Class of Genetic Adaptive Systems, PhD thesis, Ann Arbor, MI, USA, AAI7609381, 1975.

[43] Alberto V. Donati, Roberto Montemanni, Norman Casagrande, Andrea E. Rizzoli, Luca M. Gambardella, Time dependent vehicle routing problem with a multi ant colony system, European Journal of Operational Research 185 (3) (2008) 1174–1191.

[44] Marco Dorigo, Optimization, Learning and Natural Algorithms, PhD thesis, Politecnico di Milano, Italy, 1992.

[45] Mohammad Bagher Dowlatshahi, Hossein Nezamabadi-pour, Mashaallah Mashinchi, A discrete gravitational search algorithm for solving combinatorial optimization problems, Information Sciences 258 (2014) 94–107.

[46] John H. Drake, Nikolaos Kililis, Ender Özcan, Generation of VNS components with grammatical evolution for vehicle routing, in: Krzysztof Krawiec, Alberto Moraglio, Ting Hu, A. Şima Etaner-Uyar, Bin Hu (Eds.), Genetic Programming, Springer, Berlin, Heidelberg, 2013, pp. 25–36.

[47] Christophe Duhamel, Philippe Lacomme, Christian Prins, Caroline Prodhon, A GRASPxELS approach for the capacitated location-routing problem, Computers & Operations Research 37 (11) (2010) 1912–1923, Metaheuristics for Logistics and Vehicle Routing.

[48] Abdellah El-Fallahi, Christian Prins, Roberto Wolfler Calvo, A memetic algorithm and a tabu search for the multi-compartment vehicle routing problem, Computers & Operations Research 35 (5) (2008) 1725–1741.

[49] Özlem Ergun, James B. Orlin, Abran Steele-Feldman, Creating very large scale neighborhoods out of smaller ones by compounding moves, Journal of Heuristics 12 (1) (Mar 2006) 115–140.

[50] Ana Laura Vázquez-Esquivel, Jesús Aguirre-García, Andrea Isabel Cervantes-Payan, Edgar Osvaldo Escobedo-Hernández, Ernesto Liñán-García, Helue I. De la Barrera-Gómez, Luis A. López-Alday, Solving vehicle routing problem with multi-phases simulated annealing algorithm, 2018.

[51] Thomas A. Feo, Mauricio G.C. Resende, Greedy randomized adaptive search procedures, Journal of Global Optimization 6 (2) (Mar 1995) 109–133.

[52] Merrill M. Flood, The traveling-salesman problem, Operations Research 4 (1) (1956) 61–75.

[53] B.R. Fox, M.B. McMahon, Genetic operators for sequencing problems, 1990.

[54] Bernd Freisleben, Peter Merz, A genetic local search algorithm for solving symmetric and asymmetric traveling salesman problems, in: International Conference on Evolutionary Computation, 1996, pp. 616–621.

[55] Guenther Fuellerer, Karl F. Doerner, Richard F. Hartl, Manuel Iori, Ant colony optimization for the two-dimensional loading vehicle routing problem, Computers & Operations Research 36 (3) (2009) 655–673.

[56] Y. Gajpal, P. Abad, Saving-based algorithms for vehicle routing problem with simultaneous pickup and delivery, The Journal of the Operational Research Society 61 (10) (2010) 1498–1509.

[57] Yuvraj Gajpal, Prakash Abad, An ant colony system (ACS) for vehicle routing problem with simultaneous delivery and pickup, Computers & Operations Research 36 (12) (2009) 3215–3223, New developments on hub location.

[58] L.M. Gambardella, E. Taillard, G. Agazzi, MACS-VRPTW: a Multiple Ant Colony System for Vehicle Routing Problems With Time Windows, Technical report, 1999.

[59] Xiaobing Gan, Lijiao Liu, Ben Niu, L.J. Tan, F.F. Zhang, J. Liu, SRBFOs for solving the heterogeneous fixed fleet vehicle routing problem, in: De-Shuang Huang, Kang-Hyun Jo, Abir Hussain (Eds.), Intelligent Computing Theories and Methodologies, Springer International Publishing, Cham, 2015, pp. 725–732.

[60] Pablo Garrido, Carlos Castro, Stable solving of CVRPs using hyperheuristics, in: Proceedings of the 11th Annual Conference on Genetic and Evolutionary Computation, GECCO'09, ACM, New York, NY, USA, 2009, pp. 255–262.

[61] Pablo Garrido, Carlos Castro, Eric Monfroy, Towards a flexible and adaptable hyperheuristic approach for VRPs 1 (01 2009) 311–317.

[62] Michel Gendreau, Alain Hertz, Gilbert Laporte, New insertion and postoptimization procedures for the traveling salesman problem, Operations Research 40 (1086–1094) (12 1992).

[63] Michel Gendreau, Alain Hertz, Gilbert Laporte, A tabu search heuristic for the vehicle routing problem, Management Science 40 (10) (1994) 1276–1290.

[64] Keivan Ghoseiri, Seyed Farid Ghannadpour, Multi-objective vehicle routing problem with time windows using goal programming and genetic algorithm, Applied Soft Computing 10 (4) (2010) 1096–1107, Optimisation Methods & Applications in Decision-Making Processes.

[65] Fred Glover, Tabu search – part I, INFORMS Journal on Computing 1 (3) (1989) 190–206.

[66] Fred Glover, Ejection chains, reference structures and alternating path methods for traveling salesman problems, Discrete Applied Mathematics 65 (1) (1996) 223–253, First International Colloquium on Graphs and Optimization.

[67] Rajeev Goel, Raman Maini, A hybrid of ant colony and firefly algorithms (HAFA) for solving vehicle routing problems, Journal of Computational Science 25 (2018) 28–37.

[68] Fatma Pinar Goksal, Ismail Karaoglan, Fulya Altiparmak, A hybrid discrete particle swarm optimization for vehicle routing problem with simultaneous pickup and delivery, Computers & Industrial Engineering 65 (1) (2013) 39–53, Intelligent Manufacturing Systems.

[69] David E. Goldberg, Genetic Algorithms in Search, Optimization and Machine Learning, 1st edition, Addison-Wesley Longman Publishing Co., Inc., Boston, MA, USA, 1989.

[70] David E. Goldberg, Robert Lingle Jr., Alleles, loci, and the traveling salesman problem.

[71] Martina Gorges-Schleuter, Genetic Algorithms and Population Structures: a Massively Parallel Algorithm, PhD thesis, Uni Dortmund, 1991.

[72] Chrysanthos E. Gounaris, Panagiotis P. Repoussis, Christos D. Tarantilis, Wolfram Wiesemann, Christodoulos A. Floudas, An adaptive memory programming framework for the robust capacitated vehicle routing problem, Transportation Science 50 (4) (2016) 1239–1260.
[73] Philippe Grangier, Michel Gendreau, Fabien Lehuédé, Louis-Martin Rousseau, An adaptive large neighborhood search for the two-echelon multiple-trip vehicle routing problem with satellite synchronization, European Journal of Operational Research 254 (1) (2016) 80–91.
[74] Lahcene Guezouli, Mohamed Bensakhria, Samir Abdelhamid, Efficient golden-ball algorithm based clustering to solve the multi-depot VRP with time windows, International Journal of Applied Evolutionary Computation 9 (2018) 1–16.
[75] Vera C. Hemmelmayr, Jean-Francois Cordeau, Teodor Gabriel Crainic, An adaptive large neighborhood search heuristic for two-echelon vehicle routing problems arising in city logistics, Computers & Operations Research 39 (12) (2012) 3215–3228.
[76] Seda Hezer, Yakup Kara, Solving vehicle routing problem with simultaneous delivery and pick-up using an algorithm based on bacterial foraging optimization, Journal of the Faculty of Engineering and Architecture of Gazi University 28 (06 2013) 373–382.
[77] William Ho, George T.S. Ho, Ping Ji, Henry C.W. Lau, A hybrid genetic algorithm for the multi-depot vehicle routing problem, Engineering Applications of Artificial Intelligence 21 (4) (2008) 548–557.
[78] John H. Holland, Adaptation in Natural and Artificial Systems: An Introductory Analysis With Applications to Biology, Control and Artificial Intelligence, MIT Press, Cambridge, MA, USA, 1992.
[79] Ali Asghar Rahmani Hosseinabadi, Maryam Kardgar, Mohammad Shojafar, Shahab Shamshirband, Ajith Abraham, Gravitational search algorithm to solve open vehicle routing problem, in: Václav Snášel, Ajit Abraham, Pavel Krömer, Millie Pant, Azah Kamilah Muda (Eds.), Innovations in Bio-Inspired Computing and Applications, Springer International Publishing, Cham, 2016, pp. 93–103.
[80] Aliasghar Rahmani Hosseinabadi, Mohammad Yazdanpanah, Ali Shokouhi Rostami, A new search algorithm for solving symmetric traveling salesman problem based on gravity, 2013.
[81] Wenbin Hu, Huanle Liang, Chao Peng, Bo Du, Qi Hu, A hybrid chaos-particle swarm optimization algorithm for the vehicle routing problem with time window, Entropy [electronic only] 4 (04 2013).
[82] G.K. Jati, H.M. Manurung, Suyanto, Discrete cuckoo search for traveling salesman problem, in: 2012 7th International Conference on Computing and Convergence Technology, ICCCT, Dec 2012, pp. 993–997.
[83] Gilang Kusuma Jati, Suyanto, Evolutionary discrete firefly algorithm for travelling salesman problem, in: Abdelhamid Bouchachia (Ed.), Adaptive and Intelligent Systems, Springer, Berlin, Heidelberg, 2011, pp. 393–403.
[84] Liu Jie, Lin Teng, Shoulin Yin, An improved discrete firefly algorithm used for traveling salesman problem, in: Ying Tan, Hideyuki Takagi, Yuhui Shi (Eds.), Advances in Swarm Intelligence, Springer International Publishing, Cham, 2017, pp. 593–600.
[85] Wan-Rong Jih, J. Yung-Jen Hsu, Dynamic vehicle routing using hybrid genetic algorithms, in: Proceedings 1999 IEEE International Conference on Robotics and Automation, Cat. No. 99CH36288C, vol. 1, May 1999, pp. 453–458.
[86] A.S. Joshi, Omkar Kulkarni, G.M. Kakandikar, V.M. Nandedkar, Cuckoo search optimization – a review, Materials Today: Proceedings 4 (8) (2017) 7262–7269, International Conference on Advancements in Aeromechanical Materials for Manufacturing (ICAAMM-2016): Organized by MLR Institute of Technology, Hyderabad, Telangana, India.
[87] J.W. Joubert, S.J. Claasen, A sequential insertion heuristic for the initial solution to a constrained vehicle routing problem, ORiON 22 (06 2006).
[88] Soonchul Jung, Byung-Ro Moon, The natural crossover for the 2D Euclidean TSP, in: Proceedings of the 2nd Annual Conference on Genetic and Evolutionary Computation, GECCO'00, Morgan Kaufmann Publishers Inc., San Francisco, CA, USA, 2000, pp. 1003–1010.

[89] Soonchul Jung, Byung-Ro Moon, A hybrid genetic algorithm for the vehicle routing problem with time windows, in: Proceedings of the 4th Annual Conference on Genetic and Evolutionary Computation, GECCO'02, Morgan Kaufmann Publishers Inc., San Francisco, CA, USA, 2002, pp. 1309–1316.

[90] Can B. Kalayci, Can Kaya, An ant colony system empowered variable neighborhood search algorithm for the vehicle routing problem with simultaneous pickup and delivery, Expert Systems with Applications 66 (2016) 163–175.

[91] Karaboga Dervis, An Idea Based on Honey Bee Swarm for Numerical Optimization, Technical report – tr06, Erciyes University, 01 2005.

[92] James Kennedy, Russell Eberhart, Particle swarm optimization 4 (12 1995) 1942–1948.

[93] S. Kirkpatrick, C.D. Gelatt, M.P. Vecchi, Optimization by simulated annealing, Science 220 (4598) (1983) 671–680.

[94] Gözde Kizilateş, Fidan Nuriyeva, On the nearest neighbor algorithms for the traveling salesman problem, in: Dhinaharan Nagamalai, Ashok Kumar, Annamalai Annamalai (Eds.), Advances in Computational Science, Engineering and Information Technology, Springer International Publishing, Heidelberg, 2013, pp. 111–118.

[95] George Kontoravdis, Jonathan Bard, A grasp for the vehicle routing problem with time windows, INFORMS Journal on Computing 7 (02 1995) 10–23.

[96] Slavko Krajcar, Davor Skrlec, Branko Pribicevic, Snjezana Blagajac, Ga approach to solving multiple vehicle routing problem, in: Carlos Pinto-Ferreira, Nuno J. Mamede (Eds.), Progress in Artificial Intelligence, Springer, Berlin, Heidelberg, 1995, pp. 473–481.

[97] Marek Kubiak, Systematic construction of recombination operators for the vehicle routing problem 29 (01 2004).

[98] S.G. Varun Kumar, Ramasamy Panneerselvam, A study of crossover operators for genetic algorithms to solve VRP and its variants and new sinusoidal motion crossover operator, International Journal of Computational Intelligence Research 13 (06 2017) 1717–1733.

[99] Sharad N. Kumbharana, Gopal M. Pandey, Solving travelling salesman problem using firefly algorithm, International Journal for Research in Science & Advanced Technologies (March–April 2013).

[100] R.J. Kuo, Ferani E. Zulvia, Kadarsah Suryadi, Hybrid particle swarm optimization with genetic algorithm for solving capacitated vehicle routing problem with fuzzy demand – a case study on garbage collection system, Applied Mathematics and Computation 219 (5) (2012) 2574–2588.

[101] Jari Kytöjoki, Teemu Nuortio, Olli Bräysy, Michel Gendreau, An efficient variable neighborhood search heuristic for very large scale vehicle routing problems, Computers & Operations Research 34 (9) (2007) 2743–2757.

[102] Nacima Labadi, Christian Prins, Mohamed Reghioui, A memetic algorithm for the vehicle routing problem with time windows, RAIRO - Operations Research 42 (3) (2008) 415–431.

[103] G. Laporte, F. Semet, Classical heuristics for the capacitated VRP, in: The Vehicle Routing Problem, Society for Industrial and Applied Mathematics, Philadelphia, PA, USA, 2001, pp. 109–128.

[104] Bo Li, Chen Guo, Tao Ning, Yingqi Wei, Solve VRPPD with improved bacteria optimization algorithm, 01 2017.

[105] Xiangyong Li, Peng Tian, Y.P. Aneja, An adaptive memory programming metaheuristic for the heterogeneous fixed fleet vehicle routing problem, Transportation Research Part E: Logistics and Transportation Review 46 (6) (2010) 1111–1127.

[106] F-H. Liu, S-Y. Shen, The fleet size and mix vehicle routing problem with time windows, Journal of the Operational Research Society 50 (7) (Jul 1999) 721–732.

[107] R.M. Brady, Optimization strategies gleaned from biological evolution, Nature 317 (10 1985) 804–806.

[108] O. Maeda, M. Nakamura, B.M. Ombuki, K. Onaga, A genetic algorithm approach to vehicle routing problem with time deadlines in geographical information systems, in: IEEE SMC'99 Conference Proceedings. 1999 IEEE International Conference on Systems, Man, and Cybernetics, Cat. No. 99CH37028, vol. 4, Oct 1999, pp. 595–600.

[109] Yannis Marinakis, Multiple phase neighborhood search-grasp for the capacitated vehicle routing problem, Expert Systems with Applications 39 (8) (2012) 6807–6815.

[110] Yannis Marinakis, Magdalene Marinaki, Georgios Dounias, A hybrid particle swarm optimization algorithm for the vehicle routing problem, Engineering Applications of Artificial Intelligence 23 (4) (2010) 463–472.

[111] Rafael Mart, Gerhard Reinelt, The Linear Ordering Problem: Exact and Heuristic Methods in Combinatorial Optimization, 1st edition, Springer Publishing Company, Incorporated, 2011.

[112] Peter Merz, A comparison of memetic recombination operators for the traveling salesman problem, in: William B. Langdon, Erick Cantú-Paz, Keith E. Mathias, Rajkumar Roy, David Davis, Riccardo Poli, Karthik Balakrishnan, Vasant G. Honavar, Günter Rudolph, Joachim Wegener, Larry Bull, Mitchell A. Potter, Alan C. Schultz, Julian F. Miller, Edmund K. Burke, Natasa Jonoska (Eds.), GECCO, Morgan Kaufmann, 2002, pp. 472–479.

[113] Peter Merz, Bernd Freisleben, Memetic algorithms for the traveling salesman problem, Complex Systems 13 (4) (2002).

[114] S.A. MirHassani, N. Abolghasemi, A particle swarm optimization algorithm for open vehicle routing problem, Expert Systems with Applications 38 (9) (2011) 11547–11551.

[115] Nenad Mladenovic, Pierre Hansen, Variable neighborhood search, Computers & Operations Research 24 (11) (1997) 1097–1100.

[116] Jaromír Mlejnek, Jiří Kubalik, Evolutionary hyperheuristic for capacitated vehicle routing problem, in: Proceedings of the 15th Annual Conference Companion on Genetic and Evolutionary Computation, GECCO'13 Companion, ACM, New York, NY, USA, 2013, pp. 219–220.

[117] L. Moccia, J-F. Cordeau, G. Laporte, An incremental tabu search heuristic for the generalized vehicle routing problem with time windows, Journal of the Operational Research Society 63 (2) (February 2012) 232–244.

[118] Pablo Moscato, On evolution, search, optimization, genetic algorithms and martial arts – towards memetic algorithms, 1989.

[119] Pablo Moscato, Carlos Cotta, A Gentle Introduction to Memetic Algorithms, Springer US, Boston, MA, 2003, pp. 105–144.

[120] Pablo Moscato, Fernando Tinetti, Blending Heuristics With a Population-Based Approach: a "Memetic" Algorithm for the Traveling Salesman Problem, Report 92-12, Universidad Nacional de La Plata, C.C. 75, 1900 La Plata, 1994.

[121] H. Mühlenbein, M. Gorges-Schleuter, O. Krämer, Evolution algorithms in combinatorial optimization, Parallel Computing 7 (1) (1988) 65–85.

[122] Yuichi Nagata, Edge assembly crossover for the capacitated vehicle routing problem, in: Carlos Cotta, Jano van Hemert (Eds.), Evolutionary Computation in Combinatorial Optimization, Springer, Berlin, Heidelberg, 2007, pp. 142–153.

[123] Yuichi Nagata, Olli Bräysy, Edge assembly-based memetic algorithm for the capacitated vehicle routing problem, Networks 54 (4) (December 2009) 205–215.

[124] Yuichi Nagata, Olli Bräysy, Wout Dullaert, A penalty-based edge assembly memetic algorithm for the vehicle routing problem with time windows, Computers & Operations Research 37 (4) (2010) 724–737.

[125] Yuichi Nagata, Shigenobu Kobayashi, Edge assembly crossover: a high-power genetic algorithm for the travelling salesman problem, in: ICGA, 1997.

[126] Yuichi Nagata, Shigenobu Kobayashi, A memetic algorithm for the pickup and delivery problem with time windows using selective route exchange crossover, in: Robert Schaefer, Carlos Cotta, Joanna Kolodziej, Günter Rudolph (Eds.), PPSN (1), in: Lecture Notes in Computer Science, vol. 6238, Springer, 2010, pp. 536–545.

[127] Jakub Nalepa, Miroslaw Blocho, Adaptive memetic algorithm for minimizing distance in the vehicle routing problem with time windows, Soft Computing 20 (6) (2016) 2309–2327.

[128] K.K.H. Ng, C.K.M. Lee, S.Z. Zhang, Kan Wu, William Ho, A multiple colonies artificial bee colony algorithm for a capacitated vehicle routing problem and re-routing strategies under time-dependent traffic congestion, Computers & Industrial Engineering 109 (2017) 151–168.

[129] Phuong Khanh Nguyen, Teodor Gabriel Crainic, Michel Toulouse, A tabu search for time-dependent multi-zone multi-trip vehicle routing problem with time windows, European Journal of Operational Research 231 (1) (2013) 43–56.
[130] Ben Niu, Hong Wang, Li-Jing Tan, Li Li, Jing-Wen Wang, Vehicle routing problem with time windows based on adaptive bacterial foraging optimization, in: De-Shuang Huang, Jianhua Ma, Kang-Hyun Jo, M. Michael Gromiha (Eds.), Intelligent Computing Theories and Applications, Springer, Berlin, Heidelberg, 2012, pp. 672–679.
[131] I.M. Oliver, D.J. Smith, J.R.C. Holland, A study of permutation crossover operators on the traveling salesman problem, in: ICGA, 1987.
[132] Alfredo Olivera, Omar Viera, Adaptive memory programming for the vehicle routing problem with multiple trips, Computers & Operations Research 34 (1) (2007) 28–47.
[133] Beatrice Ombuki, Brian J. Ross, Franklin Hanshar, Multi-objective genetic algorithms for vehicle routing problem with time windows, Applied Intelligence 24 (1) (Feb 2006) 17–30.
[134] İlhan Or, Traveling Salesman-Type Combinatorial Problems and Their Relation to the Logistics of Regional Blood Banking, Ph. D., NorthWestern University, Illinois, 1967.
[135] E. Osaba, F. Diaz, E. Onieva, Golden ball: a novel meta-heuristic to solve combinatorial optimization problems based on soccer concepts, Applied Intelligence 41 (1) (Jul 2014) 145–166.
[136] Eneko Osaba, Roberto Carballedo, Pedro Lopez-Garcia, Fernando Diaz, Comparison between golden ball meta-heuristic, evolutionary simulated annealing and tabu search for the traveling salesman problem, in: Proceedings of the 2016 on Genetic and Evolutionary Computation Conference Companion, GECCO'16 Companion, ACM, New York, NY, USA, 2016, pp. 1469–1470.
[137] Eneko Osaba, Roberto Carballedo, Xin-She Yang, Fernando Diaz, An Evolutionary Discrete Firefly Algorithm With Novel Operators for Solving the Vehicle Routing Problem With Time Windows, Springer International Publishing, Cham, 2016, pp. 21–41.
[138] Eneko Osaba, Roberto Carballedo, Xin-She Yang, Iztok Fister Jr., Pedro Lopez-Garcia, Javier Del Ser, On Efficiently Solving the Vehicle Routing Problem With Time Windows Using the Bat Algorithm With Random Reinsertion Operators, Springer International Publishing, Cham, 2018, pp. 69–89.
[139] Eneko Osaba, X.S. Yang, Fernando Díaz, Enrique Onieva, Antonio D. Masegosa, Asier Perallos, A bio-inspired discrete firefly algorithm to solve a rich vehicle routing problem modeling a newspaper distribution problem with recycling policy, 2015.
[140] Ibrahim Hassan Osman, Metastrategy simulated annealing and tabu search algorithms for the vehicle routing problem, Annals of Operations Research 41 (4) (Dec 1993) 421–451.
[141] Aziz Ouaarab, Belaïd Ahiod, Xin-She Yang, Random-key cuckoo search for the travelling salesman problem, Soft Computing 19 (4) (Apr 2015) 1099–1106.
[142] Xinxin Ouyang, Yongquan Zhou, Qifang Luo, Huan Chen, A novel discrete cuckoo search algorithm for spherical traveling salesman problem, 2013.
[143] Fengshan Pan, Chunming Ye, Kefeng Wang, Jiangbo Cao, Research on the vehicle routing problem with time windows using firefly algorithm, Journal of Computers 8 (09 2013).
[144] D.C. Paraskevopoulos, P.P. Repoussis, C.D. Tarantilis, G. Ioannou, G.P. Prastacos, A reactive variable neighborhood tabu search for the heterogeneous fleet vehicle routing problem with time windows, Journal of Heuristics 14 (5) (Oct 2008) 425–455.
[145] K.M. Passino, Biomimicry of bacterial foraging for distributed optimization and control, IEEE Control Systems Magazine 22 (3) (June 2002) 52–67.
[146] Tantikorn Pichpibul, Ruengsak Kawtummachai, New enhancement for Clarke–Wright savings algorithm to optimize the capacitated vehicle routing problem, European Journal of Scientific Research 78 (06 2012).
[147] Telmo Pinto, C. Alves, José Carvalho, Ana Moura, An insertion heuristic for the capacitated vehicle routing problem with loading constraints and mixed linehauls and backhauls 43 (01 2015) 311–318.
[148] Michael Polacek, Richard F. Hartl, Karl Doerner, Marc Reimann, A variable neighborhood search for the multi depot vehicle routing problem with time windows, Journal of Heuristics 10 (6) (Dec 2004) 613–627.

[149] Riccardo Poli, James Kennedy, Tim Blackwell, Particle swarm optimization, Swarm Intelligence 1 (1) (Jun 2007) 33–57.

[150] Jean-Yves Potvin, Evolutionary algorithms for vehicle routing, 2007.

[151] Jean-Yves Potvin, Samy Bengio, The vehicle routing problem with time windows – part II: genetic search, INFORMS Journal on Computing 8 (2) (1996) 165–172.

[152] Jean-Yves Potvin, Tanguy Kervahut, Bruno-Laurent Garcia, Jean-Marc Rousseau, The vehicle routing problem with time windows – part I: tabu search, INFORMS Journal on Computing 8 (2) (1996) 158–164.

[153] Jean-Yves Potvin, Jean-Marc Rousseau, An exchange heuristic for routing problems with time windows, Journal of the Operational Research Society 46 (12) (Dec 1995) 1433–1446.

[154] Christian Prins, A GRASP × Evolutionary Local Search Hybrid for the Vehicle Routing Problem, Springer, Berlin, Heidelberg, 2009, pp. 35–53.

[155] Esmat Rashedi, Hossein Nezamabadi-pour, Saeid Saryazdi, GSA: a gravitational search algorithm, Information Sciences 179 (13) (2009) 2232–2248, Special Section on High Order Fuzzy Sets.

[156] Martin Reed, Aliki Yiannakou, Roxanne Evering, An ant colony algorithm for the multi-compartment vehicle routing problem, Applied Soft Computing 15 (2014) 169–176.

[157] P.P. Repoussis, C.D. Tarantilis, Solving the fleet size and mix vehicle routing problem with time windows via adaptive memory programming, Transportation Research Part C: Emerging Technologies 18 (5) (2010) 695–712, Applications of Advanced Technologies in Transportation: Selected papers from the 10th AATT Conference.

[158] Glaydston Mattos Ribeiro, Gilbert Laporte, An adaptive large neighborhood search heuristic for the cumulative capacitated vehicle routing problem, Computers & Operations Research 39 (3) (2012) 728–735.

[159] Yves Rochat, Éric D. Taillard, Probabilistic diversification and intensification in local search for vehicle routing, Journal of Heuristics 1 (1) (Sep 1995) 147–167.

[160] Stefan Ropke, David Pisinger, An adaptive large neighborhood search heuristic for the pickup and delivery problem with time windows, Transportation Science 40 (4) (2006) 455–472.

[161] Robert A. Russell, An effective heuristic for the m-tour traveling salesman problem with some side conditions, Operations Research 25 (3) (1977) 517–524.

[162] K. Ruttanateerawichien, W. Kurutach, T. Pichpibul, A new efficient and effective golden-ball-based technique for the capacitated vehicle routing problem, in: 2016 IEEE/ACIS 15th International Conference on Computer and Information Science, ICIS, June 2016, pp. 1–5.

[163] Kanjana Ruttanateerawichien, Werasak Kurutach, Tantikorn Pichpibul, An improved golden ball algorithm for the capacitated vehicle routing problem, in: Linqiang Pan, Gheorghe Păun, Mario J. Pérez-Jiménez, Tao Song (Eds.), Bio-Inspired Computing – Theories and Applications, Springer, Berlin, Heidelberg, 2014, pp. 341–356.

[164] S. Salhi, G. Nagy, A cluster insertion heuristic for single and multiple depot vehicle routing problems with backhauling, Journal of the Operational Research Society 50 (10) (Oct 1999) 1034–1042.

[165] Martin W.P. Savelsbergh, The vehicle routing problem with time windows: minimizing route duration, INFORMS Journal on Computing 4 (1992) 146–154.

[166] Fatima Sayoti, Mohammed Riffi, Random-keys golden ball algorithm for solving traveling salesman problem, International Review on Modelling and Simulations (IREMOS) 8 (02 2015).

[167] B. Schutz, Gravity From the Ground up, December 2003.

[168] Davoud Sedighizadeh, Houman Mazaheripour, Optimization of multi objective vehicle routing problem using a new hybrid algorithm based on particle swarm optimization and artificial bee colony algorithm considering precedence constraints, Alexandria Engineering Journal 57 (4) (2018) 2225–2239.

[169] Frédéric Semet, Éric D. Taillard, Solving real-life vehicle routing problems efficiently using tabu search, Annals of Operations Research 41 (4) (1993) 469–488.

[170] Paul Shaw, Using constraint programming and local search methods to solve vehicle routing problems, in: Michael Maher, Jean-Francois Puget (Eds.), Principles and Practice of Constraint Programming, CP98, Springer, Berlin, Heidelberg, 1998, pp. 417–431.

[171] Paulo Vitor Silvestrin, Marcus Ritt, An iterated tabu search for the multi-compartment vehicle routing problem, Computers & Operations Research 81 (C) (May 2017) 192–202.

[172] Marius M. Solomon, Algorithms for the vehicle routing and scheduling problems with time window constraints, Operations Research 35 (2) (April 1987) 254–265.

[173] Marius M. Solomon, Jacques Desrosiers, Time window constrained routing and scheduling problems, Transportation Science 22 (1) (1988) 1–13.

[174] Mara Soto, Marc Sevaux, Andr Rossi, Andreas Reinholz, Multiple neighborhood search, tabu search and ejection chains for the multi-depot open vehicle routing problem, Computers & Industrial Engineering 107 (C) (May 2017) 211–222.

[175] Gilbert Syswerda, in: Lawrence Davis (Ed.), Handbook of Genetic Algorithms, Van Nostrand Reinhold, 1991.

[176] W.Y. Szeto, Yongzhong Wu, Sin C. Ho, An artificial bee colony algorithm for the capacitated vehicle routing problem, European Journal of Operational Research 215 (1) (2011) 126–135.

[177] A. Taha, M. Hachimi, A. Moudden, A discrete bat algorithm for the vehicle routing problem with time windows, in: 2017 International Colloquium on Logistics and Supply Chain Management, LOGISTIQUA, April 2017, pp. 65–70.

[178] Éric Taillard, Phillipe Badeau, Michel Gendreau, François Guertin, Jean-Yves Potvin, A tabu search heuristic for the vehicle routing problem with soft time windows, Transportation Science 31 (2) (1997) 170–186.

[179] Éric D. Taillard, Parallel iterative search methods for vehicle routing problems, Networks 23 (8) (1993) 661–673.

[180] Li Jing Tan, Fu Yong Lin, Hong Wang, BFO optimization algorithms for vehicle routing problem with time windows, in: Vehicle, Mechatronics and Information Technologies II, in: Applied Mechanics and Materials, vol. 543, Trans Tech Publications, 6 2014, pp. 1884–1887.

[181] Lijing Tan, Fuyong Lin, Hong Wang, Adaptive comprehensive learning bacterial foraging optimization and its application on vehicle routing problem with time windows, Neurocomputing 151 (2015) 1208–1215.

[182] Christos D. Tarantilis, Solving the vehicle routing problem with adaptive memory programming methodology, Computers & Operations Research 32 (2005) 2309–2327.

[183] Christos D. Tarantilis, Chris T. Kiranoudis, Boneroute: an adaptive memory-based method for effective fleet management, Annals of Operations Research 115 (1–4) (2002) 227–241.

[184] Christos D. Tarantilis, F. Stavropoulou, Panagiotis P. Repoussis, A template-based tabu search algorithm for the consistent vehicle routing problem, Expert Systems with Applications 39 (2012) 4233–4239.

[185] Ehsan Teymourian, Vahid Kayvanfar, G.H.M. Komaki, M. Zandieh, Enhanced intelligent water drops and cuckoo search algorithms for solving the capacitated vehicle routing problem, Information Sciences 334–335 (2016) 354–378.

[186] S.R. Thangiah, K.E. Nygard, P.L. Juell, Gideon: a genetic algorithm system for vehicle routing with time windows, in: [1991] Proceedings. The Seventh IEEE Conference on Artificial Intelligence Application, vol. i, Feb 1991, pp. 322–328.

[187] Sam R. Thangiah, Ibrahim H. Osman, Tong Sun, Hybrid genetic algorithm, simulated annealing and tabu search methods for vehicle routing problems with time windows, 1993.

[188] Sebastian B. Thrun, Efficient Exploration in Reinforcement Learning, Technical report, Pittsburgh, PA, USA, 1992.

[189] Om Verma, Rashmi Jain, Vindhya Chhabra, Solution of travelling salesman problem using bacterial foraging optimisation algorithm, International Journal of Swarm Intelligence 1 (01 2014) 179–192.

[190] Thibaut Vidal, Teodor Gabriel Crainic, Michel Gendreau, Nadia Lahrichi, Walter Rei, A hybrid genetic algorithm for multidepot and periodic vehicle routing problems, Operations Research 60 (3) (2012) 611–624.

[191] Thibaut Vidal, Teodor Gabriel Crainic, Michel Gendreau, Christian Prins, A hybrid genetic algorithm with adaptive diversity management for a large class of vehicle routing problems with time-windows, Computers & Operations Research 40 (1) (2013) 475–489.
[192] Daniele Vigo, A heuristic algorithm for the asymmetric capacitated vehicle routing problem, European Journal of Operational Research 89 (1) (1996) 108–126.
[193] Juan G. Villegas, Christian Prins, Caroline Prodhon, Andrés L. Medaglia, Nubia Velasco, A grasp with evolutionary path relinking for the truck and trailer routing problem, Computers & Operations Research 38 (9) (2011) 1319–1334.
[194] Chao Wang, Fu Zhao, Dong Mu, John W. Sutherland, Simulated annealing for a vehicle routing problem with simultaneous pickup-delivery and time windows, in: Vittal Prabhu, Marco Taisch, Dimitris Kiritsis (Eds.), Advances in Production Management Systems. Sustainable Production and Service Supply Chains, Springer, Berlin, Heidelberg, 2013, pp. 170–177.
[195] Chao Wang, Shengchuan Zhou, Yang Gao, Chao Liu, A self-adaptive BAT algorithm for the truck and trailer routing problem, Engineering Computations 35 (01 2018).
[196] Junling Wang, Arun Kumar Ranganathan Jagannathan, Xingquan Zuo, Chase C. Murray, Two-layer simulated annealing and tabu search heuristics for a vehicle routing problem with cross docks and split deliveries, Computers & Industrial Engineering 112 (2017) 84–98.
[197] Shuihua Wang, Zeyuan Lu, Ling Wei, Genlin Ji, Jiquan Yang, Fitness-scaling adaptive genetic algorithm with local search for solving the multiple depot vehicle routing problem, Simulation 92 (7) (2016) 601–616.
[198] Lijun Wei, Zhenzhen Zhang, Defu Zhang, Stephen C.H. Leung, A simulated annealing algorithm for the capacitated vehicle routing problem with two-dimensional loading constraints, European Journal of Operational Research 265 (3) (2018) 843–859.
[199] Lijun Wei, Zhenzhen Zhang, Defu Zhang, Andrew Lim, A variable neighborhood search for the capacitated vehicle routing problem with two-dimensional loading constraints, European Journal of Operational Research 243 (3) (2015) 798–814.
[200] Shuyuan Yang, Min Wang, Licheng Jiao, A quantum particle swarm optimization, in: Proceedings of the 2004 Congress on Evolutionary Computation, IEEE Cat. No. 04TH8753, vol. 1, 2004, pp. 320–324.
[201] Xin-She Yang, Nature-Inspired Metaheuristic Algorithms, Luniver Press, 2008.
[202] Xin-She Yang, A New Metaheuristic Bat-Inspired Algorithm, Springer, Berlin, Heidelberg, 2010, pp. 65–74.
[203] Baozhen Yao, Ping Hu, Mingheng Zhang, Shuang Wang, Artificial bee colony algorithm with scanning strategy for the periodic vehicle routing problem, Simulation 89 (6) (2013) 762–770.
[204] Baozhen Yao, Qianqian Yan, Mengjie Zhang, Yunong Yang, Improved artificial bee colony algorithm for vehicle routing problem with time windows, PLoS ONE 12 (9) (09 2017) 1–18.
[205] Baozhen Yao, Bin Yu, Ping Hu, Junjie Gao, Mingheng Zhang, An improved particle swarm optimization for carton heterogeneous vehicle routing problem with a collection depot, Annals of Operations Research 242 (2) (Jul 2016) 303–320.
[206] Peng-Yeng Yin, Ya-Lan Chuang, Adaptive memory artificial bee colony algorithm for green vehicle routing with cross-docking, Applied Mathematical Modelling 40 (21) (2016) 9302–9315.
[207] Vincent F. Yu, A.A.N. Perwira Redi, Yosi Agustina Hidayat, Oktaviyanto Jimat Wibowo, A simulated annealing heuristic for the hybrid vehicle routing problem, Applied Soft Computing 53 (C) (April 2017) 119–132.
[208] Emmanouil E. Zachariadis, Christos D. Tarantilis, Chris T. Kiranoudis, A guided tabu search for the vehicle routing problem with two-dimensional loading constraints, European Journal of Operational Research 195 (3) (2009) 729–743.
[209] Shuzhu Zhang, C.K.M. Lee, K.L. Choy, William Ho, W.H. Ip, Design and development of a hybrid artificial bee colony algorithm for the environmental vehicle routing problem, Transportation Research Part D: Transport and Environment 31 (2014) 85–99.
[210] Hongqiang Zheng, Yong-Quan Zhou, Qifang Luo, A hybrid cuckoo search algorithm-grasp for vehicle routing problem, Journal of Convergence Information Technology 8 (02 2013) 821–828.

[211] Yongquan Zhou, Qifang Luo, Jian Xie, Hongqing Zheng, A Hybrid BAT Algorithm With Path Relinking for the Capacitated Vehicle Routing Problem, Springer International Publishing, Cham, 2016, pp. 255–276.
[212] Marcin Woch, Piotr Lebkowski, Sequential simulated annealing for the vehicle routing problem with time windows, Decision Making in Manufacturing and Services 3 (2009).
[213] V. Černý, Thermodynamical approach to the traveling salesman problem: an efficient simulation algorithm, Journal of Optimization Theory and Applications 45 (1985) 41–51.

Chapter 5

Hybrid algorithms for rich vehicle routing problems: a survey

Rajeev Kr. Goel[a] and Sandhya Rani Bansal[b]

[a] *C.S Deptt., Govt. College, Naraingarh, India,* [b] *C.S.E. Deptt., M.M. Deemed University, Mullana, India*

5.1 Introduction

Transportation is the "lifeline" of a nation. Efficient and effective utilization of transportation is essential for economic development of a nation [59]. Goods transportation directly affects the education sector, industry, business, inventory, logistic, distribution, and many more areas [58,60]. In India, it is a means of integration. Transport has recorded a substantial growth over the years both in terms of the length and output of the system [61]. However, road transportation is associated with some costs like congestion, pollution, etc. Keeping all these points in view, it becomes necessary to develop some new efficient methods of transportation.

One of the operational research problems in transportation that handles this issue is Vehicle Routing Problem (VRP). The problem was first formulated by Dantzig and Ramser in 1959 [62]. In a VRP, a group of vehicles having identical capacity departs from a central station (depot) for serving geographically scattered customers having certain demands such that the total cost of service is minimized. The costs represent total traveled distance, traveled time, number of vehicles used, or a concatenation of all these. Therefore, the goal of VRP is to optimize the route plan process from depot to customers such that customer demands are satisfied without violating any problem-specific constraints [63,64]. Recent developments in communication technologies, for example, massive use of Global Positioning System (GPS), hand handled smart phones, and Internet computing technologies have opened new conceivable outcomes for optimizing the planning process for smart delivery transportation [63]. On account of these advancements, many variants of VRP have been proposed. All the basic variants of VRP have been mentioned in the literature [64–68]. Even though the problem has been studied for years, the research around it is still very active. The new tendency is mainly focused on applying VRP study to real-life problems. Because

Smart Delivery Systems. https://doi.org/10.1016/B978-0-12-815715-2.00011-7

of this trend, the rich VRP comprises combining multiple constraints for tackling realistic problems. Some studies have also considered specific combinations of real-life constraints to define the emerging Rich VRP scopes. Rich VRP deals with multiobjective optimization functions, stochastic or fuzzy nature of problem, dynamism along with many real-life constraints like use of heterogeneous vehicles, competitive environment, integrating with inventory and scheduling problems, environmental and energy issues, and many more.

In spite of its simple nature, in terms of complexity theory the basic VRP and its different flavors are nonpolynomial hard problems (NP-Hard) [63,69]. This implies that the problem cannot be solved by an algorithm in finite number of steps [63,67], and in practice it is not feasible to guarantee the optimal solution. Therefore, different methods to solve VRP have been explored ranging from exact methods to metaheuristics [64,65,67,70]. Exact methods are pure optimization methods that are applicable to small-size problems. For an overview of exact methods, see [71,67,68,72]. But as the problem size increases, they fail to find an optimal solution within a reasonable time, and use of metaheuristics comes into picture. Metaheuristics are the approaches that can solve large instances of complex problems within reasonable time. For different meta heuristics used in such problems, see [73–75]. In recent years' combination of these two methods and fusion of metaheuristics have been tried to tackle RVRP [63,76–78]. In this chapter, we give a survey of commonly used hybrid algorithm for solving rich VPR and for real-life VRPs.

5.1.1 Methodology and contribution of this chapter

The literature surveyed in this chapter has been mainly selected from academic databases, bibliographies of surveys, and linking papers addressed in mined literature. In Rich VRP, we have selected the papers that focus on dynamic and stochastic VRP and on periodic and multidepot VRP. In real-life VRP, papers under the category of food distribution, newspaper distribution, fuel distribution, and waste collection are considered. The contribution of this chapter is to present an evolution of RVRP with an extensive literature review on its solution techniques using hybrid methods along with the cooperation techniques used by hybrid methods as proposed in [76]. We have also presented a critically analyzed annotated bibliography in the field of RVRP and pointed out the possible future research directions. For academic purposes, this chapter presents future scopes in the field of RVRP using hybrid methodologies. In particular, the evolution of RVRP from classical VRP and its relation with other variants are also summarized, which may inspire researchers to extend their own work toward more realistic problems.

5.1.2 Structure of the chapter

The rest of the chapter is organized as follows. Section 5.2 presents a mathematical model for traditional CVRP. Section 5.3 presents variants of CVRP and a model classification of rich VRP. Section 5.4 presents different approaches used for solving RVRP along with the cooperation schemes. In Section 5.5, we discuss the literature related to hybrid algorithms. Section 5.6 presents the concluding observations and future trends in RVRP.

5.2 Mathematical model for traditional CVRP

Let us consider a routing network represented by a directed graph $G = (C, E)$, where C is the set of vertices representing the customers $C = \{c_0, c_1, c_2, \ldots, c_n\}$, and E is the set of arcs $\{(i, j)\,/\,i, j \in C, i \neq j\}$ connecting customers c_i and c_j. Each customer is assigned with a fixed load w_i. A set of homogenous vehicles $V = \{v_1, v_2, \ldots, v_n\}$ with known capacity Q are located at depot c_0 is required to accomplish the delivery task. Every vehicle originates its services from depot and returns back to the same central station. So, the tour of vehicle v consists of customers $\{c_0, \ldots, c_i, c_{i+1}, \ldots, c_0\}$ connected by edges starting and ending at depot c_0. A cost matrix D_{ij} is associated with every vehicle v. Furthermore, it is assumed that the triangular inequality $D_{ik} + D_{kj} \geq D_{ij}$ holds for the cost matrix. The objective of CVRP is to minimize the total cost of performing the services to all the customer nodes so that following constraints are preserved:

- Each customer is to be served once and only once.
- Demand of each tour should not exceed the truck capacity.
- Each tour must start and end at depot within specified time slot.

Two types of cost functions are associated with the objective function: the minimum distance traveled and minimum number of vehicles used.

The mathematical model for VRPTW is as shown in the next paragraph. A thorough explanation is made thereafter.

5.2.1 Objective function

The overall objective is often to minimize the number of vehicles and total traveling distance. The parameter C_{ij} stands for the fixed cost of using vehicle v, and the binary variable x_{ij}^{v} becomes equal to one only if vehicle v is used by customers c_i and c_j.

5.2.2 Problem constraints

Assignment of a customer to a single vehicle:

Eq. (5.2) states that each customer must be served by exactly one vehicle. Splitting of deliveries is forbidden.

5.2.3 Flow constraint

Eqs. (5.3) and (5.4) ensure the flow constraint of vehicle, that is, each vehicle must start and end its tour at depot.

5.2.4 Capacity constraint

Eq. (5.5) states that the demand served by the vehicle v should not exceed its capacity.

5.2.5 The mathematical model of classical VRP

$$\text{Minimize } F = \sum_{i=0}^{N} \sum_{j=0}^{N} \sum_{K=1}^{V} D_{ij} x_{ij}^{v} \tag{5.1}$$

subject to:

$$\sum_{v=1}^{V} \sum_{j=1}^{N} x_{ij}^{v} \leq V \text{ for } i = 0, \tag{5.2}$$

$$\sum_{v=1}^{V} x_{ij}^{v} = \sum_{j=1}^{N} x_{ji}^{v} \leq 1 \quad \text{for } i = 0 \text{ and } v \in \{1, \ldots, V\}, \tag{5.3}$$

$$\sum_{v=1}^{V} \sum_{j=1}^{N} x_{0j}^{v} = 1, \tag{5.4}$$

$$\sum_{v=1}^{V} \sum_{i=1}^{N} x_{i0}^{v} = 1, \tag{5.5}$$

$$\sum_{i=0}^{N} c_i \sum_{j=0}^{N} x_{ij}^{v} \leq q_v \text{ for } v \in \{1, \ldots, V\}. \tag{5.6}$$

Here:

q_i is total servings for customer i,

q_v upper limit for capacity of vehicle v

Here the objective function (5.1) is to be optimized satisfying constraints (5.2)–(5.6). It corresponds to minimization of the total traveled distance.

The first constraint (5.2) ensures that all tours must be completed with at most V vehicles. Beginning and completion of the tour at central depot is ensured by (5.3). Constraints (5.4) and (5.5) restrict the partial servings, that is, every location must be visited by exactly one vehicle. Constraint (5.6) ensures that the net demand on every route must be within vehicle's capacity.

5.3 From traditional VRP to rich VRP

VRP are mainly classified into four levels according to the degree of realism present in the associated models:

Traditional VRP, Traditional Advanced VRP, Rich VRP, and Real-life VRP are shown in Fig. 5.1.

5.3.1 Traditional VRP

These are the theoretical VRPs represented by academic models. They are of great interest to researchers for developing mathematical model and exact solution techniques. Some of the common examples are Capacitated VRP (CVRP),

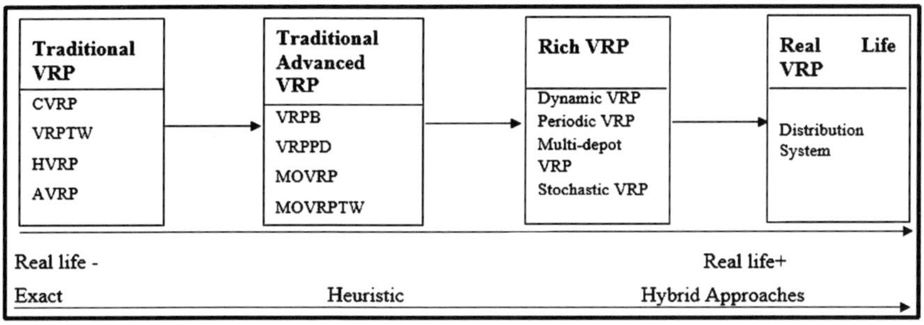

FIGURE 5.1 Classification of VRP.

VRP with Time Windows (VRPTW), Heterogeneous VRP (HVRP), and Asymmetric VRP (AVRP).

5.3.2 Traditional advanced VRP

They have slightly more realism than traditional VRP. This level includes large-size VRPs, hybrid VRPs (VRP with Back Hauls (VRPB), VRPs with simultaneous pickup and delivery (VRPPD), and multiobjective VRPs (MOVRP). These types of problems are generally solved by meta heuristic approaches.

5.3.3 Rich VRP & real-life VRP

These are the practical VRPs, and hence they are more accurate models of real-life distribution systems. These models contain VRP with complex constraints and stochastic and dynamic problems that reflect the particularities of real-world applications. The solutions obtained for these models are of great importance because they can be directly applied to real-life problems. Problems under this class are often solved by hybrid algorithms.

5.3.4 Rich VRP definition

Motivated by gap between the classical VRP and realistic scenarios of VRP, many real-life VRPs are proposed in the literature, so the definition of RVRP is vague, and hence no universally accepted definition of Rich VRP exists. Some definitions are as follows:

- In [67], "models of symmetric and asymmetric CVRP may be adapted to model some variants of the basic versions."
- In [84], "non-idealised models that represent the application at hand in an adequate way by including all important optimization criteria, constraints, and preferences."
- In [85], "real-world problems are generally characterized by several interacting attributes, which describe their feasibility and optimality structures."

- In [86], "Hence, research has turned to more specific and rich variants of the CVRP. The family of these problems is identified as rich vehicle routing problems. In order to model RVRPs, the basic CVRP must be extended by considering additional constraints or different objective functions."
- In [83], "an RVRP extends the academic variants of the VRP in the different decision levels by considering additional strategic and tactical aspects in the distribution system (4 or more) and including several daily restrictions related to the Problem Physical Characteristics (6 or more) [pure routing or operational]. Therefore, a RVRP is either a VRP that incorporates many strategic and tactical aspects and/or a VRP that reflects the complexities of the real-life context by various challenges revealed daily. The state of the art of RVRP has changed since 2006. Now studies incorporate more complex aspects of reality. Therefore, some variants described as rich by their authors in 2006 may not be considered as such anymore."
- In [63], "an RVRP reflects, as a model, most of the relevant attributes of a real-life vehicle-routing distribution system. These attributes might include dynamism, stochasticity, heterogeneity, multiperiodicity, integration with other related activities (e.g., vehicle packing, inventory management, etc.), diversity of users and policies, legal and contractual issues, environmental issues, and more. Thus, as a model, an RVRP is an accurate representation of a real-life distribution system and, therefore, the solutions obtained for the RVRP should be directly applicable to a real-life scenario".
- In [44], "Rich Vehicle Routing Problems are vehicle routing problems (VRPs) that deal with additional constraints, which aim to better take into account the particularities of real-world applications."

5.4 Solution approaches for RVRPs

As VRP is an NP-Hard problem, there is difficulty in solving it, and no polynomial-time algorithm exists for its solution. Different VRP methodologies have been explored. Depending upon the richness and size of the instances, the solution techniques can be categorized into three categories: Exact, Approximate, and Hybrid methods. The taxonomy of solution methods is shown in Fig. 5.2.

For small-size (up to 75 customers) with 2 or 3 constraints and idealistic VRP, researchers usually use exact approaches. These approaches find the optimal solution and guarantee their optimality [71,76]. This class includes numerous methods like Branch and X, Dynamic approach, Mixed Linear Programming (MILP), Column Generation, Set Partitioning, and Constraint Programming (CP). But as the size of problem and constraints grows, an exact solution becomes computationally expensive. For large-size problems, we have to resort to Approximate Methods [79–82]. These methods are again categorized into two categories, Approximate algorithms and Heuristics. Approximation algorithms can obtain acceptable solutions in acceptable computational time and guarantee their approximate solutions. Under this class, we have tour splitting and set covering approaches. On the other hand, heuristics in [71,73,75] do not guarantee achieving optimality.

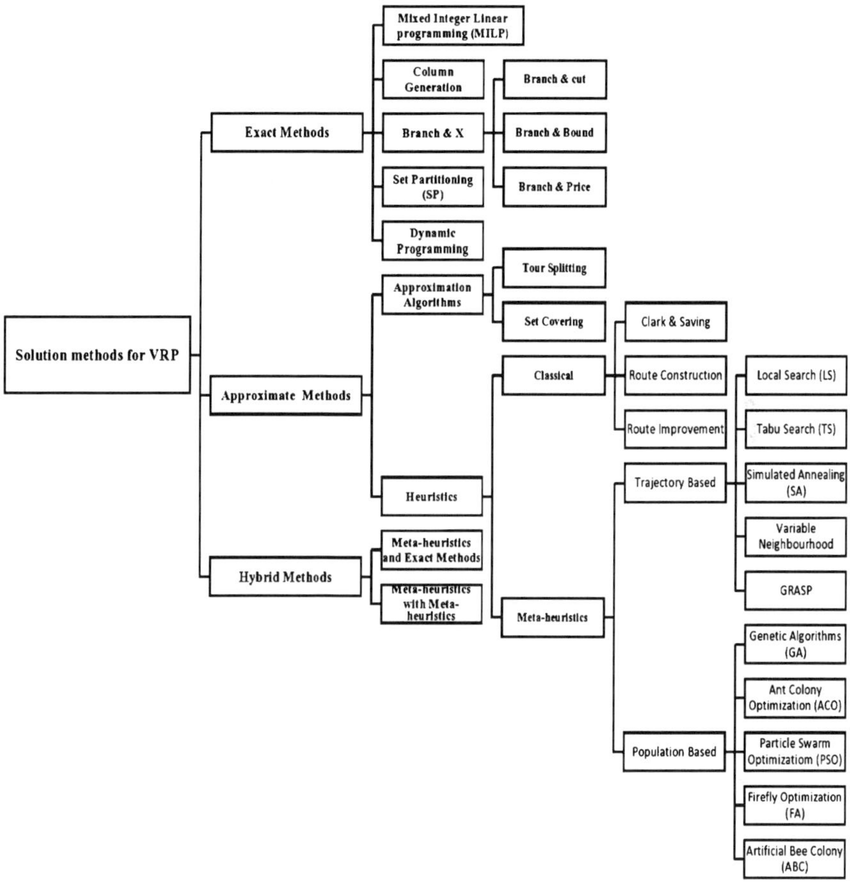

FIGURE 5.2 Taxonomy of Vehicle Routing Solution Techniques.

Heuristics are again classified into two categories, Classical Heuristics and Metaheuristics. Classical Heuristics are problem-specific heuristics, whereas Metaheuristics are the general-purpose upper-level methodologies that can be applied to all optimization problems. There are two popular categories of meta-heuristics, trajectory based and population based. The former includes VNS, LS, SA, TS, and the latter includes ACO [52], GA, PSO, FA, etc. However, as the richness of constraints in real-life problems grows, even metaheuristics are not able to generate solutions. To address real-life problems of VRP (RVRPs), during the recent years, hybrid approaches are being used. According to De la Cruz et al. [39], "A skilled combination of concepts from different optimization techniques can provide a more efficient behavior and a higher flexibility when dealing with real-world and large scale problems. Hybrid meta-heuristics are such techniques for optimization that combine different meta-heuristics or integrate artificial intelligence/operational research techniques into the procedure." These approaches hybridize the advantages of exact methods and metaheuristics. The best results have been obtained by using these approaches for RVRPs because they exploit simultaneously the advantages of good merging approaches. In [76], various cooperation schemes were presented for these hybrid methods. In this paper, we have also categorized the mined literature of RVRP according to these schemes. In [76], four different cooperation schemes were proposed for hybrid methods. These are:

a. Low-Level Relay Hybrid (LRH): This class corresponds to the algorithms that use heuristics to improve the results of exact methods. In case of hybridization between two metaheuristics the most planned approach is to run an evolutionary algorithm and then apply LS to intensify the search.
b. Low-Level Teamwork Hybrid (LTH): This scheme is used to improve performance of metaheuristic. Under this scheme, an element of a specified method is replaced by another method. For example, a mutation operator of GA by LS operator.
c. High-Level Relay Hybrid (HRH): Under this category, different methods are considered autonomous and executed sequentially.
d. High-Level Teamwork Hybrid (HTH): Under this category, different methods are autonomous but are executed in parallel.

In the next section, we present a review of hybrid approaches for RVRP with cooperation schemes and certain visible research gaps.

5.5 Literature review of hybrid approaches for VRPs

5.5.1 Real-life VRP (distribution system)

5.5.1.1 Food industry

Paper [1] presents an application of VRP in frozen food distribution industry taking into account the time windows, loading weight, and loading volume constraints. The objective of the problem is to minimize the delivery cost for a set of homogenous vehicles. The problem is solved using hybridization of various LS operators and Genetic Algorithm (GA). The effectiveness of the proposed method is demonstrated by using a real data set from a company working in Beijing. In this case, every customer receives all varieties of its required goods of different weights or volumes on a single unit. However, different sizes of products occupy different space on the trucks, so it should also be addressed from the point of view of the final number of required trucks.

In [2], Simulated Annealing (SA), Goal Programming and Local Searches are used for the optimization of biobjective VRP of perishable food distribution. Two objectives are considered for optimization: the maximization of freshness and the minimization of total traveled distance. Goal Programming is used for modeling the problem, and then SA with local searches is used for finding the optimized route. Solomon dataset is used as the benchmark data set. No realistic data set is considered.

In [3] a generalization of Asymmetric CVRP with split delivery has been addressed. The work considers the distribution of two types of products-fresh/dry and frozen food. The main goal of the problem is to minimize the total traveling cost. Integration of Mixed Linear Programming (MLP) and a two-step heuristic (local search improvement) have been used for solving the problem. The results are validated on a real application of Italian company holding food markets along the national highway network. However, it is not guaranteed that every customer will receive the product at the required time.

In [4], delivery of catering food has been optimized by shortening the time interval between production and delivery. Aggregation of MILP and large neighborhood search have been used for solving the problem. A real setting of a catering company located in Denmark has been used to demonstrate the effectiveness of the proposed approach. However, the proposed approach fails to handle situation with multiple producers.

In [5] a hybridization of Particle Swarm Optimization (PSO) and Variable Neighborhood Search (VNS) has been proposed to solve a two-Echelon Location–Routing VRPTW (2E-LRPTW) for optimization of sustainable supply chain network of perishable food. Two objectives are pursed: (i) minimization of vehicle routing cost and (ii) minimization of emissions due to shipments. Soft time windows are considered.

In [6], scheduling of refrigerator cars for fresh product distribution is addressed. The problem is solved by dividing it into two echelon subproblems. Furthermore, the scheduling is done in two stages. In the first stage an MIP model is formed, and then in the second stage, integration of PSO and VNS is used to solve the problem. However, the proposed approach is tested only on numerical experiments. No realistic dataset is used.

In [7] a rich VRP with lunch break has been solved using LS and MIP. The problem arises in furniture delivery. In this problem the driver rest period and time windows have been considered. However, balance of workload among drivers has not been considered.

In [8] a variant of vehicle allocation problem that arises in food bank distribution is addressed. The problem has been solved by hybridization of LS and perturbation schemes. However, the constraint of multiperiod with demand allocation has not been addressed.

5.5.1.2 *Waste collection management*

In [9] a hybridization of ABC and VNS is applied for solving waste collection problem with the consideration of dynamic disposal trips and minimization of carbon emission. The effectiveness of the algorithm has been demonstrated by numerical experiments. No real data set has been used to demonstrate the effectiveness.

In [10], use of SA and LS has been proposed to solve a Roll-on Roll-off waste collection VRP. The proposed problem involves a large container that collects waste generated by shopping malls and construction site. The initial solution is improved by hybrid SA and LS. However, only single recycle center and homogenous vehicles are considered.

In [11] a variant of VRP is solved by using Lagrangian relaxation and sub-gradient optimization algorithm. In this variant, waste collection sites are visited within a prescribed time windows. The proposed model has been tested on Solomon data set and randomly generated instances. Although the algorithm works well for small-scale problems, it loses its robustness as the number of customer increases.

In [12] a hybridization of PSO and LS has been applied for finding efficient waste collection and route optimization solutions. The main goal of the study is to reduce different costs involved in waste collection and environmental effects. The proposed approach is tested on randomly generated datasets, but no real data set is used. Moreover, the number of constraints specified is less in comparison to actual constraints of problems in real scenarios.

In [13], route optimization for Municipal solid waste collection has been considered using ARGIS Network Analyst and the Ant Colony System (ACS). These two approaches are based on GIS. Position of bins, road traffic conditions, road network, and other relevant parameters are first collected using GIS. Then an optimal solution is obtained by ACS. However, the performance of algorithm heavily depends upon the number of stops and the complexity of the data set.

Integration of VNS, local improvement, and exact methods have been used for node routing-based solid waste collection problems [14]. A real application of the proposed algorithm has been shown in a decision support system of a consulting company in Austria. A real data set has been used for validation of results. However, the proposed approach puts some unrealistic assumptions on fleet.

In [15], use of VNS and Tabu Search (TS) has been used for route optimization of commercial waste collection involving time windows, driver rest period, and multiple disposal sites. A dataset having 2092 customers and 19 waste disposal facilities has been used for validation of results.

A MIP and clustering approach have been presented in [16] for domestic solid waste collection. The work considers vehicle capacity, maximum work day duration, and planning horizon for two days. However, the accuracy of the results obtained depends upon the method used.

5.5.1.3 *Oil, fuel, and gas*

In [17] an extension of VRP in the field of fuel has been solved by the hybridization of VNS and simulation. The problem has been considered as rich because it considers multiproduct and multiperiod with multicompartment homogenous vehicle assumptions.

In [18] an integration of purchasing and routing for a propone gas supply chain problem has been addressed. Hybridization of applied set partitioning and TS are applied for solving the problem. The objective of the problem is to minimize the combined cost of purchasing and transportation. To test the effectiveness of the proposed approach, real data from Illinois and Michigan dispatch areas were used. In addition, hybridization of exact methods and mathematical model have also been used in [56].

5.5.1.4 News paper delivery

In [19] the first application of Discrete Firefly Algorithm (FA) has been shown for solving newspaper delivery problem with recycling policy. The problem has been modeled as RVRP, which can be considered as an extension of asymmetric VRP with simultaneous pick and deliveries. The performance of the proposed algorithm has been compared with evolutionary algorithm and evolutionary simulated annealing algorithm on real data set collected from Bizakaia, Spain. The main goal of the paper is to develop a model that is applicable to all the companies that follow the same structure. However, parameterized analysis should have also been done as it is applicable in other companies as well.

In [20] a formulation and different heuristic approaches as well as a hybrid method are used to find an effective distribution plan to deliver free newspapers from a production plant to number of points (subways, bus, or tram stations). The formulated problem combines the aspects of VRPTW, inventory routing problem, and additional constraints related to the production schedule. The main goal of the problem is to minimize the number of vehicle trips needed and to minimize the time needed to deliver all newspapers. The problem has been solved in two phases: the first phase creates an efficient delivery plan by using several constructive heuristics, and in the second phase an effective delivery plan is made by using various constructive heuristics and VNS. For computational experiments, a test set from company working in Vienna has been created. The computational time can further be reduced if both phases work in parallel.

In [21] a reactive TS-based metaheuristic has been developed for integrating the production and distribution of newspapers from plant to bulk delivery location. The problem has been modeled as an open VRP with additional constraints of zones, time windows, and multiproduct. The objective of the problem is to minimize the number of vehicles used and total traveled distance. The proposed methodology is applied to a company in a major metropolitan area. The work considers centralized loading site, but in real world, multiple loading sites are used. Chaing et al. [22] have extended this work by examining stochastic aspect of supply chain for midsize US newspapers. Stochastic phenomena include travel time, start times, production time, and loading and unloading times.

In [23] a newspaper distribution problem is modeled as SDVRP. The objective of the problem is to find optimal assignment of newspaper agents, optimal routes, and schedule for newspaper delivery in Seoul, South Korea, with 400 local agents. The problem was solved using various heuristics and digital map in two phases. The first phase allocates customers, and the second phase generates optimal routes. Savings of 15% in the cost and 40% in the delay time were obtained. Instead of several heuristics, using a single heuristic is much more beneficial.

In [24] the various newspaper distribution scenarios observed in India have been optimized by GA. The proposed work focuses on the distribution of newspaper from vendor to customers. The objective of the paper is to minimize the total delivery time. In addition, Berger [25] also aims in optimizing the delivery routes of newspaper and magazines.

In [26] a unique VRP problem called VRPTWCD has been addressed, which combines the VRPTW and cluster-dependent tour starts. The objective of the problem is to minimize the transport cost and maximize the customer satisfaction. The formulated problem has been solved by hybridizing ACO and TS. The result of hybridization has been validated on a real-life application case of one of the largest newspaper companies in Germany. No comparison of proposed approach with other hybrid heuristics is made.

5.5.2 Rich VRP

In [27] a real-world VRPTW in the area of city of Zagreb has been addressed and solved by two hybrid metaheuristics: a hybrid Iterated Local Search and a hybrid Simulated Annealing. The problem has been considered because the information of distance and travel time between customers depends on road networks. The mechanism of dynamic parameter adaptation has not been explained. In addition to this work, [54,55] can be cited.

In [28] a hybridization of GRASP and VNS has been used to solve a real-world VRP. The problem incorporates heterogeneous fleets, multiple commodities, and multiple vehicle compartments with an objective of minimizing cost routes. The diversification mechanism of GRASP and intensification of VNS have been used in relative fashion. However, the problem uses a large number of local search operators.

In [29] a practical VRPTW that occurs in delivery process has been addressed. The problem has been tackled by two hybrid metaheuristics SA and ILS. Both algorithms improve the initial solution obtained by constructive heuristics. Both benchmark and real-life data have been used for testing the effectiveness of the proposed approach. A comparison between the number of vehicles used and loss of money should be also done in case of time window violation.

In [30], hybridization of ACO and LS has been used to solve real-world VRPs. Application of ACO in time-dependent VRP and the online VRP have been addressed under the category of dynamic VRP.

In [31] a problem arising in collection of recyclable waste form large containers over a fixed planning horizon in Geneva, Switzerland, using heterogeneous vehicles has been addressed. The problem lies in the category of stochastic inventory routing problem. The problem has been solved by the hybridization of adaptive large neighborhood search and mixed-integer nonlinear program. Stochastic travel time and service need also to be considered.

The study [32] constructs models and solution approaches for three practical vehicle routing problems. The presented models are so general that they can be applied to any VRP in any distribution company. All three problems are motivated by their cooperation with industrial companies. In [33] a vehicle routing problem with cross-docking option is considered, in which goods are picked from suppliers by vehicles, combined at base station and immediately delivered to customers by the same set of vehicles. The problem has been considered rich as the consolidation decisions are made at the depot and their decisions interact with the planning of pick-up and delivery routes. MILP and TS are used for solving the problem. The performance of the proposed approach is validated on real-life data set up to 200 pairs of customers and suppliers. The author has made the assumption that orders are not splitable. However, in real life, orders are splitable.

In [34] an integrated vehicle routing and driver scheduling problem arising at the largest fresh meat producer in Denmark is addressed using MILP and large VNS. The problem has been considered as rich VRP because of additional constraints like heterogeneous vehicles and predefined work regulation of drivers. The objective of the problem is to minimize the total delivery cost. In addition to this, in [35], MILP and three heuristics are used to solve dynamic multi-period vehicle routing problem. TS is used in the third phase for improvement of routes. The problem deals with distribution of orders encountered by company operating in Sweden to a set of customers over a multiperiod time horizon. The proposed method improves the company performance in terms of objectives like travel time, customer waiting time, and balance of daily work load. However, a policy must be determined to predict the best serving time, which can further decrease the computational complexity of the algorithm.

In [36] hybridization of LS and GA (HGSADC) has been applied to solve a large class of time constrained VRP: multiple depots (MDVRPTW), multi-periods (PVRPTW), and vehicle-site dependencies (SDVRPTW). The proposed work is an extension of [57]. The addressed variants are called RVRP because of the additional constraints of route-duration, planning periods, and vehicle-customer dependencies. The proposed algorithm HGSADC uses a different approach to population diversity management. However, the efficiency of the GA heavily depends upon the giant tour algorithm. In [37], GA is proposed to solve four different variants of VRP. The proposed approach integrates set-partitioning and GA to solve the problem.

In [38], GNVS with TS is used to solve RVRP. This problem takes into account various constraints and complexities that are encountered in real life. Among these, there are time windows, capacity constraints, not returning to depot, pickup and delivery dependent orders, compatible orders, and the number of vehicles. The proposed algorithm solves capillary distribution of goods originating in Spain. The main objective of the problem is to reduce the cost by minimizing the distance and number of vehicles used, increasing customer satisfaction, and balancing load among vehicles. More real-life constraints and complexities need also to be addressed.

In [39] a practical variant of VRP known as HVRPTWMP has been solved by the combination of ACS and TS. The variant considers the heterogeneous vehicles with limited transport capacity and delivers multiproducts to customers without violating time windows. The objective of the problem is to minimize the cost compromising travel time and travel distance. The proposed approach uses regency and frequency distribution of TS to improve the quality of solution provided by ACS. No real-life data set is used for demonstrating the efficiency of the proposed approach. The demand of customer is still consistent.

In [40] a combination of GRASP and LS is used for modeling and solving an RVRP for delivery of goods in Urban areas. For modeling, MILP is used. The problem takes into account cost of vehicles, multiple trips, and time limitations for the circulation of large vehicles in city centers. More constraints need to be added to match with real-world problems.

In [41] a new algorithm based on ACS and LS is used for handling RVRP problems. The proposed algorithm is able to solve the uncertainty, heterogeneous fleet, and split delivery constraints of the problem. The main objective of the problem is to optimize the responsiveness along with the traditional objective functions. A real case in the field of air transportation and sets of well-known benchmarks are used for testing the validation of algorithm. Time windows of customers have not been included in the study.

In [42] a hybridization of ILS, SP, and VNS has been proposed for solving a broad class of rich VRPs with heterogeneous fleets. Along with this, other attributes that are considered are time windows, backhauls, multiple depots, split deliveries, duration limits, and open routes. A combination of ILS and VNS is used for solution improvement, and SP is used for exploiting memory of the past search. No real-life data set has been used. In addition to this work, [45–51,53] in the same area can be also cited.

In [43], RVRPs having multiple depots and heterogeneous fleets of vehicles have been optimized by using Monte Carlo simulation (MCS) embedded in local Search (2-opt). MCS algorithm helps in the transformation of deterministic heuristic into a probabilistic heuristic. Local search with geometric distribution is used for generation of high-quality results. Real test beds are used for validation of the approach. However, the problem is not able to handle large data sets.

In [44] an adaptive solution based on GVNS and LS has been proposed for solving real-world VRPTW. The attributes that make the problem RVRP are heterogeneous fleets, soft and multiple TW, and customer priorities. The proposed approach has been embedded into a fleet management company situated in Canary Island. The objective of the problem is to minimize the travel distance and achieve a balance between time and distance. Additional attributes like fleet cost are also important.

TABLE 5.1 Annoted Survey of Real Life VRP.

Author	Year	Journal	Product	Constraint	Objectives	Real Data Set	Hybrid Approach	Cooperation Scheme	RESEARCH GAP
Song Sung Hun et al.	2002	CIE, Elsevier	Newspaper Logistics	Constraints of SDVRP	optimal assignment of newspaper agents, optimal routes and schedule for newspaper delivery	YES	Several Heuristic + Digital Map	HRH	Several Heuristics are used
Chiang Wen-Chyuan, and Robert A. Russell	2004	EJOR, Elsevier	Propone Gas	–	Minimize cost of purchasing and transportation	YES	Set partitioning, TS with route improvement strategies	LRH	–
Ambrosino Daniela and Anna Sciomachen	2007	IMA Journal of Mgmt. Math, Oxford Academic	Fresh/Dry and Frozen Food	Operative and Customer Requirement	Minimize TC	YES	MILP + LS	LRH	Time Windows should be considered
Karadimas Nikolaos V., et al.	2008	WSEAS Transactions on Computers	Waste Collection	–	Optimal routes	NO	ARGIS + ACS	LRH	Performance of algorithm heavily depends upon the number of stops and the complexity of the data set
Russell Robert et al.	2008	COR, Elsevier	Newspaper Logistics	OVRP, zoning, time windows	Minimize TD, NV	YES	Parallel construction + TS	LRH	Centralized Sites

continued on next page

TABLE 5.1 *(continued)*

Author	Year	Journal	Product	Constraint	Objectives	Real Data Set	Hybrid Approach	Cooperation Scheme	RESEARCH GAP
Chiang Wen-Chyuan, et al.	2009	IJPE, Elsevier	Newspaper Distribution	Stochastic VRP, zoning, time windows	Minimize TD, NV	YES	TS	LRH	---
Bohnlein Dominik et al.	2009	IEEE Proceedings	Newspaper Distribution	Heterogeneous editions per vehicle and cluster-dependent tour starts	Minimize transport cost and maximize the customer satisfaction	YES	ACO + TS	HTH	No comparison with other approaches
Benjamin Aida Mauziah, and J. E. Beasley	2010	COR, Elsevier	Waste Collection	Time windows, driver rest period, multiple	Minimize total travel distance and no. of vehicles used	YES	VNS + TS	HRH	Complex Approach
Poorya Farahani et al.	2012	FSM, Springer	Catering Food	Various Constraints	Minimize Food Decay and Transportation Cost	YES	MILP + LNS	LTH	Fails to handle situation with multiple producers
Popović Dražen et al.	2012	ESA, Elsevier	Fuel Delivery	Multiproduct, multiperiod, multicompart-ment	Minimize cost	NO	VNS + simulation	LRH	Homogenous Vehicles
Hemmel-mayr Vera, et al.	2013	JH, Springer	Waste Collection	Long Planning period, Assignment of waste to IFs	Minimize TC	YES	VNS + LS + Exact methods	LTH	unrealistic assumptions on fleet

continued on next page

TABLE 5.1 (*continued*)

Author	Year	Journal	Product	Constraint	Objectives	Real Data Set	Hybrid Approach	Cooperation Scheme	RESEARCH GAP
Archetti Claudia et al.	2013	EJOR, Elsevier	Newspaper Delivery	Time window, constraints of inventory problem, production	No. of trips and time of delivery	YES	Constructive Heuristic + VNS	LTH	Sequential working of phases
Y. Zhang and X. D. Chen	2014	JARCT, Elsevier	Frozen Food	Time windows, loading weight and loading volume	Minimize the delivery cost	YES	GA + LS	LTH	No. of trucks required as objective should be considered
Govindan Kannan et al.	2014	IJPE, Elsevier	Perishable food	–	Minimize routing cost + emissions	NO	PSO + VNS	HRH	Soft time windows are considered
Ufuk Yapar, Fulya Altiparmak	2016	Proceedings	Perishable food	–	Maximization of freshness and minimization TD	NO	SA + GP + LS	LRH	No realistic data set
Coelho Leandro C. et al.	2016	JORS, Springer	Furniture	Driver Rest Period, Time Windows	Minimize TD, NV, TC	YES	MILP + LS	LRH	Balance of workload among drivers should also be considered
Rabbani M. et al.	2016	IJAOR, Springer	Waste Collection	Type of Trip	Minimize TD	NO	SA + LS	LTH	Single Recycle center and homogenous vehicles

continued on next page

TABLE 5.1 *(continued)*

Author	Year	Journal	Product	Constraint	Objectives	Real Data Set	Hybrid Approach	Cooperation Scheme	RESEARCH GAP
Adewumi Aderemi Oluyinka, and Olawale Joshua Adeleke	2016	AMS, Hikari	Waste Collection	Time Windows and road attributes	Minimize TD, Maximize visited customers	NO	Lagrangian relaxation and subgradient	LRH	Less Robust
Patiño Chirva Johana Andrea et al.	2016	Revista Ingeniería	Waste Collection	Vehicle capacity, workday duration, planning horizon	Maximize the amount of waste recollected	YES	MILP + clustering	LRH	Accuracy of result depends upon method used
Hu Hongtao et al.	2017	IJPR, Taylor & Francis	Fresh Products	–	Minimize Total Cost of Cars	NO	MILP + PSO + VNS	HTH	No realistic data set
Wei Qu et al.	2017	Proceedings, IEEE	Waste Collection	Dynamic Disposal trips	Minimize total carbon emission	NO	ABC + VND	HRH	No realistic data set
Osaba Eneko, et al.	2017	Soft Computing, Springer	Newspaper Delivery	Flow for collection, delivery demands	Minimize Cost	YES	Discrete FireFly	–	Absence of Parameterized Analysis
Reihaneh, Mohammad & Ahmed Ghoniem	2018	JORS, Springer	Food	–	Minimize vehicle routing, customer assignment	NO	LS + Permutation	LTH	Constraint of multiperiod with demand allocation should also be addressed

TABLE 5.2 Annotated Survey of RVRP.

Author	Year	Journal	Problem	Why RVRP (Constraint or Objective)	Real Data Set	Hybrid Approach	Cooperation Scheme	RESEARCH GAP
Repoussis Panagiotis P et al.	2006	Springer Proceedings	Fleet Size and Mix VRPTW	Heterogeneous fleets, multiple commodities, multiple vehicle compartments	NO	GRASP + VNS	HRH	Use of large number of local search operators
Carić Tonči, et al.	2007	Springer Proceedings	VRPTW	Minimize transport cost	NO	ILS + SA	LTH	Mechanism of dynamic parameter adaptation is not explained
Rizzoli Andrea Emilio, et al.	2007	Swarm Intell, Springer	Dynamic VRP	Minimize TD, NV	YES	ACO + LS	LTS	–
Wen Min, et al.	2010	JOSR, Springer	VRP with Cross Docking	Consolidated decisions at depot	YES	MILP + TS	LRH	Nonsplitable orders
Wen Min, et al.	2010	COR, Elsevier	Dynamic multi-period VRP	Multiperiod time horizon	YES	MILP + Three Heuristics	LRH	Policy to predict best time
Wen Min, et al.	2011	Networks, Wiley	VRP with Driver Scheduling	Heterogeneous vehicles, predefined work regulation of drivers	YES	MILP + VNS	LRH	–
Taner Filip et al.	2012	PROMET-Traffic & Transportation	Delivery Process	Minimize TD	YES	SA + ILS	LTH	No comparison between loss of money and number of vehicle in case of time window violation

continued on next page

TABLE 5.2 (*continued*)

Author	Year	Journal	Problem	Why RVRP (Constraint or Objective)	Real Data Set	Hybrid Approach	Cooperation Scheme	RESEARCH GAP
Vidal Thibaut, et al.	2013	COR, Elsevier	MDVRPTW, PVRPTW, SDVRPTW	Route Duration, planning Period, Vehicle customer Dependencies	NO	LS + GA	LTH	Efficiency of the GA heavily depends upon the giant tour algorithm
Jair J., et al.	2013	J. of Heuristic, Springer	HVRPTWMP	Heterogeneous vehicles, multiproducts, limited transport capacity	NO	ACS + TS	HRH	No real data set
Schyns Michael	2015	EJOR-Elsevier	DVRPTWSD	Uncertainty, heterogeneous fleet, split Delivery, Optimize Responsiveness	YES	ACS + LS	LTS	No Time Windows
de Armas Jesica, et al.	2015	Elsevier	RVRPTW	Heterogeneous Fleets, soft and multiple TW, customer priorities	YES	GVNS + LS	LTS	Less real-life attributes
Sicilia Juan Antonio, et al.	2016	JCAM, Elsevier	Realistic VRP	Non return to depot, compatible orders, increase customer satisfaction, balancing load among vehicles	YES	GNVS + TS	LTH	Less real-life constraints
Souza Neto et al.	2016	Pesquisa Operacional	Delivery of goods	Multitrips, avoiding large vehicle in city centers	YES	GRASP + LS	LTH	Less real life constraints
Alemany Gabriel, et al.	2016	Conference, IEEE	Multi depot RVRP	Multi depot, Heterogeneous Fleets, External Replenishment	YES	Monte Carlo + LS	LTH	Not able to solve large data set

continued on next page

TABLE 5.2 *(continued)*

Author	Year	Journal	Problem	Why RVRP (Constraint or Objective)	Real Data Set	Hybrid Approach	Cooperation Scheme	RESEARCH GAP
Markov Iliya Dimitrov	2017	Dissertation	Stochastic inventory problem	Heterogeneous vehicles	YES	MILP + LNS	LTH	Stochastic travel time and service should also be considered
Penna Puca Huachi Vaz, et al.	2017	Annals of Operations Research	HRVRP	Backhauls, multiple depots, split deliveries, site dependency, open routes, duration limits, and time windows	NO	ILS + VNS + Set Partitioning	LTH	No real data set
Repoussis Panagiotis P et al.	2006	Springer Proceedings	Fleet Size and Mix VRPTW	Heterogeneous fleets, multiple commodities, multiple vehicle compartments	NO	GRASP + VNS	HRH	Use of large number of local search operators
Carić Tonči, et al.	2007	Springer Proceedings	VRPTW	Minimize transport cost	NO	ILS + SA	LTH	Mechanism of dynamic parameter adaptation is not explained
Rizzoli Andrea Emilio, et al.	2007	Swarm Intell, Springer	Dynamic VRP	Minimize TD, NV	YES	ACO + LS	LTS	–
Wen Min, et al.	2010	JOSR, Springer	VRP with Cross Docking	Consolidated decisions at depot	YES	MILP + TS	LRH	Non splitable orders
Wen Min, et al.	2010	COR, Elsevier	Dynamic multi-period VRP	Multiperiod time horizon	YES	MILP + Three Heuristics	LRH	Policy to predict best time

continued on next page

TABLE 5.2 (*continued*)

Author	Year	Journal	Problem	Why RVRP (Constraint or Objective)	Real Data Set	Hybrid Approach	Cooperation Scheme	RESEARCH GAP
Wen Min, et al.	2011	Networks, Wiley	VRP with Driver Scheduling	Heterogeneous vehicles, predefined work regulation of drivers	YES	MILP + VNS	LRH	–
Taner Filip et al.	2012	PROMET-Traffic & Transportation	Delivery Process	Minimize TD	YES	SA + ILS	LTH	No comparison between loss of money and number of vehicle in case of time window violation
Vidal Thibaut, et al.	2013	COR, Elsevier	MDVRPTW, PVRPTW, SDVRPTW	Route Duration, planning Period, Vehicle customer Dependencies	NO	LS + GA	LTH	Efficiency of the GA heavily depends upon the giant tour algorithm
Jair J., et al.	2013	J. of Heuristic, Springer	HVRPTWMP	Heterogeneous vehicles, multiproducts, limited transport capacity	NO	ACS + TS	HRH	No real data set
Schyns Michael	2015	EJOR-Elsevier	DVRPTWSD	Uncertainty, heterogeneous fleet, split Delivery, Optimize Responsiveness	YES	ACS + LS	LTS	No Time Windows
de Armas Jesica, et al.	2015	Elsevier	RVRPTW	Heterogeneous Fleets, soft and multiple TW, customer priorities	YES	GVNS + LS	LTS	Less real life attributes

continued on next page

TABLE 5.2 *(continued)*

Author	Year	Journal	Problem	Why RVRP (Constraint or Objective)	Real Data Set	Hybrid Approach	Cooperation Scheme	RESEARCH GAP
Sicilia Juan Antonio, et al.	2016	JCAM, Elsevier	Realistic VRP	Non return to depot, compatible orders, increase customer satisfaction, balancing load among vehicles	YES	GNVS + TS	LTH	Less real life constraints
Souza Neto et al.	2016	Pesquisa Operacional	Delivery of goods	Multitrips, avoiding large vehicle in city centers	YES	GRASP + LS	LTH	Less real life constraints
Alemany Gabriel, et al.	2016	Conference, IEEE	Multi depot RVRP	Multi depot, Heterogeneous Fleets, External Replenishment	YES	Monte Carlo + LS	LTH	Not able to solve large data set
Markov Iliya Dimitrov	2017	Dissertation	Stochastic inventory problem	Heterogeneous vehicles	YES	MILP + LNS	LTH	Stochastic travel time and service should also be considered
Penna Puca Huachi Vaz, et al.	2017	Annals of Operations Research	HRVRP	Backhauls, multiple depots, split deliveries, site dependency, open routes, duration limits, and time windows	NO	ILS + VNS + Set Partitioning	LTH	No real data set

In Tables 5.1 and 5.2, we present a critically analyzed annotated survey of the mined literature, with research gaps and the cooperation scheme used by the hybrid method. In Table 5.1 the literature of real-life VRP has been summarized. The first column of the table indicates the author, and the second and third columns indicate the year and journal of publication, respectively. The fourth column indicates the product of distribution. The fifth and sixth columns indicate the constraints and objectives of the problem. The seventh column of the table indicates whether the considered problem validates its approach on real data sets or not. The next two columns of the table indicate respectively the hybrid approach used and the cooperation scheme used. Finally, the last column indicates the research gap in the considered problem. TD, NV, and TC are used as abbreviations for the total travel distance, number of vehicles, and total cost, respectively. Similarly to Table 5.1, Table 5.2 summarizes the literature on RVRP.

5.6 Conclusion and future directions

In this chapter, we presented a review of real-life complex problems called RVRP. As these problems have many constraints and objectives, they cannot be solved by exact methods or by approximate methods alone. In fact, they are solved by integration of these two approaches. Even in some cases the heuristics are also combined with DSS and GIS software. In this chapter, we presented a survey on the hybrid methods for RVRPs. The chapter reviews the hybrid methodologies for rich VRP and for real VRP. Their constraints, objectives, hybrid solution techniques, cooperation schemes, and research gaps are summarized in Tables 5.1 and 5.2. From the review of literature it is clear that most of the papers use LTH and LRH schemes. Very few of them focus on HTH and HRH schemes. So in future, cooperative schemes that focus on parallel implementation of exact metaheuristics need be proposed so as to improve the algorithmic efficiency. Secondly, a general framework for solving RVRPs needs also to be proposed. Finally, benchmark problems for RVRPs should also be developed. Some papers do not include any real-life scenario for validating their developed hybrid methods.

References

[1] Y. Zhang, X.D. Chen, An optimization model for the vehicle routing problem in multi-product frozen food delivery, Journal of Applied Research and Technology 12 (2) (2014) 239–250.

[2] Yapar Ufuk, Fulya Altiparmak, Bi-objective optimization of perishable food distribution: a heuristic approach, in: Proceedings of the IRES 25th International Conference, Istanbul, Turkey, 24th January 2016, ISBN 978-81-925751-3-1.

[3] Daniela Ambrosino, Anna Sciomachen, A food distribution network problem: a case study, IMA Journal of Management Mathematics 18 (1) (2007) 33–53.

[4] Poorya Farahani, Martin Grunow, H-O. Günther, Integrated production and distribution planning for perishable food products, Flexible Services and Manufacturing Journal 24 (1) (2012) 28–51.

[5] K. Govindan, A. Jafarian, R. Khodaverdi, K. Devika, Two-echelon multiple-vehicle location–routing problem with time windows for optimization of sustainable supply chain network of perishable food, International Journal of Production Economics 152 (2014) 9–28.
[6] Hongtao Hu, Ye Zhang, Lu Zhen, A two-stage decomposition method on fresh product distribution problem, International Journal of Production Research 55 (16) (2017) 4729–4752.
[7] C. Coelho Leandro, Jean-Philippe Gagliardi, Jacques Renaud, Angel Ruiz, Solving the vehicle routing problem with lunch break arising in the furniture delivery industry, Journal of the Operational Research Society 67 (5) (2016) 743–751.
[8] Mohammad Reihaneh, Ahmed Ghoniem, A multi-start optimization-based heuristic for a food bank distribution problem, Journal of the Operational Research Society (2018) 1–16.
[9] Qi Liu Wei Qu, Zhaoxia Guo, An intelligent optimization approach for waste collection with dynamic disposal trips, in: 2017 IEEE International Conference on Industrial Engineering and Engineering Management, IEEM, IEEE, 2017.
[10] M. Rabbani, F. Haeri Tabrizi, H. Farrokhi-Asl, A hybrid metaheuristic algorithm for solving a roll-on roll-off waste collection vehicle routing problem considering waste separation and recycling center, International Journal of Applied 6 (2) (2016) 19–31.
[11] Aderemi Oluyinka Adewumi, Olawale Joshua Adeleke, A new model for optimizing waste disposal based on customers' time windows and road attributes, Applied Mathematical Sciences 10 (42) (2016) 2051–2063.
[12] M.A. Hannan, et al., Capacitated vehicle-routing problem model for scheduled solid waste collection and route optimization using PSO algorithm, Waste Management 71 (2018) 31–41.
[13] Nikolaos V. Karadimas, et al., Routing optimization heuristics algorithms for urban solid waste transportation management, WSEAS Transactions on Computers 7 (12) (2008) 2022–2031.
[14] Vera Hemmelmayr, et al., A heuristic solution method for node routing based solid waste collection problems, Journal of Heuristics 19 (2) (2013) 129–156.
[15] Aida Mauziah Benjamin, J.E. Beasley, Metaheuristics for the waste collection vehicle routing problem with time windows, driver rest period and multiple disposal facilities, Computers & Operations Research 37 (12) (2010) 2270–2280.
[16] Johana Andrea Patiño Chirva, Yesica Xiomara Daza Cruz, Eduyn Ramiro Lopez-Santana, A hybrid mixed-integer optimization and clustering approach to selective collection services problem of domestic solid waste, Ingeniería 21 (2) (2016) 235–257.
[17] Dražen Popović, Milorad Vidović, Gordana Radivojević, Variable neighborhood search heuristic for the inventory routing problem in fuel delivery, Expert Systems with Applications 39 (18) (2012) 13390–13398.
[18] Wen-Chyuan Chiang, Robert A. Russell, Integrating purchasing and routing in a propane gas supply chain, European Journal of Operational Research 154 (3) (2004) 710–729.
[19] Eneko Osaba, et al., A discrete firefly algorithm to solve a rich vehicle routing problem modelling a newspaper distribution system with recycling policy, Soft Computing 21 (18) (2017) 5295–5308.
[20] Claudia Archetti, Karl F. Doerner, Fabien Tricoire, A heuristic algorithm for the free newspaper delivery problem, European Journal of Operational Research 230 (2) (2013) 245–257.
[21] Robert Russell, Wen-Chyuan Chiang, David Zepeda, Integrating multi-product production and distribution in newspaper logistics, Computers & Operations Research 35 (5) (2008) 1576–1588.
[22] Wen-Chyuan Chiang, et al., A simulation/metaheuristic approach to newspaper production and distribution supply chain problems, International Journal of Production Economics 121 (2) (2009) 752–767.
[23] Sung Hun Song, Kwan Suk Lee, Gi Sub Kim, A practical approach to solving a newspaper logistics problem using a digital map, Computers & Industrial Engineering 43 (1–2) (2002) 315–330.
[24] Sachin Suresh Kamble, Rakesh Raut, Shradha Ashok Gawankar, Optimizing the newspaper distribution scenarios using genetic algorithm: a case study of India, American Journal of Applied Sciences 14 (4) (2017) 478–495.

[25] Berndt Burger, Optimizing delivery routes for newspapers and magazines using vehicle routing problem, Research Report, http://hdl.handle.net/2263/21414.
[26] Dominik Bohnlein, Christian Gahm, Axel Tuma, A hybrid meta-heuristic for the VRPTW with cluster-dependent tour starts in the newspaper industry, in: 42nd Hawaii International Conference on System Sciences, 2009, HICSS'09, IEEE, 2009.
[27] Tonči Carić, et al., Empirical analysis of two different metaheuristics for real-world vehicle routing problems, in: International Workshop on Hybrid Metaheuristics, Springer, Berlin, Heidelberg, 2007.
[28] Panagiotis P. Repoussis, Christos D. Tarantilis, George Ioannou, A hybrid metaheuristic for a real life vehicle routing problem, in: International Conference on Numerical Methods and Applications, Springer, Berlin, Heidelberg, 2006.
[29] Filip Taner, Ante Galić, Tonči Carić, Solving practical vehicle routing problem with time windows using metaheuristic algorithms, PROMET-Traffic & Transportation 24 (4) (2012) 343–351.
[30] Andrea Emilio Rizzoli, et al., Ant colony optimization for real-world vehicle routing problems, Swarm Intelligence 1 (2) (2007) 135–151.
[31] Iliya Dimitrov Markov, Rich Vehicle and Inventory Routing Problems With Stochastic Demands, Diss., École Polytechnique Fédérale de Lausanne, 2017.
[32] Min J. Clausen Wen, J. Larsen, Rich Vehicle Routing Problems and Applications, DTU Management, 2010.
[33] Min Wen, et al., Vehicle routing with cross-docking, Journal of the Operational Research Society 60 (12) (2009) 1708–1718.
[34] Min Wen, et al., A multilevel variable neighborhood search heuristic for a practical vehicle routing and driver scheduling problem, Networks 58 (4) (2011) 311–322.
[35] Min Wen, et al., The dynamic multi-period vehicle routing problem, Computers & Operations Research 37 (9) (2010) 1615–1623.
[36] Thibaut Vidal, et al., A hybrid genetic algorithm with adaptive diversity management for a large class of vehicle routing problems with time-windows, Computers & Operations Research 40 (1) (2013) 475–489.
[37] Ismail Yusuf, Mohd Sapiyan Baba, Nur Iksan, An optimal approach to solve rich vehicle routing problem, International Journal of Computer Science and Electronics Engineering (IJCSEE) 1 (1) (2013), ISSN 2320–4028 (online).
[38] Juan Antonio Sicilia, et al., An optimization algorithm for solving the rich vehicle routing problem based on variable neighborhood search and tabu search metaheuristics, Journal of Computational and Applied Mathematics 291 (2016) 468–477.
[39] J. Jair, et al., A two-pheromone trail ant colony system—tabu search approach for the heterogeneous vehicle routing problem with time windows and multiple products, Journal of Heuristics 19 (2) (2013) 233–252.
[40] Neto Souza, José de Ferreira, Vitória Pureza, Modeling and solving a rich vehicle routing problem for the delivery of goods in urban areas, Pesquisa Operacional 36 (3) (2016) 421–446.
[41] Michael Schyns, An ant colony system for responsive dynamic vehicle routing, European Journal of Operational Research 245 (3) (2015) 704–718.
[42] Puca Huachi Vaz Penna, et al., A hybrid heuristic for a broad class of vehicle routing problems with heterogeneous fleet, Annals of Operations Research (2017) 1–70.
[43] Gabriel Alemany, et al., Combining Monte Carlo simulation with heuristics to solve a rich and real-life multi-depot vehicle routing problem, in: Proceedings of the 2016 Winter Simulation Conference, IEEE Press, 2016.
[44] Jesica de Armas, et al., GVNS for a real-world rich vehicle routing problem with time windows, Engineering Applications of Artificial Intelligence 42 (2015) 45–56.
[45] Hugo Tsugunobu Yoshida Yoshizaki, Scatter search for a real-life heterogeneous fleet vehicle routing problem with time windows and split deliveries in Brazil, European Journal of Operational Research 199 (3) (2009) 750–758.

[46] Farah Belmecheri, et al., Particle swarm optimization algorithm for a vehicle routing problem with heterogeneous fleet, mixed backhauls, and time windows, Journal of Intelligent Manufacturing 24 (4) (2013) 775–789.
[47] Berghida Meryem, Boukra Abdelmadjid, Resolving a vehicle routing problem with heterogeneous fleet, mixed backhauls and time windows using cuckoo behavior approach, International Journal of Operational Research 24 (2) (2015) 132–144.
[48] José Brandão, A deterministic tabu search algorithm for the fleet size and mix vehicle routing problem, European Journal of Operational Research 195 (3) (2009) 716–728.
[49] José Brandão, A tabu search algorithm for the heterogeneous fixed fleet vehicle routing problem, Computers & Operations Research 38 (1) (2011) 140–151.
[50] Jose Caceres-Cruz, et al., A savings-based randomized heuristic for the heterogeneous fixed fleet vehicle routing problem with multi-trips, Journal of Applied Operational Research 6 (2) (2014) 69–81.
[51] Jose Bernal, John Willmer Escobar, Rodrigo Linfati, A granular tabu search algorithm for a real case study of a vehicle routing problem with a heterogeneous fleet and time windows, Journal of Industrial Engineering and Management 10 (4) (2017) 646.
[52] Zhang Xiao, Wang Jiang-qing, Hybrid ant algorithm and applications for vehicle routing problem, Physics Procedia 25 (2012) 1892–1899.
[53] Liu Qiang, Xu Jiuping, A study on vehicle routing problem in the delivery of fresh agricultural products under random fuzzy environment, International Journal of Information and Management Sciences 19 (4) (2008) 673–690.
[54] Hitoshi Kanoh, Souichi Tsukahara, Solving real-world vehicle routing problems with time windows using virus evolution strategy, International Journal of Knowledge-Based and Intelligent Engineering Systems 14 (3) (2010) 115–126.
[55] Tonči Carić, et al., A modelling and optimization framework for real-world vehicle routing problems, in: Vehicle Routing Problem, InTech, 2008.
[56] Rahma Lahyani, et al., A multi-compartment vehicle routing problem arising in the collection of olive oil in Tunisia, Omega 51 (2015) 1–10.
[57] Thibaut Vidal, et al., A hybrid genetic algorithm for multidepot and periodic vehicle routing problems, Operations Research 60 (3) (2012) 611–624.
[58] Saugato Datta, The impact of improved highways on Indian firms, Journal of Development Economics 99 (1) (2012) 46–57.
[59] Abhijit Banerjee, Esther Duflo, Nancy Qian, On the Road: Access to Transportation Infrastructure and Economic Growth in China, No. w17897, National Bureau of Economic Research, 2012.
[60] S. Gibbons, Kurt Van Dender, Road transport: the effects on firms, Paris School of Economics, available online: http://www.parisschoolofeconomics.eu/IMG/pdf/Gibbons-Lyytikainen-Overman-RUES-2012june.pdf, 2012.
[61] Kumares C. Sinha, Samuel Labi, Transportation Decision Making: Principles of Project Evaluation and Programming, John Wiley & Sons, 2011.
[62] George B. Dantzig, John H. Ramser, The truck dispatching problem, Management Science 6 (1) (1959) 80–91.
[63] Jose Caceres-Cruz, et al., Rich vehicle routing problem: survey, ACM Computing Surveys (CSUR) 47 (2) (2015) 32.
[64] Rajeev Goel, Raman Maini, Vehicle routing problem and its solution methodologies: a survey, International Journal of Logistics Systems and Management 28 (4) (2017) 419–435.
[65] Bruce L. Golden, Subramanian Raghavan, Edward A. Wasil (Eds.), The Vehicle Routing Problem: Latest Advances and New Challenges, vol. 43, Springer Science & Business Media, 2008.
[66] Suresh Nanda Kumar, Ramasamy Panneerselvam, A survey on the vehicle routing problem and its variants, Intelligent Information Management 4 (03) (2012) 66.
[67] Paolo Toth, Daniele Vigo (Eds.), The Vehicle Routing Problem, Society for Industrial and Applied Mathematics, 2002.
[68] Gilbert Laporte, Fifty years of vehicle routing, Transportation Science 43 (4) (2009) 408–416.

[69] Jan Karel Lenstra, A.H.G. Kan, Complexity of vehicle routing and scheduling problems, Networks 11 (2) (1981) 221–227.
[70] Thibaut Vidal, et al., Heuristics for multi-attribute vehicle routing problems: a survey and synthesis, European Journal of Operational Research 231 (1) (2013) 1–21.
[71] Gilbert Laporte, The vehicle routing problem: an overview of exact and approximate algorithms, European Journal of Operational Research 59 (3) (1992) 345–358.
[72] Paul Shaw, Using constraint programming and local search methods to solve vehicle routing problems, in: International Conference on Principles and Practice of Constraint Programming, Springer, Berlin, Heidelberg, 1998.
[73] Gilbert Laporte, et al., Classical and modern heuristics for the vehicle routing problem, International Transactions in Operational Research 7 (4–5) (2000) 285–300.
[74] Bruno De Backer, et al., Solving vehicle routing problems using constraint programming and metaheuristics, Journal of Heuristics 6 (4) (2000) 501–523.
[75] Vijay Kumar Sandhya, Issues in solving vehicle routing problem with time window and its variants using meta heuristics – a survey, International Journal of Engineering and Technology 3 (6) (2013) 668–672.
[76] Laetitia Jourdan, Matthieu Basseur, E-G. Talbi, Hybridizing exact methods and metaheuristics: a taxonomy, European Journal of Operational Research 199 (3) (2009) 620–629.
[77] Karl F. Doerner, Verena Schmid, Survey: matheuristics for rich vehicle routing problems, in: International Workshop on Hybrid Metaheuristics, Springer, Berlin, Heidelberg, 2010.
[78] Yves Caseau, Glenn Silverstein, Francois Laburthe, Learning hybrid algorithms for vehicle routing problems, arXiv preprint, arXiv:cs/0405092, 2004.
[79] Cristina Bazgan, Refael Hassin, Jérôme Monnot, Approximation algorithms for some vehicle routing problems, Discrete Applied Mathematics 146 (1) (2005) 27–42.
[80] Sven O. Krumke, et al., Approximation algorithms for a vehicle routing problem, Mathematical Methods of Operations Research 68 (2) (2008) 333.
[81] R. Nallusamy, et al., Optimization of multiple vehicle routing problems using approximation algorithms, arXiv preprint, arXiv:1001.4197, 2010.
[82] Devanshu Pandey, Vehicle Routing: A Survey of Approximation Algorithm Techniques, 2013.
[83] Rahma Lahyani, Mahdi Khemakhem, Frédéric Semet, Rich vehicle routing problems: from a taxonomy to a definition, European Journal of Operational Research 241 (1) (2015) 1–14.
[84] Geir Hasle, Arne Løkketangen, Silvano Martello, Rich models in discrete optimization: formulation and resolution (ECCO XVI), European Journal of Operational Research (2006) 1752–1753.
[85] Teodor Gabriel Crainic, et al., A concurrent evolutionary approach for rich combinatorial optimization, in: Proceedings of the 11th Annual Conference Companion on Genetic and Evolutionary Computation Conference: Late Breaking Papers, ACM, 2009.
[86] Julia Rieck, Jürgen Zimmermann, A new mixed integer linear model for a rich vehicle routing problem with docking constraints, Annals of Operations Research 181 (1) (2010) 337–358.

Chapter 6

Parallel algorithms for solving rich vehicle routing problems

Miroslaw Blocho
ABB, Krakow, Poland

The main requirement of good algorithms from practical point of view is obtaining high-quality solutions in a reasonable time. In case of small-size problems, this requirement can be met by exact algorithms, but in most of the cases, metaheuristics have to be used. The major challenge in most of the optimization problems is finding balance between available CPU, problem size, processing time required, and the quality of solution. In case the problem size is larger, the parallelization approaches can help in reducing time required to get high-quality solutions [18]. In this section, we will discuss various parallelization approaches and different cooperative strategies focusing on solutions proposed for the family of the Vehicle Routing Problems.

6.1 Parallelism ideas and taxonomies

Parallel Computing or Distributed Computing means that a problem instance is solved by several processes working simultaneously on some processors [18]. The total computational load dedicated to solve a given optimization problem is decomposed into tasks and distributed to available processors, thus attempting to make the base algorithm more "robust" against the increasing problem size.

The decomposition of the load may concern the algorithm, the search space, or the problem structure. Decomposition of the algorithm is associated with the so-called *functional parallelism,* where different tasks are allocated to different processors, and they work on the same data cooperating each other, if needed. Decomposition of the search space refers to *data parallelism,* where the problem domain is decomposed, and a specific solution methodology is used to address the problem on each resulting component. Decomposition of the problem structure refers to decomposing the problem along sets of attributed, and the generated tasks work either on subproblems related to particular sets of attributes or combine subproblem solutions into one complete solution to the problem [18]. *Fine-grained* and *coarse-grained* parallelisms are related to *small* and *large* tasks resulting from the decomposition of the problem.

Smart Delivery Systems. https://doi.org/10.1016/B978-0-12-815715-2.00010-5

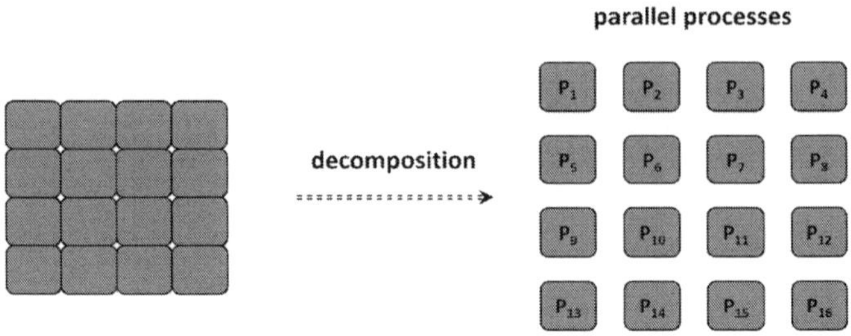

FIGURE 6.1 Example of parallel decomposition.

In general, there are two major approaches to partition the search space, described in the literature as *domain decomposition* and *multisearch*. The *domain decomposition* strategy explicitly partitions the search space, whereas the *multisearch* strategy implicitly divides it through concurrent exploration by so-called *search threads*. An example of the parallel decomposition is shown in Fig. 6.1. In case of *multisearch* method a common problem is with nonoverlapping exploration of the search space, which can be reduced however by application of different search strategies [18]. Looking from an algorithmic point of view, the simplest source of parallelism comes from concurrent execution of inner loop iterations with synchronization parts, which can yield however significant delays.

Crainic and Nourredine [17] presented in 2005 a classification for grouping various parallel strategies for metaheuristics. They introduced three dimensions of the classification indicating how global process is controlled and how data are exchanged between processes and solution techniques used in the search process. The first dimension is called *Search Control Cardinality,* and it specifies whether a single process (*1C; 1-control*) or multiple processes (*pC; p-control*) control the global search. The second dimension is called *Search Control and Communications,* and it describes the issue of data exchanges between processes. To reflect the quantity and quality of data exchanged, four classes of either synchronous or asynchronous communication schemes between processes are indicated: *Rigid Synchronization* (*RS*), *Knowledge Synchronization* (*KS*), *Collegial* (*C*), and *Knowledge Collegial* (*KC*). The third dimension is called *Search Differentiation,* and it describes whether search processes start from the same or different solution and indicates if they use the same or different search methods. Four cases were identified and named SPSS (*Same Initial Point/Population, Same Search Strategy*), SPDS (*Same Initial Point/Population, Different Search Strategy*), MPSS (*Multiple Initial Point/Population, Same Search Strategy*), and MPDS (*Multiple Initial Point/Population, Different Search Strategy*).

Most parallel exact or heuristic methods belong to *1C/RS/SPSS* indicating search process controlled by one master process, where exploration is initiated from a single solution/population and executed according to a single search strategy, containing rigid synchronization between master process and all others. These methods are typically implemented following the classical master–slave parallel programming model, where no communication takes place between slave processes. They are also called *Low-Level 1-Control Parallelization Strategies* as their only objective is to accelerate the search process, not to modify the algorithm logic or the search space.

A bit different approach takes place in the so-called *Neighborhood-based 1C/RS/SPSS*, where neighborhood search is parallelized among the processes. Generating a neighbor solution and evaluating it can be done independently due to no data dependency, and thus parallelization of this step can be implemented and carried out relatively easily. Such mechanism can be implemented also using the classical master–slave mechanism, where each slave examines the neighboring solution, and the master process gathers all the results, selects the best neighbor solution, and continues with the search.

6.2 Cooperative search strategies

The Independent Multi-Search (IMS) is one of the most popular *p-Control* parallelization strategy comprising performing multiple searches simultaneously on the whole search space, initialized from the same or different initial populations, selecting one best solution at the end with no co-operation between the processes during the search. Despite its simplicity, the IMS proved to be very powerful and effective by application to various optimization problems such as traveling salesman problem [37], vehicle routing problem [24], quadratic assignment problem [5], or job scheduling problem [56]. In terms of taxonomy the IMS belongs to the *pC/RS* class using multiple initial point/population and the same search strategy (MPSS) in most cases.

The Cooperative Search strategies go one step further compared to the IMS methods as they contain some mechanism to share and exchange data among processes while the search is in progress, not only at the end [18]. The exchange-data mechanism usually allows for obtaining even better final solutions than in case of still powerful independent multisearch techniques. The data-sharing cooperation mechanism is used to specify how different search threads (associated with possible different metaheuristics) interact with each other. In detail, the fundamental design parameters that need to be taken into account while designing cooperative parallel strategies are: what information is shared and between which processes, when and how it is exchanged, and how this data is used by the other processes [57]. For example, a cooperative mechanism comprising a set of independent metaheuristics that periodically restart from the globally best solution indicates when the processes interact (periodically), between which processes (all processes), what data is shared (best solution from each process and overall best solution), and how this information is used (each process restarts from the overall best solution) [18]. Examples of the Ring cooperation, Knowledge synchronization, and Randomized cooperation schemes between processes are shown in Figs. 6.2, 6.3 and 6.4, respectively.

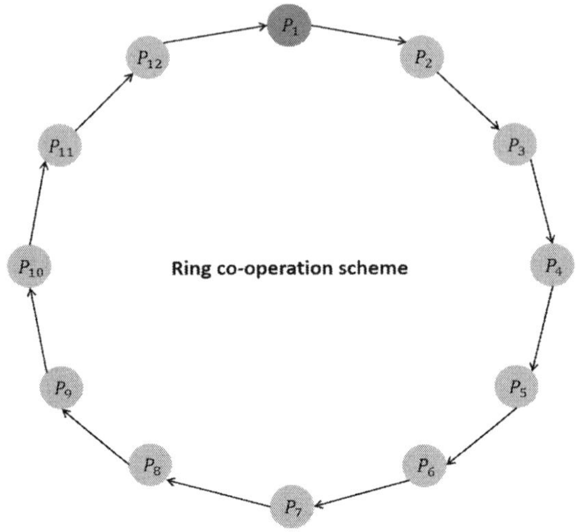

FIGURE 6.2 Ring co-operation scheme.

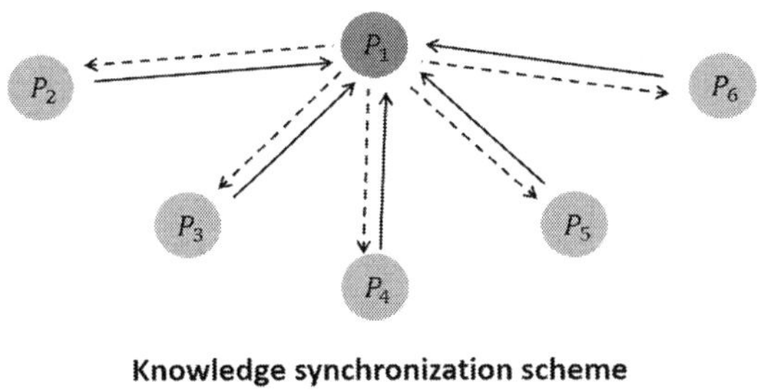

FIGURE 6.3 Knowledge synchronization scheme.

The search processes may exchange data directly between particular processes or indirectly usually through the socalled *blackboard*. The *blackboards* being independent structures store information coming from all the processes and at the same time can be used as a source of information achievable by all processes at any time. Cooperation types are split into synchronous and asynchronous communications. In case of synchronous cooperation when some knowledge about solutions must be synchronized among processes at a specified time common to all processes, it is referred to as *p-Control Knowledge Synchronization* (*pC/KS*) strategy. In case of asynchronous cooperation that is fully distributed with cooperation actions initialized independently by search processes, it is referred to as *p-Control Collegial* (*pC/C*) strategy. In turn, in case *pC/C* communication is performed not only to exchange data but additionally used to alter the knowledge of particular processes, it is referred to as *p-Control Knowledge Collegial* (*pC/KC*) strategy. Some advanced *pC/KC* strategies can

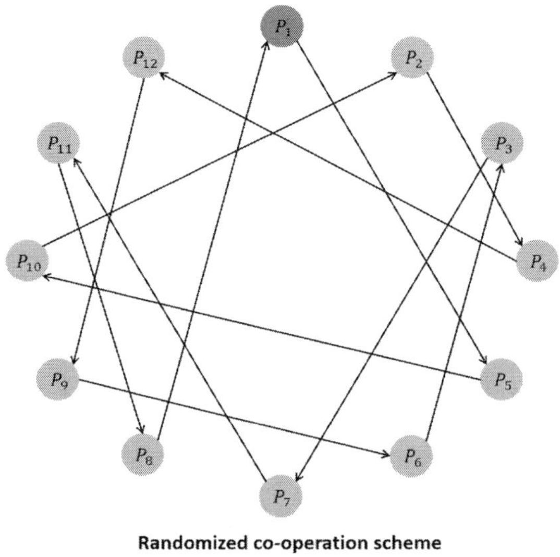

FIGURE 6.4 Randomized cooperation scheme.

even generate completely new solutions during co-operation (see Section 6.5 for some examples).

The parallel algorithms for the traveling salesman problem and for the family of the vehicle routing problems are in the interest of the researchers for almost three decades. In the following sections, we will describe various applications of the cooperative search strategies grouped by metaheuristic classes applied to different routing problems.

6.3 Parallel tabu search

Tabu-Search metaheuristics were one of the most popular algorithms taken into account for the parallelization purposes. In 1994, Garcia et al. [26] applied a neighborhood-based *1C/RS/SPSS* strategy to a tabu search for the vehicle routing problem with time windows. In this approach a master process was grouping neighbor solutions into the appropriate number of tasks assigned later to slave processes. Each slave was examining the neighbor solution, and the results were sent back to the master. Fiechter proposed in 1994 a Knowledge Synchronization approach in a parallel tabu search algorithm for large Traveling Salesman Problem (TSP) instances, in which a full metaheuristic was executed on subsets of the search space, periodically changing the partition and restarting the search process. Such an approach proved to be very successful and competitive especially for larger TSP test instances. General split of Parallel Tabu Search algorithms is shown in Fig. 6.5.

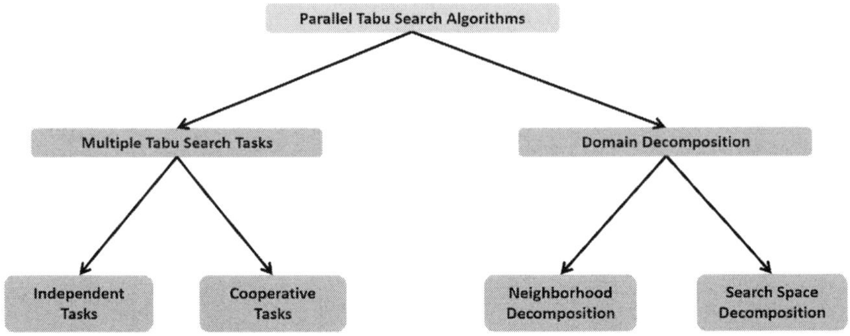

FIGURE 6.5 General split of Parallel Tabu Search algorithms.

The approach of search space decomposition between multiple processes with Knowledge Synchronization schemes was proposed in 1993 by Taillard [55]. This approach was applied for the VRP, where all customers were partitioned with vehicles assigned to particular regions, and each subproblem was solved in parallel by independent tabu search algorithm. The knowledge synchronization was conducted periodically every certain number of iterations by exchanging customers or vehicles between neighboring regions.

The Multiple Population Same Search strategy combined with Rigid Synchronization was undertaken in 1996 by Rego and Roucairol [52], who proposed a parallel tabu search for the VRP. In the neighborhood search the suggested algorithm used compound moves generated by an ejection chain process, whereas parallel processing was used to explore the solution space more extensively and to accelerate the search process.

In 1997, Badeau et al. [2] proposed a parallel tabu search combined with adaptive memory approach applied to the VRP with time windows. They proved that standard parallelizing of the sequential tabu search does not reduce the solution quality and thus provides substantial speedups in practice.

Gendreau et al. [29] developed in 1999 a parallel tabu search applied to the dynamic problem of the Real Time Vehicle Routing and Dispatching problem. They chose a master–slave mechanism, where the master process was used to manage the adaptive memory and to construct new initial solutions for slave processes, which were applying tabu search procedures to improve the quality of particular solutions.

In 2005, Bouthillier and Crainic [14] proposed a cooperative parallel metaheuristic for the VRPTW based on the solution warehouse strategy, in which several search threads cooperate by asynchronously exchanging information on the best solutions found to date. The asynchronous data exchanges were carried out through so-called solution warehouses used to hold and manage a pool of solutions. All independent processes run either tabu search procedure or an evolutionary algorithm to further improve the quality of solutions. In the same year, Crainic et al. [17] did an outstanding overview of various parallel tabu search methods applied to the TSP, VRP, Quadratic Assignment Problem, and other optimization problems.

More recently, in 2012, Cordeau and Maischberger [16] developed the parallel iterated tabu search heuristic for the classical VRP, periodic VRP, multidepot VRP, and site-dependent VRP. The parallel implementation of *pC/KS/MPDS* type comprised iterated local search framework, tabu search combined with a simple perturbation mechanism to ensure a broad exploration of the search space. The control of the algorithm was shared between different parallel processes, knowledge about solutions was synchronized among processes, and the search processes started from different points using different parameters.

Another parallel multineighborhood cooperative tabu search for Capacitated VRP was proposed in 2012 by Jin et al. [32] utilizing several different neighborhood structures. A single neighborhood or neighborhood combinations were encapsulated in tabu search threads, and they cooperated through a solution pool to exploit their joint power.

6.4 Parallel genetic and evolutionary algorithms

The group of various parallel genetic and evolutionary algorithms becomes more and more popular, mostly because that parallelization of population-based approaches gives substantial improvements in solution quality. Most of the evolutionary-genetic parallelization techniques use an initial population of size n separated into n/p groups for p available processors; however, based on multiple researches, there should be always found a proper balance between population size and the speed of crossover and fitness evaluation to get highest quality solutions possible [18]. Other approaches consider running full-sized population on several available processors, which reduce however the problem of small-size populations, but the convergence time becomes longer. Various approaches and cooperative strategies for parallel genetic and evolutionary algorithms are described in the remaining part of this section.

Muhlenbein [40] proposed in 1992 one of the first parallel genetic algorithms for the traveling salesman problem. In this algorithm, selection for mating was distributed among available processes, and thus selection of a mate was done independently by each individual in its neighborhood. Moreover, each individual was able to improve its fitness during its lifetime. This parallel algorithm was run on MIMD parallel computers using small-sized populations comparing to sequential genetic algorithm, which used full-size populations. Based on computational experiments, the author proved the power of parallelized version, which was able to find high-quality solutions for very large problems for the TSP.

In 1999, Gehring and Homberger [27] developed a powerful parallel hybrid evolutionary metaheuristic for the VRP with Time Windows. The parallelization followed the concept of cooperative autonomy, where several autonomous sequential solution procedures cooperate by exchanging solutions. Gehring and

Homberger [28] further extended this approach by implementing a parallel two-phase metaheuristic for the VRPTW with independent multisearch threads running the classical master–slave model. Each slave process performs certain number of iteration, and then it sends a custom control signal back to master. The master process terminates the concurrent search process after control signals are received back from all slaves and gathers best solutions found by all slave processes.

Berger and Barkaoui [6] presented in 2004 a parallel hybrid genetic algorithm for the VRPTW using route-directed hybrid genetic approach based on the simultaneous evolution of two populations of solutions focusing on separate objectives subject to temporal constraint relaxation. The aim of first population was evolving individuals to minimize the total traveled distance, and the aim of the second population was to minimize constraint violations to generate a feasible solution. The parallel procedure followed a classical master–slave message-passing paradigm, in which the master process controlled the execution of the algorithm, coordinated genetic operations, and handled parent selection, whereas the slave elements concurrently executed reproduction and mutation operators. Their approach proved to be very competitive allowing for multiple new best-known solutions.

Borovska [13] developed in 2006 a parallel genetic algorithm to solve the TSP on multicomputer cluster, in which functional decomposition of the parallel applications was done as the base for parallel algorithm design. Performance estimation of the parallel version was calculated based on testing the MPI implementation of this algorithm.

In 2013, Wang et al. [60] presented a comparative study of five different coarse-grained parallel genetic algorithms (PGAs) for the TSP. All versions were implemented on the same parallel machine, tested on the same problem instances, and started from the same set of initial populations. The version of the PGA that combines a new subtour technique with a migration approach was identified as the best one.

6.5 Parallel memetic algorithms

The Memetic Algorithms (MAs) are population-based hybrid genetic algorithms hybridized with local search procedures. First Parallel Memetic Algorithms (PMAs) were developed just a few years after the MAs, for example, in 1992 by Moscato and Norman [38], who proposed a parallel memetic approach for the TSP on MIMD Message-Passing parallel computers. They used Order Crossover and Strategic Edge Crossover operators for recombination purposes, and the local search was supported by a hybrid of Monte Carlo simulated annealing. The parallel version was implemented on a MIMD distributed memory machine, where all individuals were mapped to different processors, and cooperation between them was carried out by interprocessor communication through message passing system. An example of the parallel memetic algorithm is shown in Fig. 6.6.

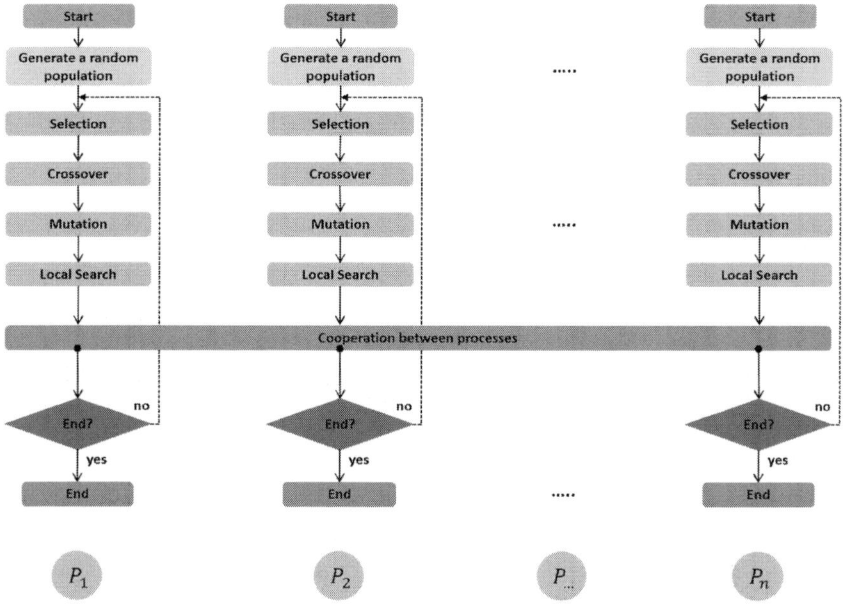

FIGURE 6.6 Example of the Parallel Memetic Algorithm.

Nalepa and Czech [48] developed in 2012 a parallel memetic algorithm for the VRP with Time Windows, where the influence of the population diversification and child generation on the accuracy was analyzed together with the speedups for the memetic algorithm in the second phase. Blocho and Czech [7] suggested a parallel EAX-based algorithm for minimizing the number of routes in the PMA for the VRPTW, in which a novel approach concerning the usage of the edge assembly crossover (EAX) operator during exchanging the best solutions between the processes was implemented. Their approach allowed them to find eight new world's best solutions for the Gehring and Homberger's benchmark problem instances. One year later, in 2013, Blocho and Czech [8] developed a parallel memetic algorithm for the VRPTW comprising a novel randomized cooperation scheme between the processes. In the randomized cooperation scheme the best solution found so far was transferred through all the processes in a ring, but the order of processes was generated randomly in each iteration. Such an approach proved to be very powerful and allowed them to find 171 out of 300 world's best solutions for the Gehring and Homberger's (GH) benchmarking tests for the VRPTW.

Nalepa et al. [47] further investigated various cooperation schemes for this PMA for the VRPTW to find out how cooperation schemes influence the search convergence and solutions quality. They investigated independent runs (no cooperation between processes), Pool (each process sends certain number of its best solutions to the master process managing pool of best solutions from all processes), Pool with EAX (pool approach with additional solutions enhanced by EAX recombinations), Ring (migration topology as a ring), Randomized EAX (order of processes in the ring is randomized, with solutions additionally enhanced by EAX recombinations), Knowledge Synchronization (the master process synchronizes the knowledge acquired by all process during the search by distributing the best solutions among all other processes). Based on extensive computational experiments, they found out that selection of the best cooperation scheme depends strongly on cooperation frequency and subclass of test instances. Based on gathered results, Nalepa and Blocho [41] carried out extensive computational experiments on 1000-customer Gehring and Homberger's (GH) benchmark tests, which gave a detailed insight into the PMA algorithm performance and search capabilities. They reported 19 out of 60 new world's best solutions to 1000 GH test instances, and they gave consistent guidelines on how to select a proper cooperation scheme in PMA algorithm based on the test characteristics. In cases of C1 and C2 test instances, Randomized EAX and Knowledge Synchronization cooperation schemes give best results, respectively. In cases of R1, R2, RC1, and RC2, best cooperation schemes vary between Knowledge Synchronization and Ring depending on whether execution time is a priority or not, respectively.

The impact of population size and the number of generated child solutions on PMA algorithm efficacy was investigated in the further studies by Blocho and Nalepa [9]. They proved that improper selection of the parameters may easily jeopardize the search, and they showed that larger populations converge to high-quality solutions in a smaller number of subsequent generations and that creating more children helps exploiting parent solutions. The studies on the PMA for the VRPTW were further extended in 2016 by Nalepa and Blocho [43] by investigating temporally adaptive cooperation schemes as an approach of changing cooperation scheme during execution between Ring scheme and Knowledge Synchronization scheme with Search-Space partition [42].

Similar studies were carried out also on the Pickup and Delivery Problem with Time Windows. In 2015, Blocho and Nalepa [10] developed a parallel algorithm for minimizing the first stage in PDPTW by taking advantage of asynchronous Ring cooperation scheme, being able to get 10 new world's best solutions to the Li and Lim's benchmark. An interesting modification of that algorithm by introducing Search-Space partition between parallel processes was introduced by Nalepa and Blocho [42] in 2016, who were able to get 10 other new world's best solutions by applying this new technique. A full parallel memetic algorithm for the PDPTW comprising described parallel algorithm for minimizing the fleet size and the memetic approach for the minimizing total traveled distance was developed by Nalepa and Blocho [44] in 2017. They took advantage of the Ring cooperation scheme and the LCS-SREX (Longest Common Subsequence-Selective Route Exchange Crossover) developed previously by Blocho and Nalepa [12]. They performed additional studies to analyze the complexity [11] and verification of correctness [46] of this parallel memetic algorithm.

In 2018, Nalepa and Blocho [45] proposed a novel approach of dynamic cooperation schemes in the PMA for the PDPTW comprising alternately setting either exploitative or explorative scheme. Considered cooperation schemes were Knowledge Synchronization and Ring. Balancing the exploration and exploitation of the solution space by means of the adaptive cooperation proved to have tremendous influence on the convergence capabilities of the parallel memetic algorithm and on the quality of solutions, which were significantly boosted when the adaptive scheme was utilized.

6.6 Parallel ant colony algorithms

The parallelization approaches were considered also for the Ant Colony Algorithms (ACOs). In 1999, Bullnheimer et al. [15] proposed the *1C/RS/SPSS* parallelization strategy for the Ant Colony algorithm for the traveling salesman problem. They followed the classical master–slave model, in which master builds tasks containing several ants and distributes all of them to the available processes. All slave processes were performing searches, and the best found solutions were sent back to the master. Then the master was selecting one global best solution, which was used to update the pheromone matrix and then was sent again to all the slaves. A very similar approach was applied by Randall and Lewis [51] for the TSP and Doerner et al. [23] for the Capacitated VRP.

In 2005, Delisle et al. [22] improved this approach by a master–slave ACO implementation for the TSP in a shared memory computer. They used a global memory to replace the master process, where the global pheromone matrix and best solutions were stored. To avoid the mutual update of global information by particular slave processes, the critical regions were used. Moreover, to decrease communication overhead, the slave processes updated information only periodically, not in every iteration [49]. Based on computational experiments, they found out that increasing number of ants per slave process and decreasing number of iterations allowed to raise the performance; however, the quality of obtained solutions was reduced.

In 2010, Fu et al. [25] extended this approach even further by developing a fine-grain implementation of the ACO for the TSP on GPU. In their approach, GPU was used to hold the pheromone matrix, being updated in every step. The data of the master process was stored partially in the CPU and partially in the global memory of the GPU. Moreover, the GPU was used to generate random cities for each ant, whereas the CPU was used for managing only small pieces of data related to visited routes and cities. They got speedup values up to 30 for this GPU implementation for the TSP.

A bit different approach of parallel independent runs being application of multistart searches on multiple processes running the same ACO algorithm was tackled by Stutzle et al. [54] for the TSP. Such a model did not involve any data exchange mechanism between the processes during the search, only gathering the best results at the end. This approach was further applied for the GPU by Bai et al. [3], where each thread block was used for independent run, and each thread executed only one ant. The main algorithm was run totally on the GPU, whereas the CPU was used to initialize the solutions and to control the subsequent iterations.

Middendorf et al. [36] studied a parallel ACO implementation for the TSP following the multicolony model. The multicolony model provides cooperative search mechanisms and usually simple implementation in distributed memory systems such as clusters. While creating the best design approach, they took into account the neighborhood topology, communication frequency, type of data exchanged between colonies, and how this data was used after exchanges. They concluded that the best results can be obtained by sending the best local solution using a unidirectional ring topology, updating the best global solution accordingly, and cooperation frequency should be set properly to avoid degradation of solution quality or computational efficiency. The multicolony approach was also adapted to the multideport VRP by Yu et al. [61] in 2011, in which outstanding ants were exchanged at certain intervals using a ring connection topology. They were able to solve instances of up to 360 customers using this multicolony approach with eight computers.

Another approach for data exchange, in which each colony dynamically determines a destination colony to send the best solution using an adaptive method, was proposed in 2008 by Chen et al. [33]. The frequency of cooperation was set dynamically based on the diversity of solutions, whereas the adaptive data exchange strategy allowed for a proper balance between the convergence and quality of solutions. The further studies on multicolony ACO and computational experiments done by Twomey et al. [58] in 2010 proved that the best cooperation strategies depend on the usage of local search methods. Moreover, preventing high cooperation frequency makes ant colonies to focus more on various regions of the search space, extending exploration and improving final solutions.

A specialized multicolony savings-based ACO for the VRP was proposed by Lucka et al. [34] with an implementation in multicore environment, where each ant colony was assigned a different thread. The best solutions of the colonies were exchanged asynchronously using shared memory within the same node and shared files within different nodes. Such an approach allowed them to solve VRP instances with up to 420 customers by 32 ant colonies.

Just recently in 2016, Skinderowicz [53] proposed for the TSP three novel parallel Ant Colony System versions for GPUs, where the first two use the standard pheromone matrix, and the third uses a novel selective pheromone memory. The efficiency of proposed approach was proven by extensive experiments made on TSP instances up to 2392 cities with achieved speedup of up to 24.29.

In 2018, Zhou et al. [62] proposed for the TSP a new parallel ACO model for multicore SIMD CPU architecture, where each ant is mapped with a CPU core, and the tour construction of each ant is accelerated by vector instructions. They suggested also a new fitness proportionate selection approach named Vector-based Roulette Wheel (VRW) in the tour construction stage, in which the fitness values are grouped into SIMD lanes and the prefix sum is computed in vector-parallel mode. They tested implementation of this algorithm on TSP instances up to 4461 cities, getting speedup of up to 57.8.

Gulce et al. [30] developed in 2018 a parallel cooperative hybrid algorithm based on ant colony optimization and 3-Opt algorithm (PACO-3Opt) for solving the TSP. The PACO-3Opt comprises multiple colonies, master–slave paradigm, and 3-Opt algorithm to avoid local minima.

6.7 Parallel simulated annealing

Due to high popularity of the simulated annealing (SA) methods for the family of the vehicle routing problems, the parallelization approaches brought high researcher attention too.

In 1989, Malek et al. [35] proposed both serial and parallel implementations of simulated annealing applied to the traveling salesman problem. They used a novel approach using abbreviated cooling schedule allowed for achieving a superlinear speedup. Similar good results were obtained by Allwright and Carpenter [1] in 1989, who suggested a distributed implementation of SA for the TSP running on a linear chain of processors driven by a host processor. The role of the host processor was only to supervise the other processors, so that the bulk of processing takes place on the chain and the efficiency of the algorithm remains high as the number of processors is increased.

Jeong and Kim [31] developed in 1991 a fast parallel simulated annealing algorithm for solving the TSP on SIMD machines with linear interconnections among processing elements. They introduced also move operations executed in proportional to the time taken to broadcast a bit from one process to all others. This allows for units having broadcasting capabilities to implement move operations in constant time making whole time complexity of simulated annealing algorithm proportional only to the number of the moves. Another parallel implementation for the TSP was proposed in 1996 by Ram and Subramaniam [50], who observed superlinear speedups by running their algorithm.

In 2002, Czech and Czarnas [20] developed a parallel SA for the VRPTW, where all processes cooperate periodically every predefined constant number of steps, passing their best solution found-to-date in a ring scheme. This approach was further extended in 2003 by Czarnas et al. [19] and in 2009 by Czech et al. [21], who proposed both MPI and OpenMP implementation of the parallel SA for the VRPTW, in which the number of components cooperate periodically by exchanging their best solutions found to date.

More recently, in 2013, Banos et al. [4] proposed the parallel Multiple Temperature Pareto Simulated Annealing (MT-PSA) to cope with the VRP with time windows. The sequential MT-PSA was parallelized with MPI using the island-based model, and computational experiments indicated that the island-based parallelization produces Pareto fronts of higher quality than those obtained by the sequential versions without increasing the computational cost but also producing significant reduction in the runtimes while maintaining solution quality.

In 2015, Wang et al. [59] developed a parallel SA for the VRP with simultaneous pickup-delivery and time windows. Their parallel version was combined with a Residual Capacity and Radial Surcharge (RCRS) insertion-based heuristic allowing for 12 new best solutions.

The parallel SA for the VRP with simultaneous pickup and delivery following traditional sequential SA algorithm, combined with an integrated asynchronous and synchronous multiple Markov chain approach and incorporated master–slave structure, was developed in 2016 by Mu et al. [39].

6.8 Summary

In this chapter, we described the most important parallelization techniques, cooperative search strategies, and the most widely used parallel algorithms for solving rich vehicle routing problems. The exact methods are mostly used when an optimal solution can be found in acceptable time. The heuristics and metaheuristics techniques do not guarantee optimum but solutions with sufficient quality to meet problem objectives. Both parallelization approaches and cooperative strategies can be used for both exact and approximation methods for larger problem instances as they can significantly help reduce time required to get high-quality solutions. The parallelization methods are based on the idea of solving combinatorial problems by several processes working simultaneously on some processors. The cooperative search strategies additionally contain some mechanisms to share and exchange data among processes while the search is in progress. Due to high popularity of the metaheuristics and multiple parallelization attempts, we described the most popular parallel tabu search, parallel simulated annealing, parallel genetic and evolutionary algorithms, parallel ant colony algorithms, and very efficient parallel memetic algorithms for solving rich vehicle routing problems.

References

[1] James R.A. Allwright, D.B. Carpenter, A distributed implementation of simulated annealing for the travelling salesman problem, Parallel Computing 10 (3) (1989) 335–338.

[2] P. Badeau, M. Gendreau, F. Guertin, J.-Y. Potvin, E. Taillard, A parallel tabu search heuristic for the vehicle routing problem with time windows, Transportation Research Part C: Emerging Technologies 5 (2) (1997) 109–122.

[3] H. Bai, D. OuYang, X. Li, L. He, H. Yu, Max-min ant system on GPU with CUDA, in: 2009 Fourth International Conference on Innovative Computing, Information and Control, ICICIC, Dec 2009, pp. 801–804.

[4] Raul Baños, Julio Ortega, Consolación Gil, Antonio Fernández, Francisco de Toro, A simulated annealing-based parallel multi-objective approach to vehicle routing problems with time windows, Expert Systems with Applications 40 (5) (2013) 1696–1707.

[5] Roberto Battiti, Giampietro Tecchiolli, Parallel biased search for combinatorial optimization: genetic algorithms and tabu, Microprocessors and Microsystems 16 (7) (1992) 351–367.

[6] Jean Berger, Mohamed Barkaoui, A parallel hybrid genetic algorithm for the vehicle routing problem with time windows, Computers & Operations Research 31 (12) (2004) 2037–2053.

[7] Miroslaw Blocho, Zbigniew J. Czech, A parallel EAX-based algorithm for minimizing the number of routes in the vehicle routing problem with time windows, in: Geyong Min, Jia Hu, Lei (Chris) Liu, Laurence Tianruo Yang, Seetharami Seelam, Laurent Lefévre (Eds.), HPCC-ICESS, IEEE Computer Society, 2012, pp. 1239–1246.

[8] Miroslaw Blocho, Zbigniew J. Czech, A parallel memetic algorithm for the vehicle routing problem with time windows, in: Fatos Xhafa, Leonard Barolli, Dritan Nace, Salvatore Venticinque, Alain Bui (Eds.), 3PGCIC, IEEE, 2013, pp. 144–151.

[9] Miroslaw Blocho, Jakub Nalepa, Impact of parallel memetic algorithm parameters on its efficacy, in: Stanislaw Kozielski, Dariusz Mrozek, Pawel Kasprowski, Bozena Malysiak-Mrozek, Daniel Kostrzewa (Eds.), BDAS, in: Communications in Computer and Information Science, vol. 521, Springer, 2015, pp. 299–308.

[10] Miroslaw Blocho, Jakub Nalepa, A parallel algorithm for minimizing the fleet size in the pickup and delivery problem with time windows, in: Jack J. Dongarra, Alexandre Denis, Brice Goglin, Emmanuel Jeannot, Guillaume Mercier (Eds.), EuroMPI, ACM, 2015, pp. 15:1–15:2.

[11] Miroslaw Blocho, Jakub Nalepa, Complexity analysis of the parallel guided ejection search for the pickup and delivery problem with time windows, CoRR, arXiv:1704.06724, 2017.

[12] Miroslaw Blocho, Jakub Nalepa, LCS-based selective route exchange crossover for the pickup and delivery problem with time windows, in: Bin Hu, Manuel López-Ibáñez (Eds.), EvoCOP, in: Lecture Notes in Computer Science, vol. 10197, 2017, pp. 124–140.

[13] Plamenka Borovska, Solving the travelling salesman problem in parallel by genetic algorithm on multicomputer cluster, in: International Conference on Computer Systems and Technologies, COMPSYSTECH, 2006.

[14] Alexandre Le Bouthillier, Teodor Gabriel Crainic, A cooperative parallel meta-heuristic for the vehicle routing problem with time windows, Computers & Operations Research 32 (2005) 1685–1708.

[15] Bernd Bullnheimer, Gabriele Kotsis, Christine Strauß, Parallelization Strategies for the Ant System, Springer US, Boston, MA, 1998, pp. 87–100.

[16] Jean-Francois Cordeau, Mirko Maischberger, A parallel iterated tabu search heuristic for vehicle routing problems, Computers & Operations Research 39 (9) (2012) 2033–2050.

[17] Teodor Gabriel Crainic, Michel Gendreau, Jean-Yves Potvin, Parallel Tabu Search, John Wiley & Sons, Ltd, 2005, pp. 289–313, chapter 13.

[18] Teodor Gabriel Crainic, Michel Toulouse, Parallel Meta-Heuristics, Springer US, Boston, MA, 2010, pp. 497–541.

[19] Piotr Czarnas, Zbigniew J. Czech, Przemyslaw Gocyla, Parallel simulated annealing for bicriterion optimization problems, in: Roman Wyrzykowski, Jack Dongarra, Marcin Paprzycki, Jerzy Wasniewski (Eds.), PPAM, in: Lecture Notes in Computer Science, vol. 3019, Springer, 2003, pp. 233–240.

[20] Zbigniew J. Czech, Piotr Czarnas, Parallel simulated annealing for the vehicle routing problem with time windows, in: PDP, IEEE Computer Society, 2002, p. 376.

[21] Zbigniew J. Czech, Wojciech Mikanik, Rafal Skinderowicz, Implementing a parallel simulated annealing algorithm, in: Roman Wyrzykowski, Jack J. Dongarra, Konrad Karczewski, Jerzy Wasniewski (Eds.), PPAM (1), in: Lecture Notes in Computer Science, vol. 6067, Springer, 2009, pp. 146–155.

[22] Pierre Delisle, Marc Gravel, Michaël Krajecki, Caroline Gagné, Wilson L. Price, Comparing parallelization of an ACO: message passing vs. shared memory, in: María J. Blesa, Christian Blum, Andrea Roli, Michael Sampels (Eds.), Hybrid Metaheuristics, Springer, Berlin, Heidelberg, 2005, pp. 1–11.

[23] Karl Doerner, Richard F. Hartl, Guenter Kiechle, Mária Lucká, Marc Reimann, Parallel ant systems for the capacitated vehicle routing problem, in: Jens Gottlieb, Günther R. Raidl (Eds.), EvoCOP, in: Lecture Notes in Computer Science, vol. 3004, Springer, 2004, pp. 72–83.

[24] Karl F. Doerner, Richard F. Hartl, Siegfried Benkner, Mária Lucká, Parallel cooperative savings based ant colony optimization – multiple search and decomposition approaches, Parallel Processing Letters 16 (3) (2006) 351–370.

[25] Jie Fu, Lin Lei, Guohua Zhou, A parallel ant colony optimization algorithm with GPU-acceleration based on all-in-roulette selection, in: Third International Workshop on Advanced Computational Intelligence, 09 2010, pp. 260–264.

[26] Bruno-Laurent Garcia, Jean-Yves Potvin, Jean-Marc Rousseau, A parallel implementation of the tabu search heuristic for vehicle routing problems with time window constraints, Computers & Operations Research 21 (9) (1994) 1025–1033.

[27] Hermann Gehring, Jörg Homberger, A parallel hybrid evolutionary metaheuristic for the vehicle routing problem with time windows, in: University of Jyväskylä, 1999, pp. 57–64.

[28] Hermann Gehring, Jörg Homberger, A parallel two-phase metaheuristic for routing problems with time windows, Asia-Pacific Journal of Operational Research - APJOR 18 (05 2001).

[29] Michel Gendreau, Francois Guertin, Jean-Yves Potvin, Éric D. Taillard, Parallel tabu search for real-time vehicle routing and dispatching, Transportation Science 33 (4) (1999) 381–390.

[30] Saban Gülcü, Mostafa Mahi, Ömer Kaan Baykan, Halife Kodaz, A parallel cooperative hybrid method based on ant colony optimization and 3-opt algorithm for solving traveling salesman problem, Soft Computing 22 (5) (2018) 1669–1685.

[31] Chang-Sung Jeong, Myung Ho Kim, Fast parallel simulated annealing for traveling salesman problem on SIMD machines with linear interconnections, Parallel Computing 17 (2–3) (1991) 221–228.

[32] Jianyong Jin, Teodor Gabriel Crainic, Arne Løkketangen, A parallel multi-neighborhood cooperative tabu search for capacitated vehicle routing problems, European Journal of Operational Research 222 (3) (2012) 441–451.

[33] Shu Wang Ling Chen, Hai-Ying Sun, Parallel implementation of ant colony optimization on MPP, in: 2008 International Conference on Machine Learning and Cybernetics, vol. 2, July 2008, pp. 981–986.

[34] Maria Lucka, Piecka Stanislav, Parallel posix threads based ant colony optimization using asynchronous communication, in: Proceedings of the 8th International Conference on Applied Mathematics, 2, 01 2009.

[35] Miroslaw Malek, Mohan Guruswamy, Mihir Pandya, Howard Owens, Serial and parallel simulated annealing and tabu search algorithms for the traveling salesman problem, Annals of Operations Research 21 (1) (Dec 1989) 59–84.

[36] Martin Middendorf, Frank Reischle, Hartmut Schmeck, Multi colony ant algorithms, Journal of Heuristics 8 (3) (2002) 305–320.

[37] Mitsunori Miki, Tomoyuki Hiroyasu, Jun'ya Wako, Takeshi Yoshida, Adaptive temperature schedule determined by genetic algorithm for parallel simulated annealing, in: IEEE Congress on Evolutionary Computation (1), IEEE, 2003, pp. 459–466.

[38] Pablo Moscato, La Plata, La Plata, Michael G. Norman, A "memetic" approach for the traveling salesman problem implementation of a computational ecology for combinatorial optimization on message-passing systems, in: Proceedings of the International Conference on Parallel Computing and Transputer Applications, IOS Press, 1992, pp. 177–186.

[39] Dong Mu, Chao Wang, Fu Zhao, John Sutherland, Solving vehicle routing problem with simultaneous pickup and delivery using parallel simulated annealing algorithm, International Journal of Shipping and Transport Logistics 8 (01 2016) 81–106.

[40] Heinz Mühlenbein, Parallel genetic algorithms, population genetics, and combinatorial optimization, in: Jörg D. Becker, Ignaz Eisele, Friedhelm Mündemann (Eds.), Parallelism, Learning, Evolution, in: Lecture Notes in Computer Science, vol. 565, Springer, 1989, pp. 398–406.

[41] Jakub Nalepa, Miroslaw Blocho, Co-operation in the parallel memetic algorithm, International Journal of Parallel Programming 43 (5) (2015) 812–839.

[42] Jakub Nalepa, Miroslaw Blocho, A parallel algorithm with the search space partition for the pickup and delivery with time windows, in: Fatos Xhafa, Leonard Barolli, Fabrizio Messina, Marek R. Ogiela (Eds.), 3PGCIC, IEEE Computer Society, 2015, pp. 92–99.

[43] Jakub Nalepa, Miroslaw Blocho, Temporally adaptive co-operation schemes, in: Fatos Xhafa, Leonard Barolli, Flora Amato (Eds.), 3PGCIC, vol. 1, in: Lecture Notes on Data Engineering and Communications Technologies, Springer, 2016, pp. 145–156.
[44] Jakub Nalepa, Miroslaw Blocho, A parallel memetic algorithm for the pickup and delivery problem with time windows, in: Igor V. Kotenko, Yiannis Cotronis, Masoud Daneshtalab (Eds.), PDP, IEEE Computer Society, 2017, pp. 1–8.
[45] Jakub Nalepa, Miroslaw Blocho, Adaptive cooperation in parallel memetic algorithms for rich vehicle routing problems, International Journal of Grid and Utility Computing 9 (2) (2018) 179–192.
[46] Jakub Nalepa, Miroslaw Blocho, Verification of Correctness of Parallel Algorithms in Practice, Springer International Publishing, Cham, 2018, pp. 135–151.
[47] Jakub Nalepa, Miroslaw Blocho, Zbigniew J. Czech, Co-operation schemes for the parallel memetic algorithm, in: Roman Wyrzykowski, Jack Dongarra, Konrad Karczewski, Jerzy Wasniewski (Eds.), PPAM (1), in: Lecture Notes in Computer Science, vol. 8384, Springer, 2013, pp. 191–201.
[48] Jakub Nalepa, Zbigniew Czech, A parallel heuristic algorithm to solve the vehicle routing problem with time windows, Studia Informatica 33 (01 2012) 91–106.
[49] Martín Pedemonte, Sergio Nesmachnow, Héctor Cancela, A survey on parallel ant colony optimization, Applied Soft Computing 11 (8) (2011) 5181–5197.
[50] D. Janaki Ram, T.H. Sreenivas, K. Ganapathy Subramaniam, Parallel simulated annealing algorithms, Journal of Parallel and Distributed Computing 37 (2) (1996) 207–212.
[51] Marcus Randall, Andrew Lewis, A parallel implementation of ant colony optimization, Journal of Parallel and Distributed Computing 62 (9) (2002) 1421–1432.
[52] César Rego, Catherine Roucairol, A Parallel Tabu Search Algorithm Using Ejection Chains for the Vehicle Routing Problem, Springer US, Boston, MA, 1996, pp. 661–675.
[53] Rafal Skinderowicz, The GPU-based parallel ant colony system, Journal of Parallel and Distributed Computing 98 (2016) 48–60.
[54] Thomas Stützle, Parallelization strategies for ant colony optimization, in: A.E. Eiben, Thomas Bäck, Marc Schoenauer, Hans-Paul Schwefel (Eds.), PPSN, in: Lecture Notes in Computer Science, vol. 1498, Springer, 1998, pp. 722–731.
[55] Éric D. Taillard, Parallel iterative search methods for vehicle routing problems, Networks 23 (8) (1993) 661–673.
[56] Éric D. Taillard, Parallel taboo search techniques for the job shop scheduling problem, ORSA Journal on Computing 6 (2) (1994) 108–117.
[57] Michel Toulouse, Teodor G. Crainic, Michel Gendreau, Communication Issues in Designing Cooperative Multi-Thread Parallel Searches, Springer US, Boston, MA, 1996, pp. 503–522.
[58] Colin Twomey, Thomas Stützle, Marco Dorigo, Max Manfrin, Mauro Birattari, An analysis of communication policies for homogeneous multi-colony ACO algorithms, Information Sciences 180 (12) (2010) 2390–2404.
[59] Chao Wang, Dong Mu, Fu Zhao, John W. Sutherland, A parallel simulated annealing method for the vehicle routing problem with simultaneous pickup-delivery and time windows, Computers & Industrial Engineering 83 (2015) 111–122.
[60] Lee Wang, Anthony A. Maciejewski, Howard Jay Siegel, Vwani P. Roychowdhury, Bryce D. Eldridge, A study of five parallel approaches to a genetic algorithm for the traveling salesman problem, Intelligent Automation & Soft Computing 11 (4) (2005) 217–234.
[61] B. Yu, Z.Z. Yang, J.-X. Xie, A parallel improved ant colony optimization for multi-depot vehicle routing problem, Journal of the Operational Research Society 62 (1) (2011) 183–188.
[62] Yi Zhou, Fazhi He, Neng Hou, Yimin Qiu, Parallel ant colony optimization on multi-core SIMD CPUs, Future Generation Computer Systems 79 (2018) 473–487.

Chapter 7

Where machine learning meets smart delivery systems

Jakub Nalepa
Silesian University of Technology, Gliwice, Poland

7.1 Introduction

The amount of data produced every day grows tremendously in most real-life domains, including medical imaging, genomics, text categorization, computational biology, smart delivery systems, and many others. Although it appears beneficial at the first glance (more data could mean more possibilities of extracting and revealing useful underlying knowledge), handling massively large datasets became a challenging issue and attracts research attention, especially in the era of *big data,* where the data *volume* (the amount of generated data), *veracity* (its quality), *velocity* (the speed of data generation), and *variety* (data may come from various sources and can be of different types) are its four main pillars (Fig. 7.1). Given such data, we are aimed at extracting *knowledge* from that (the *value* is often considered as the fifth pillar of big data).

The big data revolution affected many research fields, including statistics, machine learning, parallel computing, designing and implementing smart delivery systems, and computer systems in general [28]. Conveniently, the explosion in the amount of generated data is coupled with boosted hardware processing abilities and increased storage. This unprecedented availability of data and computing resources allows us to design data-driven methods, which benefit from *experience* extracted from such historical and/or incoming data [80].

7.1.1 A gentle introduction to machine learning

Given the available computing power and the amount of generated data, the role of computers in science and engineering is being actively altered and enhanced. Learning is the most important hallmark of human intelligence, and it is equally valid in a subfield of artificial intelligence referred to as *machine learning*. Storing and analyzing the acquired historical information should allow us to predict the label of an incoming *feature vector,* which represents a real-life entity and is composed of a number of numerical and/or categorical features describing this entity. If the dependent variable (a label) is categorical, then this is

Smart Delivery Systems. https://doi.org/10.1016/B978-0-12-815715-2.00013-0

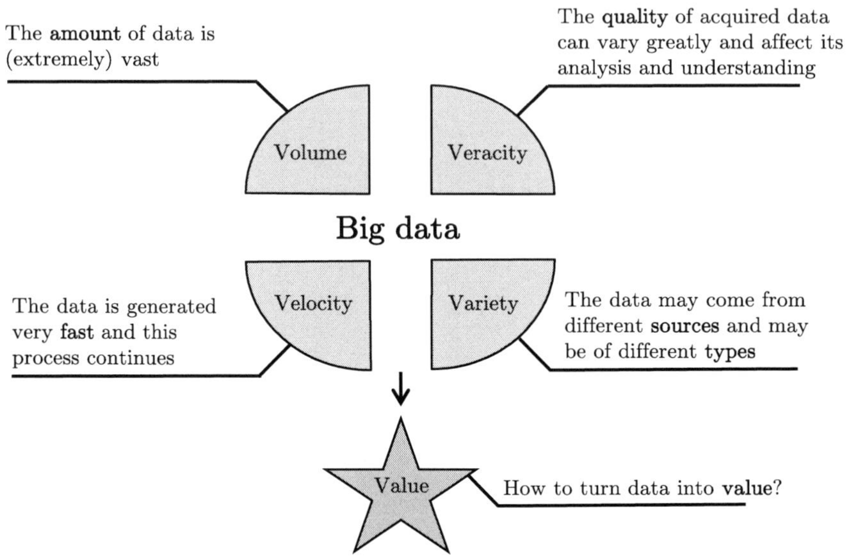

FIGURE 7.1 How to turn data into *knowledge* (value)? In the current settings, we generate a tremendous amount (volume) of data every second (velocity), it may come from different sources and may be of different types (variety), and the quality of some of its pieces may be questionable (veracity). Therefore, modern artificial intelligence (and machine learning in particular) algorithms must effectively cope with such massive data collections.

the *classification* task. Otherwise—if the labels are numerical—it is a *regression* problem. Classification may be further divided into the *binary* (two-class) classification (only two classes of vectors exist, usually denoted as the "positive" and "negative" classes; see Fig. 7.2) and the multiclass classification (the number of classes defined by the corresponding labels is greater than two). To exploit the already-seen data, the representative *model* of that information is built in the process of *learning*. The model is then utilized for either classification or regression. The learning concept encompasses *unsupervised* and *supervised* learning.

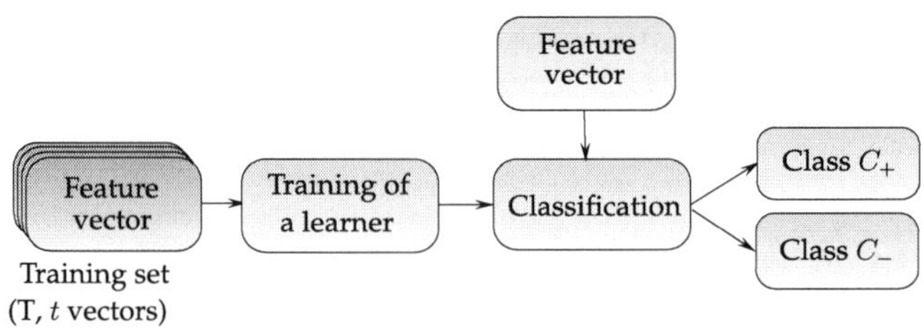

FIGURE 7.2 In supervised learning, we perform a training of a learner (based on *experience*, i.e., a set of training data with the corresponding class labels) to perform the classification of incoming (unseen) examples. Here we can appreciate a binary classification task, where only two classes (the "positive" class C_+ and the "negative" class C_-) are considered.

The unsupervised learning techniques aim at finding an internal (usually hidden) "structure" of the unlabeled data. The most representative examples of these approaches include clustering and blind signal separation (using, e.g., principal component analysis [32], singular value decomposition [85], independent component analysis [30], and many others [49]). On the other hand, in supervised learning algorithms the data examples are labeled and based on those labels the model is generated to predict the labels of the incoming feature vectors. They include a plethora of techniques, including—among others—artificial neural networks [27], nearest-neighbor algorithms [11], random forests [5], decision trees [74], various kernel methods [79], and support vector machines [10].

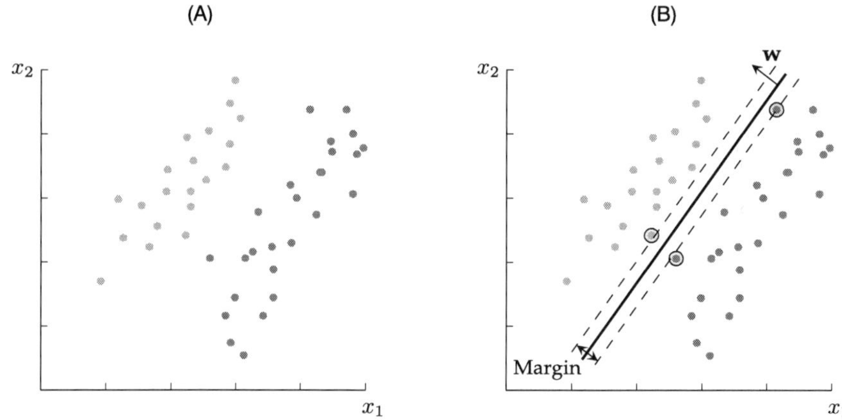

FIGURE 7.3 Example of determining a decision hyperplane (using a support vector machine) for a simple two-dimensional (i.e., each example is characterized by two features, x_1 and x_2) linearly separable classification case (as already mentioned, feature vectors encompass two features, x_1 and x_2): (A) blue (mid gray in print version) and red (dark gray in print version) dots visualize vectors belonging to two classes; (B) three vectors are annotated as support vectors and are used to determine the position of the SVM hyperplane. This figure is inspired by an excellent tutorial by Burges [6]; we highly recommend it for the readers interested in support vector machines and pattern recognition in general.

An example of a binary classification task is presented in Fig. 7.3. The examples belonging to two classes (rendered as blue (mid gray in print version) and red (dark gray in print version) dots) are classified by a support vector machine using a decision hyperplane; the margin between these classes is maximized during the support vector machine training process (three examples from the training set, encircled in black, constitute *support vectors*, which determine the position of the decision hyperplane). Finally, the decision hyperplane is exploited to classify the incoming (previously unseen) test examples.

In Fig. 7.4, we visualize an end-to-end, real-life binary-classification example, where we consider a problem of tumor detection from computed tomography (CT) images. In *conventional* machine learning the features quantifying the objects (in this case, CT images) have to be extracted manually in the process of *feature engineering*. Such features can encompass, among others, the intensity histogram or textural features (elaborated for each image), and their number can grow fairly fast. To deal with that, we often apply feature selection algorithms to find the most informative features (i.e., those that convey the most important information about the underlying objects). An obvious drawback of such conventional machine learning approaches that involve hand-crafted features is the necessity of designing feature extractors appropriate for a given task. This process is very user dependent and prone to errors. Also, it does not allow us to build data representation that encompasses features unknown for humans (and should be automatically *learnt*).

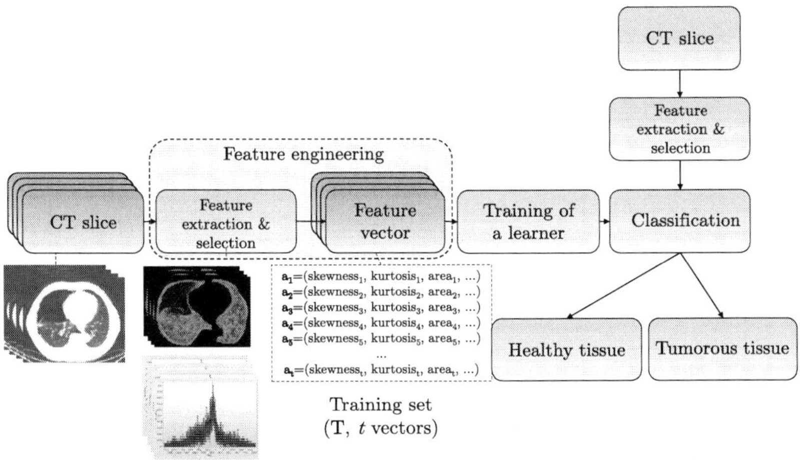

FIGURE 7.4 Example of an end-to-end classical classification example of healthy and tumorous computed-tomography scans. Here the features are extracted and selected in the *feature engineering* process (such features extracted for medical images may include, among others, intensity, shape, histogram, or textural characteristics). Then a supervised learner is trained and used to classify the incoming examples for which the same set of features is extracted.

Ensemble methods—which train multiple different *classifiers* (called the *base* classifiers) to tackle the same problem—combine these components into a single answer of the ensemble system [42]. These classifiers may be either *homogeneous* (they are of the same type) or *heterogeneous*, employing the classifiers of various types. It is worth noting that ensemble methods are able to significantly boost the performance of weak base classifiers (very often the performance of base classifiers is close to random guessing) and to provide high-quality classification [43].

The ensembles are usually divided into three types: (i) the combinations of classifiers that perform quite well on their own, (ii) ensembles of weak classifiers, and (iii) mixtures of experts (in which the parametric models are learnt and combined into a single answer following the divide-and-conquer strategy).

Although an arbitrarily large number of base classifiers can be applied in the ensembles, it is common to prune them and select best-performing and diverse base classifiers (to decrease the memory requirements and the computational time)—it became a vital issue nowadays [44,54,93]. An excellent book on ensemble methods has been published by Zhou [102]. Finally, the *No-Free-Lunch* theorem is worth mentioning. It says that it is not possible to find a learning algorithm that outperforms the others in *all* problems [94]. However, it considers the entire problem space, which is not valid in practice, where the learning algorithm is usually selected and tailored to a specific task. In this case, finding the best-performing and robust to handle any kind of data learning algorithm is perfectly valid and desirable [102].

The availability of processing power (especially graphics processing units) led to the development of *deep learning* techniques [46]. Such deep models, which are commonly composed of a number of processing layers of different types (e.g., convolutions or pooling), are able to automatically determine (i.e., to *learn*) representation of an input data at different levels of abstractions. Introducing deep neural networks (DNNs) was undoubtedly a breakthrough in artificial intelligence. DNNs have achieved remarkable success in a plethora of tasks and established the state-of-the-art in various fields—their applications include (but are not limited to):

- Text classification [26],
- Temporal and time-series analysis [17],
- Self-driving cars[1] [75,37],
- Voice generation [50],
- Music composition [45],
- Real-time analysis of human behavior [25],
- Machine translation and natural language processing [53],
- Speech recognition [99],
- Language modeling [56],
- Document summarization [35],
- Segmentation of (various kinds of) images [18,86,78,64],
- Object detection and recognition [100],
- Hand-written text generation [3], and many more.

As already mentioned, DNNs are able to learn underlying data representation in an automatic way (hence, they automatically analyze the feature space) [8]. It allows us to redesign the medical-image classification system presented in Fig. 7.4 and replace manual feature engineering with automatic representation learning (Fig. 7.5). This approach not only makes the process of designing classification (or regression) engines much easier, but also allows for extracting features, which might have been omitted (or unknown) by practitioners. It, in turn, leads to training deep models, which commonly perform on par with humans [47,57].

1. For details on the current deep-learning advances in self-driving cars, see the NVIDIA website: https://www.nvidia.com/en-us/self-driving-cars/. Access date: March 27, 2019.

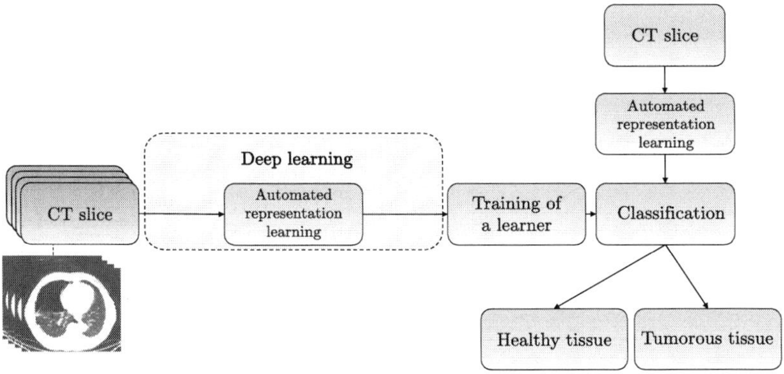

FIGURE 7.5 Example of an end-to-end deep learning-powered classification example of healthy and tumorous computed-tomography scans. We can observe that feature engineering has been replaced by automated representation learning.

In convolutional neural networks, features are extracted by the initial (e.g., convolutional and pooling) layers, and they are processed by fully connected layers to classify the data (or achieve other analysis goals). In an example rendered in Fig. 7.6, we present a basic convolutional neural network with three convolutional layers (this part of the DNN performs the feature extraction) and a fully connected network for classification (note that we did not include other types of the layers for brevity). It is interesting that the *deep features* extracted with DNNs [101] can be further selected using well-established methods and exploited by supervised learners (e.g., support vector machines). Poria et al. [71] proposed to use a deep convolutional neural net to extract visual features, which are selected using cyclic correlation feature subset selection or principal component analysis and classified with support vector machines. In the work by Sun et al. [81], the forward–backward greedy algorithm was exploited to select the deep features for face verification. Nezhad et al. [66] analyzed the demographic data for assessing the health risk factors. The features extracted using stacked autoencoders are selected with random forests and are subject to supervised learning. A different research direction is to exploit deep architectures, especially deep belief networks, for selecting the features, for example, in genomics [51] and remote sensing [103]. Kowaliw et al. [40] proposed an evolutionary deep feature extractor, which exploits genetic programming. Finally, Nalepa et al. [63] showed that the subset of deep features can be effectively evolved using a memetic algorithm (being a hybrid of an evolutionary method and various local-search procedures, which intensify the exploitation of particular regions of the entire solution space).

The literature on deep learning is growing extremely fast,[2] and there are new deep-network architectures and new deep-learning applications proposed daily.

2. Google Scholar reports almost 16,000 papers and patents (March 27, 2019) with a "deep learning" keyword in their titles and/or abstracts published just in 2019.

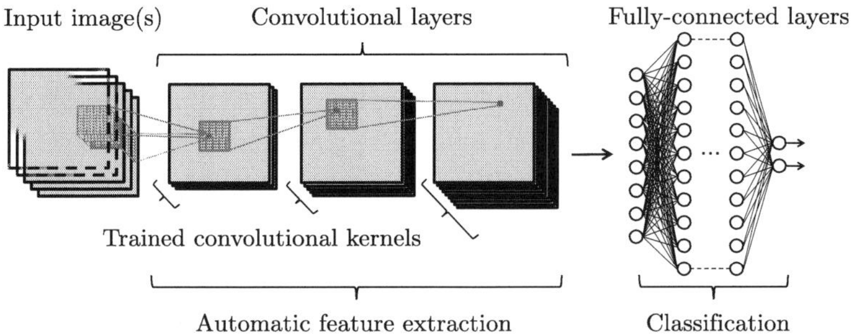

FIGURE 7.6 Example of a basic deep neural network architecture, in which the automatic (deep) feature extraction is performed with three convolutional layers, whereas the classification is performed using a fully connected network.

For a good introduction into the field, we recommend an excellent review paper by LeCun et al. [47], alongside a textbook by Goodfellow et al. [19],[3] which provides a comprehensive background not only in deep learning, but in machine learning in general.

7.1.2 Where machine learning meets smart delivery systems – an overview

In this book, we already showed that there exist various techniques for solving rich vehicle routing problems, ranging from exact to heuristic, metaheuristic, and hyperheuristic approaches (both sequential and parallel). Such techniques for solving discrete optimization problems (not only concerning smart delivery systems) are being actively enhanced by machine learning algorithms, which benefit from the available (or acquired) data. In general, machine learning can be applied in the context of smart delivery systems in the following context:

1. **Tuning hyper-parameters of existent algorithms for solving rich vehicle routing problems using machine learning**. Virtually all (meta/hyper)-heuristics for solving routing problems are heavily parameterized and require performing the process of hyper-parameters tuning. It is important that different sets of hyper-parameters are often selected for different problem instances (e.g., tests containing clustered vs. randomized customers to be served in a pickup and delivery problem). Machine learning can be effectively utilized to not only help find appropriate hyper-parameter values for a task at hand, but also to select them according to the underlying test instance characteristics.
2. **Solving rich vehicle routing problems using hybrid algorithms that exploit machine learning**. Guiding the search into the desired parts of the solution space is not an easy task by any means. We can therefore observe

3. See also the website: http://www.deeplearningbook.org/. Access date: March 27, 2019.

a new trend in the literature, in which machine learning techniques are hybridized with other algorithms for tackling discrete optimization problems. Incorporating (un)supervised learning into heuristics for rich vehicle routing problems offers new opportunities in the exploration/exploitation context (e.g., the search may be guided based on the historical information acquired during the optimization process, which is then transformed into useful knowledge, e.g., about the landscape of the solution space). The hybrid algorithms can also exploit various predictors and/or simulators that help better respond to the inherently dynamic changes in the transportation networks.

3. **Solving rich vehicle routing problems using data-driven machine learning algorithms**. The exceptional abilities of learning from experience, which are key in machine learning, allowed for initiating a new research pathway in combinatorial optimization, where the problems are entirely solved by data-driven algorithms. This exciting research area is still relatively fresh; however, the already proposed techniques were shown to be extremely powerful, and we anticipate—due to the amount of available data and the complexity of modern smart delivery systems—that it will be blooming in the nearest future.

This chapter provides an overview of the data-driven algorithms applied for solving rich vehicle routing problems (hence designing effective smart delivery systems). We review the most exciting and important approaches belonging to all of the aforementioned categories. Although we did our best to highlight the most important works in the literature, we realize that the field is developing extremely fast and the list of the approaches discussed in this chapter will be outdated fairly soon. However, we believe that it will provide a concise summary of the machine learning techniques applied in the optimization tasks and will help understand the benefits of hybridizing such approaches.

7.1.3 Structure of this chapter

The rest of this chapter is structured as follows. In Section 7.2, we show how machine learning techniques can be applied to optimize hyper-parameters of various algorithms for solving vehicle routing problems. The approaches which hybridize machine learning and (meta)heuristics are discussed in Section 7.3. The exciting data-driven algorithms for designing delivery systems in an end-to-end manner are presented in Section 7.4. Finally, Section 7.5 summarizes this chapter.

7.2 Tuning hyper-parameters of existent algorithms for solving rich vehicle routing problems using machine learning

Approximate methods for solving discrete optimization problems are often heavily parameterized, and the values of such hyper-parameters can drastically affect their performance [77]. Therefore, they should be either carefully *tuned* before the optimization or *adapted* during the search. Additionally, responding appropriately to the current optimization phase may help improve the convergence capabilities of (meta)-heuristics, especially biologically-inspired techniques [59]. Since various approximate methods for tackling rich routing problems are randomized, they should be executed multiple times to understand their performance. Thus, improperly selected hyper-parameter values drastically increase their execution time which constitutes their serious drawback that should be endured in practice. To deal with this problem, metaheuristics are commonly coupled with the process of their parameter tuning or *controlling* (adapting). In the former approach the space of hyper-parameter values is (grid) searched before the optimization in a very time-consuming process. It must be noted that most of the tuning techniques optimize one hyper-parameter at the time, which often leads to suboptimal choices, because such hyper-parameters are rarely independent [59,70].

On the other hand, the control algorithms are aimed at adapting the hyper-parameters on the fly (e.g., evolutionary approaches that utilize such techniques are often referred to as *dynamic* evolutionary algorithms). According to Eiben et al. [14], we can divide control metaheuristics (in this case, evolutionary approaches) based on the following criteria:

- **What** is being updated (i.e., which hyper-parameters).
- **How** this update is performed (is it self-adaptation? Is any feedback from the running algorithm involved in the update scheme? Is the update process deterministic and nailed down before the actual execution?).
- What is the **scope** of the change (is the entire population of solutions adapted? Is it a single individual in the population?).
- What is the **evidence** that the updating process should be initiated (is it deterministic, e.g., is it executed every *nth* generation? Is the optimization progress traced by any means? Does the lack of diversity in the population trigger the update?).

Machine learning algorithms are currently being incorporated into various heuristics to tune and/or control their hyper-parameters. As already mentioned, the characteristics of a problem instance may strongly influence the choice of appropriate hyper-parameters (or even a variant of the algorithm to be applied; some algorithmic components might be useful in certain types of the instances, e.g., encompassing customers clustered in several locations on the map [60]).

TABLE 7.1 Overview of six classes of the PDPTW problem instances.

Characteristic ↓	c1	c2	r1	r2	rc1	rc2
Structure	clustered	clustered	random	random	mixed	mixed
Time windows	tight	wide	tight	wide	tight	wide
Truck capacity	small	large	small	large	small	large

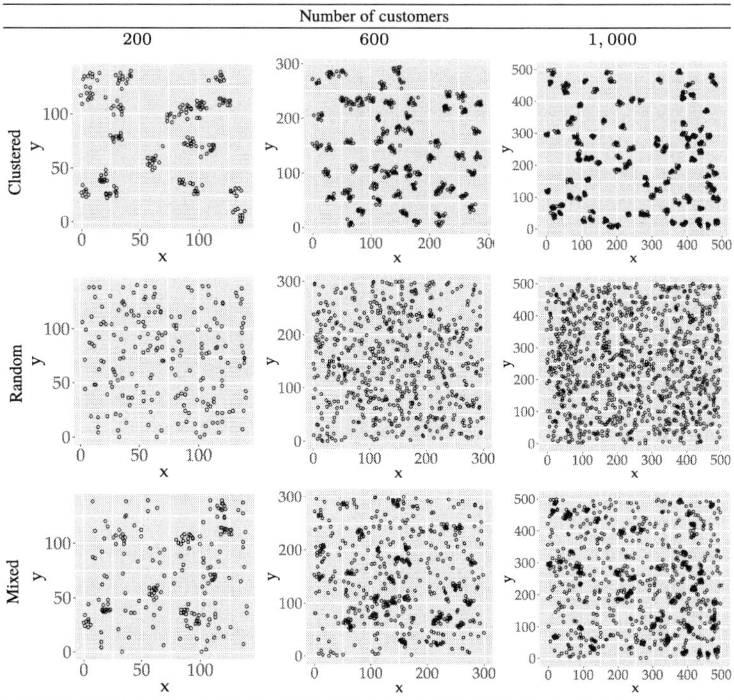

FIGURE 7.7 Example benchmark instances (with different numbers of travel points, ranging from 200 to 1000) encompassing clustered, randomized, and mixed customers. This figure is inspired by Nalepa and Blocho [61].

The underlying structure of benchmark tests for the pickup and delivery problem with time windows (PDPTW) was investigated by Nalepa and Blocho [61] to select the best variant of a guided ejection search for minimizing the number of routes in PDPTW. In the preprocessing step the authors executed the clustering and histogram-based analysis to extract features quantifying different aspects of the problem. Then the k-nearest neighbor algorithm was applied to classify this instance into one (out of six) class. These classes are gathered in Table 7.1. We can observe that they are different with respect to the problem structure, that is, the positions of the customers: clustered, random, and mixed (see the example visualizations of travel point positions rendered in Fig. 7.7), the width of time windows: tight vs. wide, and the truck capacity: small vs. large.

The structural information concerning the customer positioning was extracted using both hierarchical and nonhierarchical clustering with several aggregation (linkage) techniques (including, among others k-means, McQuitty and Ward), alongside 27 clustering validity indices[4] (exploited to find the number of

4. Such indices combine the information about the intercluster isolation with the information on the intracluster compactness.

travel-point clusters "hidden" in the analyzed problem instance), alongside the affinity propagation clustering [15]. The clustering-based analysis was coupled with the histogram-based feature extraction, in which 2D histograms show the number of travel points in various parts of the map (this map is divided into smaller cells, as presented in Table 7.2; this table is inspired by Nalepa and Blocho [61,62]).

Given the histograms of travel-points locations (quantifying the numbers of customers in each "cell" at different scales), the authors extracted various histogram features (including skewness, kurtosis, standard deviation, and entropy). Finally, the authors calculated the minimal number of trucks that can feasibly serve all transportation requests. All of the aforementioned features formed feature vectors, which were classified using the k-nearest-neighbor algorithm, and—based on the outcome of this classification—the appropriate guided ejection search variant was selected. The experiments (performed over 354 Li and Lim's problem instances in a 10-fold cross-validation setup) showed that this approach allowed for obtaining perfect classification (when all features were exploited) and hence allowed for determining the best optimization technique for a problem instance at hand.

A very similar research pathway was followed by Rasku et al. [76], who extracted a number of instance descriptors for vehicle routing problems and applied them for instance-specific algorithm configuration. For capacitated vehicle routing problems with time windows, the features encompassed node (travel-point) distribution features (e.g., the distance to the centroid, normalized cluster sizes, silhouette coefficients, or the number of outlier travel points), minimum spanning tree features (e.g., the edge cost and node degree), local-search probing features (the number of local-search steps to the local minimum, intraroute intersections, or the autocorrelation length), branch-and-cut probing features (e.g., an improvement per added cut), geometric features (e.g., squareness, i.e., the area of an enclosing rectangle or the total length of the convex-hull edges), nearest-neighbor features (e.g., the distance to the nearest neighbor, cosine similarity between the edges to two nearest neighbors, or the node input degree in the k-nearest-neighbor graph), and those that are specific to the problem characteristics (e.g., the number of travel points, location of the depot, ratio between the largest demand and the truck capacity, and so forth). Additionally, the authors exploited principal component analysis to deal with a fairly large number of extracted features (seven principal components were exploited instead of 386 extracted features). The experiments showed that every resulting hyper-parameter configuration was superior to the default one (the authors optimized various heuristics, including simulated annealing and the ejection metaheuristics).

Cooray and Rupasinghe [9] exploited machine learning to tune the hyper-parameters of their genetic algorithm for solving an energy-minimizing vehicle routing problem (EMVRP), in which the energy consumption (the product of the traveled distance and the load of a vehicle) is to be minimized as well. The authors considered the mutation rate, size of the population, and the number of

TABLE 7.2 Extracting histogram features: (A), (C) 2D histograms at various scales, (B), (D) histograms with the numbers of "cells" with different numbers of customers.

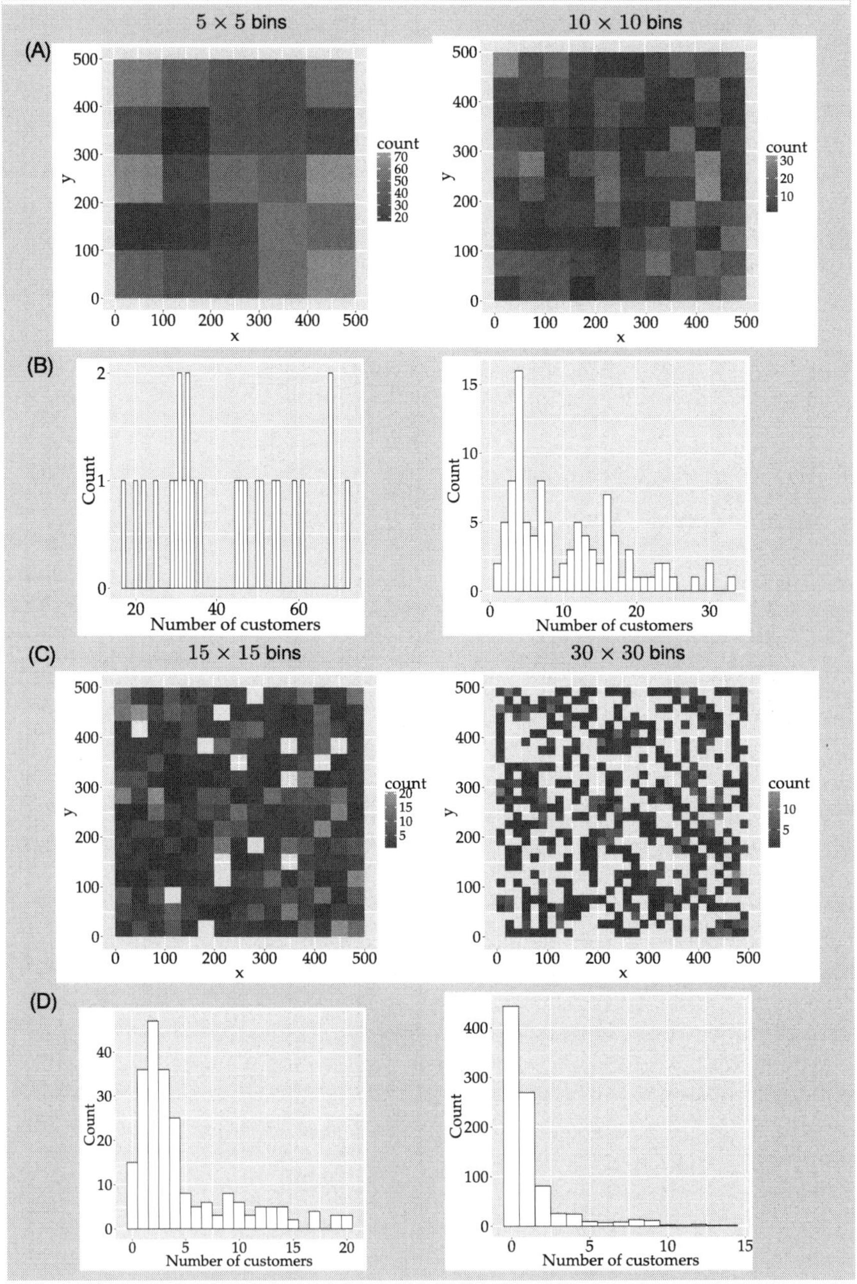

generations processed during the evolution (however, only mutation rate was controlled). These hyper-parameters were updated using a very basic version of k-means clustering (exploiting the total accumulated demand and the total number of travel points to be visited). Once the travel points are clustered, different mutation rates were applied to separate clusters. This simple machine learning-powered approach for selecting mutation rates showed potential, and the resulting instances of a genetic algorithm were competitive and could improve the performance of the baseline genetic algorithm.

7.3 Solving rich vehicle routing problems using hybrid algorithms that exploit machine learning

Designing hybrid algorithms that couple (meta/hyper)heuristics with artificial intelligence is not new. Potvin et al. [72] combined a neural network with a genetic algorithm to improve the initialization and construction phase of a parallel insertion heuristics for the vehicle routing problem with time windows almost 25 years ago.[5] In their approach a neural network is utilized to identify the seed customers (which are distributed across the map). These customers are used during the initialization, and a genetic algorithm is aimed at optimizing the hyper-parameters of the route-construction phase that follows the initialization. The authors observed that the customers located (clustered) in the same geographical area should be served by the same vehicle. To cluster the travel points, a competitive neural network was used, and for each cluster, a seed customer is selected (it is then exploited in an insertion heuristics, where each route serves a single customer at the beginning). Otherwise, if the structure of the problem instance does not reveal any travel-point groups, then a customer that is widely apart from the others is annotated as the seed customer (the motivation is to cover the entire geographic area and to create routes for different geographic regions). Once the seed points are determined, a genetic algorithm is executed to optimize the hyper-parameters of the parallel insertion heuristics. The experiments (over the Solomon's benchmark) indicated that a combination of a neural network, metaheuristics, and the insertion heuristics helped improve the quality of the final routing schedules (the numbers of routes were decreased).

Phiboonbanakit et al. [69] exploited a similar idea of grouping customers into clusters and aimed at using an unsupervised algorithm for classifying travel points into specific geographical regions on the map in the vehicle routing problem with time windows. The authors used two separate *agents* (coordinate and route agents), where one agent transfers its knowledge to the other one. Thanks to such a knowledge transfer, the optimization algorithm may use the information learnt from past experience to better adapt to changes in routing conditions, such as customer demands, possible traffic incidents, or varying travel times. Then the routes are determined within their known subregion of the map (the

5. Note that Hopfield networks were applied for optimizing decisions in combinatorial optimization problems back in 1985.

authors used k-means clustering to elaborate them), and the routes are later combined into the final solution. Although the experimental study reported in this work was fairly limited (a very small number of test instances were processed), it was observable that the machine learning-based subregion determination delivered improvement in the final quality of routing schedules when compared with those obtained using a classical genetic algorithm.

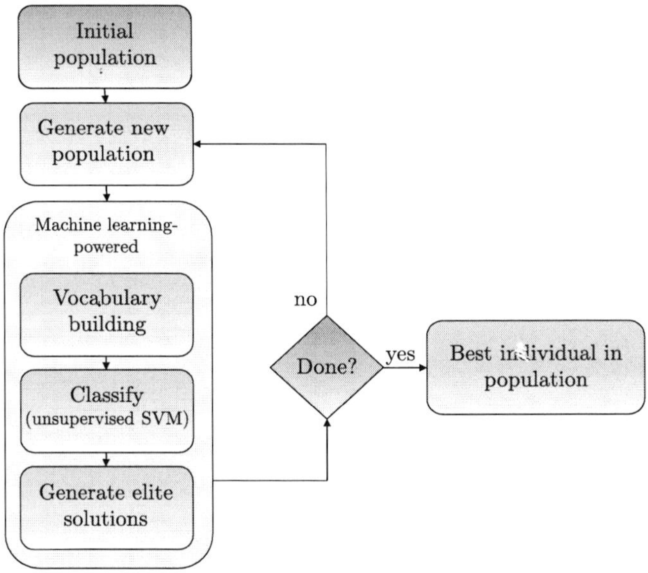

FIGURE 7.8 Zennaki and Ech-cherif [98] enhanced a population-based metaheuristics (a genetic algorithm) with kernel clustering to guide the optimization process.

Zennaki and Ech-cherif [98] investigated the possibility of using kernel clustering techniques coupled with data fusion methods for solving challenging combinatorial optimization problems (in this work the authors tackled the vehicle routing problem with time windows). Such kernel-based algorithms were incorporated into population-based metaheuristics that rely on spatial fusion of solutions. The ultimate aim of this combination was to better guide the search toward the promising parts of the solution space (which encompasses high-quality solutions). In the proposed algorithm (Fig. 7.8) a local-search heuristics is used to generate an initial population. Then the vocabulary of solutions is built (as in scatter search and path relinking), and kernel clustering of the vocabulary (in this case using unsupervised support vector machines) is executed to group similar solutions into clusters. Finally, to diversify the search, a set of elite solutions is generated—the authors used solution combination techniques known from the path relinking algorithms. The experiments (over a small subset of the Solomon's benchmark) indicated that the machine learning-powered population-based technique is competitive when compared with tabu search.

In modern smart delivery systems, transportation requests are no longer static and can be received over time and hence must be incorporated into the current routing schedule in real time. Hajjam et al. [24] identified the features that make such dynamic optimization problems different from their static counterparts and proposed a memetic algorithm in which a self-organizing map (SOM) neural network is incorporated into a population-based memetic algorithm. The authors benefit from the intermediate structure provided by the vanilla SOM and derive a set of operators to perform massive and distributed insertions of dynamic transportation requests (the objective was to simultaneously minimize the total traveled distance alongside the total waiting time of the customers). Therefore the SOM-elaborated operators enhance local search (which is also referred to as *education* in memetic algorithms). An extensive experimental study (performed for the traveling salesmen problem, vehicle routing problem, and vehicle routing problem with time windows) revealed that the hybrid algorithm outperformed other heuristics when the dynamic variant of the vehicle routing problems is considered. Importantly, the memetic SOM can be conveniently parallelized (at different granulation levels) and therefore can be potentially applied in massively large real-life transportation systems with dynamic transportation requests. For further reading on the dynamic smart delivery systems (i.e., those with dynamic inputs—the elements in either objective function or the constraints set which can vary in a predictable way) and the applications of machine learning for optimizing them, we refer to Calvet et al. [7].

Since the transportation requests in intelligent delivery systems (and the delivery systems as such) are dynamic by their nature [96], there exist approaches for predicting such dynamic events (i.e., when and where will they likely occur), based on the historical information (e.g., covered routes, GPS data, utilization of the vehicles in the fleet, and so forth). This acquired (or very often generated using various traffic simulations) knowledge can help build predictors, which, in turn, may allow for better allocation of the available vehicles and other resources. The other aspects of delivery systems which can be predicted based on the acquired knowledge include, among others:

- Travel times [82,58].
- Arrival and release times of the vehicles [16].
- Short-term travel behavior (e.g., short-term travel destinations of the vehicles within the fleet) [41].
- Vehicle routes (which may help predict various traffic situations and events directly) [34,97,2].
- Vehicle state (position, velocity, and acceleration) [52].
- Distribution of the transportation demands and requests [83,95].
- Traffic congestion [92].
- Energy consumption of electric vehicles [36,1,87].

Kim et al. [38] proposed a dynamic vehicular routing algorithm and coupled it with traffic prediction (the authors aimed at exploiting both real-time and predictive traffic information to optimize their delivery system). A microtraffic simulator was additionally introduced to investigate the performance of the dynamic routing with prediction. The authors observed that a standard metric—mean travel time (being the average time taken for all vehicles to move from the origin to the destination)—used to quantify the performance of optimization algorithms in the context of dynamic routing may be insufficient to model the preferences of individual drivers and proposed a new metric that captures the travel time distribution of all vehicles. The experiments showed that the predictive information decreases travel times and improves the road efficiency.

Traffic simulation has become an important research topic recently, since the traffic congestion and jams are a real-life problem and the number of vehicles is constantly growing, especially in agglomerations [20]. There exist numerous traffic simulators (often accompanied with various visualization tools), which help understand the real impact of traffic simulated in different parts of the transportation network. Note that the acquisition of the domain knowledge, which might be used for effective simulation, is an important research task as well. Wasilewski and Gora [89] treated traffic as a complex process having a hierarchical structure and showed that the expert knowledge is indispensable for building well-performing classifiers and predictors for such processes. The current trend in the literature involves exploiting machine learning-based techniques (often coupled with metaheuristics) for traffic simulation. Gora [21] introduced a traffic simulator that couples evolutionary algorithms and neural networks and experimentally showed that it can be useful for autonomous vehicles too. Neural networks were also used for approximating traffic simulation outcomes, such as the total waiting time. Gora and Bardonski [22] proved that their tensor-based traffic simulator[6] operates orders of magnitude faster when compared with microscopic traffic models (and that it can be trained even using a relatively small training set). Finally, Gora et al. [23] introduced a very interesting hybrid approach, in which they combined neural networks with gradient boosting models, which are later utilized as fitness functions in evolutionary algorithms (here, genetic techniques) for traffic optimization.

7.4 Solving rich vehicle routing problems using data-driven machine learning algorithms

The current deep learning advancements opened new exciting possibilities of solving complex transportation problems in a fully data-driven manner. It would ultimately allow us to replace hand-crafted (meta)heuristics designed for specific discrete optimization problems with techniques that directly operate in the solution space [4,31]. The introduction of *pointer networks* that use the attention modules to generate a permutation of an input sequence and return it as an output may be considered as a breakthrough in solving combinatorial optimization problems using deep learning [84]. The pointer networks—as their name

6. For more details, see http://tensortraffic.com/. Access date: March 29, 2019.

indicates—learn the *pointers* to positions in an input sequence and exploit attention as a pointer to select a member of the input. It is remarkable that such networks are able to generalize to test problems having more points than in the training problems (a good example where this property is extremely useful is solving the traveling salesmen problem, in which the number of travel points varies from test to test). The authors considered the planar symmetric traveling salesmen problem and showed that it can be effectively solved in a pure data-driven fashion using pointer networks (upon test time, the beam search filters out invalid tours). Nazari et al. [65] replaced the long short-term memory encoder utilized in a pointer network with elementwise projections and applied the modified model to tackle the vehicle routing problem with split deliveries (a single model finds near-optimal solutions for problem instances sampled from a given distribution, only by observing the reward signals and by following certain feasibility rules).

In a very recent work, Kool et al. [39] proposed a new model based on attention layers, which benefits from the pointer networks. It is important that the authors trained this model with the REINFORCE approach (using effective greedy rollout baseline) [91]. Also, Kool et al. [39] clearly stated that the motivation is not to outperform highly specialized heuristics for solving specific rich routing problems, but rather show the flexibility of data-driven approaches in tackling such discrete optimization problems. An extensive experimental study (encompassing different routing problems: the vehicle routing problem, the orienteering problem, and a stochastic variant of the prize collecting traveling salesmen problem) revealed that the elaborated data-driven heuristics are extremely competitive—they outperformed the baseline solutions and got very close to the highly optimized (and fairly specialized) solvers. Independently of the work by Kool et al. [39], Deudon et al. [13] presented an attention-based model[7] for solving the traveling salesmen problem to the operations research community. In their approach the travel-point coordinates are fed into the neural network (trained using reinforcement learning), and the prediction over the travel-point permutations is generated. Additionally, this technique was enhanced with a standard 2-Opt heuristics. Similarly, the experimental results (over the traveling salesmen problem test instances) reported in the paper are very promising and clearly show that the machine learning techniques can compete with specialized heuristics known from the operational research literature.

Other interesting approaches for solving rich vehicle routing problems include exploiting the models based on graph embeddings [12] and graph neural networks [67,68]. In the latter approach the deep model outputs an adjacency matrix corresponding to a route. Finally, Kaempfer and Wolf [33] tackled the multiple traveling salesmen problem in a data-driven manner. The authors encoded the travel points, depot, and the salesmen as three independent sets (their

7. The code has been made available at GitHub: https://github.com/MichelDeudon/encode-attend-navigate. Access date: March 29, 2019.

approach is actually invariant to their order) and generalized the PointNet deep architecture [73] for solving the problems that involve multiple sets. The authors experimentally proved that their deep architecture can be competitive with sophisticated heuristics crafted for this variant of the routing problem (and indicated their future plans, which involve applying this method to other versions of the multiple traveling salesmen problem, especially with the time-windows constraints).

Overall, the literature concerning the data-driven machine learning-powered algorithms for solving rich vehicle routing problems is blooming, and we anticipate that this trend will continue. Although there are still open issues that require further investigation (especially concerning sophisticated constraints that are relevant to complex variants of routing problems), the already-published papers clearly show that learning from experience and examples is extremely powerful in the context of solving discrete optimization problems—note that these techniques may be applied to other problems in the operations research field as well [29,48,90,55,88].

7.5 Summary

We witness the breakthroughs in virtually all fields of science and engineering due to the emerging data-driven algorithms benefitting from experience (i.e., acquired and data-mined information). The big data revolution already changed the way of solving (and thinking about) various tasks (especially in image analysis, computer vision, and pattern recognition) by making enormously large, heterogeneous, and constantly increasing datasets available for processing, analysis, and ultimately understanding by the extraction of useful knowledge from them. This knowledge is later used to elaborate high-quality solutions (possibly in shorter time) when compared with those obtained using hand-crafted approximate algorithms.

Designing smart delivery systems is not an exception here, and the data-driven techniques are being actively introduced at various levels of optimization of such systems. Since rich vehicle routing problems are known from their NP-hardness, this trend is not surprising at all, and we can anticipate that the machine learning-powered algorithms will become the mainstream in the operations research field.

In this chapter, we reviewed the data-driven approaches applied in the context of smart delivery systems. We divided all techniques into three groups, encompassing the algorithms which are aimed at:

- Tuning hyper-parameters of existent algorithms for solving rich vehicle routing problems using machine learning.
- Solving rich vehicle routing problems using hybrid algorithms that exploit machine learning.
- Solving rich vehicle routing problems using data-driven machine learning algorithms.

Since the field is developing at an extreme speed, the list of the machine learning techniques discussed in this chapter will be outdated before the book is actually published. However, we believe that this chapter provides a concise overview of the data-driven algorithms applied in smart delivery systems and sheds more light on the areas that may benefit from such approaches in the near future.

Acknowledgments

This work was supported by the Silesian University of Technology Grant for young researchers (BKM-556/RAU2/2018).

References

[1] R. Abousleiman, O. Rawashdeh, Energy consumption model of an electric vehicle, in: 2015 IEEE Transportation Electrification Conference and Expo, ITEC, June 2015, pp. 1–5.
[2] A.T. Akabane, R.W. Pazzi, E.R.M. Madeira, L.A. Villas, Modeling and prediction of vehicle routes based on hidden Markov model, in: 2017 IEEE 86th Vehicular Technology Conference, VTC-Fall, Sep. 2017, pp. 1–5.
[3] E. Aksan, F. Pece, O. Hilliges, Deepwriting: making digital ink editable via deep generative modeling, CoRR, arXiv:1801.08379, 2018.
[4] I. Bello, H. Pham, Q.V. Le, M. Norouzi, S. Bengio, Neural combinatorial optimization with reinforcement learning, CoRR, arXiv:1611.09940, 2016.
[5] L. Breiman, Random forests, Machine Learning 45 (1) (2001) 5–32, https://doi.org/10.1023/A:1010933404324.
[6] C.J. Burges, A tutorial on support vector machines for pattern recognition, Data Mining and Knowledge Discovery 2 (2) (1998) 121–167.
[7] L. Calvet, J. Armas, D. Masip, J. Angel, Learnheuristics: hybridizing metaheuristics with machine learning for optimization with dynamic inputs, Open Mathematics 15 (2017) 261–280, https://doi.org/10.1515/math-2017-0029.
[8] G.B. Cavallari, L.S.F. Ribeiro, M.A. Ponti, Unsupervised representation learning using convolutional and stacked auto-encoders: a domain and cross-domain feature space analysis, CoRR, arXiv:1811.00473, 2018.
[9] P. Cooray, T. Rupasinghe, Machine learning-based parameter tuned genetic algorithm for energy minimizing vehicle routing problem, Journal of Industrial Engineering (2017) 1–13, https://doi.org/10.1155/2017/3019523.
[10] C. Cortes, V. Vapnik, Support-vector networks, Machine Learning 20 (3) (1995) 273–297.
[11] T. Cover, P. Hart, Nearest neighbor pattern classification, Information Theory, IEEE Transactions 13 (1) (2006) 21–27, https://doi.org/10.1109/TIT.1967.1053964.
[12] H. Dai, E.B. Khalil, Y. Zhang, B. Dilkina, L. Song, Learning combinatorial optimization algorithms over graphs, CoRR, arXiv:1704.01665, 2017.
[13] M. Deudon, P. Cournut, A. Lacoste, Y. Adulyasak, L.-M. Rousseau, Learning heuristics for the TSP by policy gradient, in: W.-J. van Hoeve (Ed.), Integration of Constraint Programming, Artificial Intelligence, and Operations Research, Springer International Publishing, Cham, 2018, pp. 170–181.
[14] A. Eiben, R. Hinterding, Z. Michalewicz, Parameter control in evolutionary algorithms, IEEE Transactions on Evolutionary Computation 3 (2) (1999) 124–141.
[15] B.J. Frey, D. Dueck, Clustering by passing messages between data points, Science 315 (5814) (2007) 972–976, http://science.sciencemag.org/content/315/5814/972.
[16] J. Fu, L. Wang, M. Pan, Z. Zuo, Q. Yang, Bus arrival time prediction and release: system, database and Android application design, in: X.-h. Sun, W. Qu, I. Stojmenovic, W. Zhou, Z.

Li, H. Guo, G. Min, T. Yang, Y. Wu, L. Liu (Eds.), Algorithms and Architectures for Parallel Processing, Springer International Publishing, Cham, 2014, pp. 404–416.

[17] J.C.B. Gamboa, Deep learning for time-series analysis, CoRR, arXiv:1701.01887, 2017.

[18] E. Gibson, W. Li, C. Sudre, L. Fidon, D.I. Shakir, G. Wang, Z. Eaton-Rosen, R. Gray, T. Doel, Y. Hu, T. Whyntie, P. Nachev, M. Modat, D.C. Barratt, S. Ourselin, M.J. Cardoso, T. Vercauteren, Niftynet: a deep-learning platform for medical imaging, Computer Methods and Programs in Biomedicine 158 (2018) 113–122, http://www.sciencedirect.com/science/article/pii/S0169260717311823.

[19] I. Goodfellow, Y. Bengio, A. Courville, Deep Learning, The MIT Press, 2016.

[20] P. Gora, Traffic simulation framework, in: 14th International Conference on Computer Modelling and Simulation, 2012 UKSim, Cambridge, United Kingdom, March 28–30, 2012, 2012, pp. 345–349.

[21] P. Gora, Simulation-based traffic management system for connected and autonomous vehicles, in: G. Meyer, S. Beiker (Eds.), Road Vehicle Automation 4, Springer International Publishing, Cham, 2018, pp. 257–266.

[22] P. Gora, M. Bardonski, Training neural networks to approximate traffic simulation outcomes, in: 2017 5th IEEE International Conference on Models and Technologies for Intelligent Transportation Systems, MT-ITS, June 2017, pp. 889–894.

[23] P. Gora, M. Brzeski, M. Mozejko, A. Klemenko, A. Kochanski, Investigating performance of neural networks and gradient boosting models approximating microscopic traffic simulations in traffic optimization tasks, CoRR, arXiv:1812.00401, 2018.

[24] A. Hajjam, J.-C. Créput, A. Koukam, From the TSP to the Dynamic VRP: An Application of Neural Networks in Population Based Metaheuristic, Springer, Berlin, Heidelberg, 2013, pp. 309–339.

[25] J.S. Hartford, J.R. Wright, K. Leyton-Brown, Deep learning for predicting human strategic behavior, in: D.D. Lee, M. Sugiyama, U.V. Luxburg, I. Guyon, R. Garnett (Eds.), Advances in Neural Information Processing Systems 29, Curran Associates, Inc., 2016, pp. 2424–2432.

[26] A. Hassan, A. Mahmood, Efficient deep learning model for text classification based on recurrent and convolutional layers, in: 2017 16th IEEE International Conference on Machine Learning and Applications, ICMLA, Dec 2017, pp. 1108–1113.

[27] M.H. Hassoun, Fundamentals of Artificial Neural Networks, 1st edition, MIT Press, Cambridge, MA, USA, 1995.

[28] S. Haykin, S. Wright, Y. Bengio, Big data: theoretical aspects [scanning the issue], Proceedings of the IEEE 104 (1) (Jan 2016) 8–10.

[29] M.R. Hilliard, G.E. Liepins, M. Palmer, Machine learning applications to job shop scheduling, in: Proceedings of the 1st International Conference on Industrial and Engineering Applications of Artificial Intelligence and Expert Systems, vol. 2, IEA/AIE'88, ACM, New York, NY, USA, 1988, pp. 728–737.

[30] A. Hyvärinen, E. Oja, Independent component analysis: algorithms and applications, Neural Networks 13 (4–5) (2000) 411–430, https://doi.org/10.1016/S0893-6080(00)00026-5.

[31] R. Jin, Deep learning at Alibaba, in: Proceedings of the 26th International Joint Conference on Artificial Intelligence, IJCAI'17, AAAI Press, 2017, pp. 11–16, http://dl.acm.org/citation.cfm?id=3171642.3171644.

[32] I. Jolliffe, Principal Component Analysis, John Wiley & Sons, Ltd, 2005, pp. 1–488.

[33] Y. Kaempfer, L. Wolf, Learning the multiple traveling salesmen problem with permutation invariant pooling networks, CoRR, arXiv:1803.09621, 2018.

[34] A. Karbassi, M. Barth, Vehicle route prediction and time of arrival estimation techniques for improved transportation system management, in: IEEE IV2003 Intelligent Vehicles Symposium, Proceedings (Cat. No. 03TH8683), June 2003, pp. 511–516.

[35] C. Khatri, G. Singh, N. Parikh, Abstractive and extractive text summarization using document context vector and recurrent neural networks, CoRR, arXiv:1807.08000, 2018.

[36] E. Kim, J. Lee, K.G. Shin, Real-time prediction of battery power requirements for electric vehicles, in: 2013 ACM/IEEE International Conference on Cyber-Physical Systems, ICCPS, April 2013, pp. 11–20.

[37] J. Kim, J.F. Canny, Interpretable learning for self-driving cars by visualizing causal attention, in: IEEE International Conference on Computer Vision, ICCV 2017, Venice, Italy, October 22–29, 2017, 2017, pp. 2961–2969.

[38] K. Kim, M. Kwon, J. Park, Y. Eun, Dynamic vehicular route guidance using traffic prediction information, Mobile Information Systems 2016 (2016) 3727865, https://doi.org/10.1155/2016/3727865.

[39] W. Kool, H. van Hoof, M. Welling, Attention, learn to solve routing problems!, CoRR, arXiv:1803.08475, 2018.

[40] T. Kowaliw, W. Banzhaf, R. Doursat, Networks of transform-based evolvable features for object recognition, in: Proc. GECCO, ACM, USA, 2013, pp. 1077–1084.

[41] C.M. Krause, L. Zhang, Short-term travel behavior prediction with GPS, land use, and point of interest data, Transportation Research Part B: Methodological (2018), http://www.sciencedirect.com/science/article/pii/S0191261518305800.

[42] B. Krawczyk, One-class classifier ensemble pruning and weighting with firefly algorithm, Neurocomputing 150, Part B (2015) 490–500, http://www.sciencedirect.com/science/article/pii/S0925231214012375.

[43] B. Krawczyk, M. Woźniak, Untrained method for ensemble pruning and weighted combination, in: Z. Zeng, Y. Li, I. King (Eds.), Advances in Neural Networks - ISNN 2014, in: Lecture Notes in Computer Science, vol. 8866, Springer International Publishing, 2014, pp. 358–365.

[44] B. Krawczyk, M. Woźniak, Pruning ensembles with cost constraints, in: N.T. Nguyen, B. Trawinski, R. Kosala (Eds.), Intelligent Information and Database Systems, in: Lecture Notes in Computer Science, vol. 9011, Springer International Publishing, 2015, pp. 503–512.

[45] H. Kumar, B. Ravindran, Polyphonic music composition with LSTM neural networks and reinforcement learning, CoRR, arXiv:1902.01973, 2019.

[46] Y. LeCun, Deep learning hardware: past, present, and future, in: IEEE International Solid-State Circuits Conference, ISSCC 2019, San Francisco, CA, USA, February 17–21, 2019, 2019, pp. 12–19.

[47] Y. LeCun, Y. Bengio, G.E. Hinton, Deep learning, Nature 521 (7553) (2015) 436–444, https://doi.org/10.1038/nature14539.

[48] C.-Y. Lee, S. Piramuthu, Y.-K. Tsai, Job shop scheduling with a genetic algorithm and machine learning, International Journal of Production Research 35 (4) (1997) 1171–1191, https://doi.org/10.1080/002075497195605.

[49] D.D. Lee, H.S. Seung, Algorithms for non-negative matrix factorization, in: Proceedings of the NIPS International Conference, MIT Press, 2000, pp. 556–562.

[50] Y. Lee, T. Kim, S. Lee, Voice imitating text-to-speech neural networks, CoRR, arXiv:1806.00927, 2018.

[51] Y. Li, C.-Y. Chen, W.W. Wasserman, Deep feature selection: theory and application to identify enhancers and promoters, Journal of Computational Biology 23 (5) (2016) 322–336.

[52] Y. Liu, D. Cheng, Y. Wang, J. Cheng, S. Gao, A novel method for predicting vehicle state in Internet of vehicles, Mobile Information Systems 2018 (2018) 9728328, https://doi.org/10.1155/2018/9728328.

[53] Y. Liu, J. Zhang, Deep Learning in Machine Translation, Springer, Singapore, 2018, pp. 147–183.

[54] R. Łysiak, M. Kurzynski, T. Wołoszynski, Optimal selection of ensemble classifiers using measures of competence and diversity of base classifiers, Neurocomputing 126 (2014) 29–35, http://www.sciencedirect.com/science/article/pii/S092523121300698X.

[55] Y. Martínez, A. Nowé, J. Suárez, R. Bello, A reinforcement learning approach for the flexible job shop scheduling problem, in: C.A.C. Coello (Ed.), Learning and Intelligent Optimization, Springer, Berlin, Heidelberg, 2011, pp. 253–262.

[56] T. Mikolov, M. Karafiát, L. Burget, J. Cernocký, S. Khudanpur, Recurrent neural network based language model, in: INTERSPEECH, 2010.

[57] M.R. Minar, J. Naher, Recent advances in deep learning: an overview, CoRR, arXiv:1807.08169, 2018.

[58] G. Musolino, A. Polimeni, C. Rindone, A. Vitetta, Travel Time Forecasting and Dynamic Routes Design for Emergency Vehicles, Procedia – Social and Behavioral Sciences, vol. 87, sIDT Scientific Seminar, 2013, pp. 193–202, http://www.sciencedirect.com/science/article/pii/S1877042813040482, 2012.

[59] J. Nalepa, M. Blocho, Adaptive memetic algorithm for minimizing distance in the vehicle routing problem with time windows, Soft Computing 20 (6) (Jun 2016) 2309–2327, https://doi.org/10.1007/s00500-015-1642-4.

[60] J. Nalepa, M. Blocho, Enhanced guided ejection search for the pickup and delivery problem with time windows, in: Intelligent Information and Database Systems - 8th Asian Conference, ACIIDS 2016, Da Nang, Vietnam, March 14–16, 2016, Proceedings, Part I, 2016, pp. 388–398.

[61] J. Nalepa, M. Blocho, Adaptive guided ejection search for pickup and delivery with time windows, Journal of Intelligent and Fuzzy Systems 32 (2) (2017) 1547–1559, https://doi.org/10.3233/JIFS-169149.

[62] J. Nalepa, M. Blocho, Parameter-less (meta)heuristics for vehicle routing problems, in: Proceedings of the Genetic and Evolutionary Computation Conference Companion, GECCO 2018, Kyoto, Japan, July 15–19, 2018, 2018, pp. 27–28.

[63] J. Nalepa, G. Mrukwa, M. Kawulok, Evolvable deep features, in: Applications of Evolutionary Computation – 21st International Conference EvoApplications 2018, Parma, Italy, April 4–6, 2018, Proceedings, 2018, pp. 497–505.

[64] J. Nalepa, M. Myller, M. Kawulok, Validating hyperspectral image segmentation, IEEE Geoscience and Remote Sensing Letters (2019) 1–5, https://doi.org/10.1109/LGRS.2019.2895697.

[65] M. Nazari, A. Oroojlooy, L.V. Snyder, M. Takác, Deep reinforcement learning for solving the vehicle routing problem, CoRR, arXiv:1802.04240, 2018.

[66] M.Z. Nezhad, D. Zhu, X. Li, K. Yang, P. Levy, Safs: a deep feature selection approach for precision medicine, in: Proc. IEEE BIBM, IEEE, 2016, pp. 501–506.

[67] A. Nowak, S. Villar, A.S. Bandeira, J. Bruna, A note on learning algorithms for quadratic assignment with graph neural networks, CoRR, arXiv:1706.07450, 2017.

[68] A. Nowak, S. Villar, A.S. Bandeira, J. Bruna, Revised note on learning quadratic assignment with graph neural networks, in: 2018 IEEE Data Science Workshop, DSW 2018, Lausanne, Switzerland, June 4–6, 2018, 2018, pp. 229–233.

[69] T. Phiboonbanakit, T. Horanont, T. Supnithi, V.-N. Huynh, Knowledge-based learning for solving vehicle routing problem, in: Proceedings of the 2018 ACM International Joint Conference and 2018 International Symposium on Pervasive and Ubiquitous Computing and Wearable Computers, UbiComp'18, ACM, New York, NY, USA, 2018, pp. 1103–1111.

[70] N. Pholdee, S. Bureerat, A comparative study of eighteen self-adaptive metaheuristic algorithms for truss sizing optimisation, KSCE Journal of Civil Engineering 22 (8) (Aug 2018) 2982–2993, https://doi.org/10.1007/s12205-017-0095-y.

[71] S. Poria, E. Cambria, A. Gelbukh, Deep convolutional neural network textual features and multiple kernel learning for utterance-level multimodal sentiment analysis, in: Proc. EMNLP, 2015, pp. 2539–2544.

[72] J.-Y. Potvin, D. Dubé, C. Robillard, A hybrid approach to vehicle routing using neural networks and genetic algorithms, Applied Intelligence 6 (3) (Jul 1996) 241–252, https://doi.org/10.1007/BF00126629.

[73] C.R. Qi, H. Su, K. Mo, L.J. Guibas, PointNet: deep learning on point sets for 3D classification and segmentation, CoRR, arXiv:1612.00593, 2016.

[74] J.R. Quinlan, Induction of decision trees, Machine Learning 1 (1) (1986) 81–106, https://doi.org/10.1023/A:1022643204877.

[75] S. Ramos, S.K. Gehrig, P. Pinggera, U. Franke, C. Rother, Detecting unexpected obstacles for self-driving cars: fusing deep learning and geometric modeling, in: IEEE Intelligent Vehicles Symposium, IV 2017, Los Angeles, CA, USA, June 11–14, 2017, 2017, pp. 1025–1032.

[76] J. Rasku, T. Kärkkäinen, N. Musliu, Feature extractors for describing vehicle routing problem instances, in: OASICS (50), 2016.
[77] J. Rasku, N. Musliu, T. Kärkkäinen, On automatic algorithm configuration of vehicle routing problem solvers, Journal on Vehicle Routing Algorithms (Feb 2019), https://doi.org/10.1007/s41604-019-00010-9.
[78] H.R. Roth, C. Shen, H. Oda, M. Oda, Y. Hayashi, K. Misawa, K. Mori, Deep learning and its application to medical image segmentation, CoRR, arXiv:1803.08691, 2018.
[79] J. Shawe-Taylor, N. Cristianini, Kernel Methods for Pattern Analysis, Cambridge University Press, New York, NY, USA, 2004.
[80] O. Simeone, A very brief introduction to machine learning with applications to communication systems, CoRR, arXiv:1808.02342, 2018, 1–20.
[81] Y. Sun, Y. Chen, X. Wang, X. Tang, Deep learning face representation by joint identification-verification, in: Proc. NIPS, 2014, pp. 1988–1996.
[82] D. Tiesyte, C.S. Jensen, Similarity-based prediction of travel times for vehicles traveling on known routes, in: Proceedings of the 16th ACM SIGSPATIAL International Conference on Advances in Geographic Information Systems, GIS'08, ACM, USA, 2008, pp. 14:1–14:10.
[83] N. Van Son, B. Babaki, A. Dries, P.Q. Dung, N. Xuan, Prediction-based optimization for online people and parcels share a ride taxis, in: 2017 9th International Conference on Knowledge and Systems Engineering, KSE, Oct 2017, pp. 42–47.
[84] O. Vinyals, M. Fortunato, N. Jaitly, Pointer networks, in: Advances in Neural Information Processing Systems 28: Annual Conference on Neural Information Processing Systems 2015, December 7–12, 2015, Montreal, Quebec, Canada, 2015, pp. 2692–2700, http://papers.nips.cc/paper/5866-pointer-networks.
[85] M. Wall, A. Rechtsteiner, L. Rocha, Singular value decomposition and principal component analysis, A Practical Approach to Microarray Data Analysis (2003) 91–109.
[86] G. Wang, W. Li, M.A. Zuluaga, R. Pratt, P.A. Patel, M. Aertsen, T. Doel, A.L. David, J. Deprest, S. Ourselin, T. Vercauteren, Interactive medical image segmentation using deep learning with image-specific fine tuning, IEEE Transactions on Medical Imaging 37 (7) (July 2018) 1562–1573.
[87] J. Wang, I. Besselink, H. Nijmeijer, Battery electric vehicle energy consumption prediction for a trip based on route information, Proceedings of the Institution of Mechanical Engineers, Part D: Journal of Automobile Engineering 232 (11) (2018) 1528–1542, https://doi.org/10.1177/0954407017729938.
[88] B. Waschneck, A. Reichstaller, L. Belzner, T. Altenmuller, T. Bauernhansl, A. Knapp, A. Kyek, Optimization of global production scheduling with deep reinforcement learning, Procedia CIRP 72 (2018) 1264–1269, 51st CIRP Conference on Manufacturing Systems, http://www.sciencedirect.com/science/article/pii/S221282711830372X.
[89] P. Wasilewski, P. Gora, Traffic-related knowledge acquired by interaction with experts, in: Proceedings of the 23th International Workshop on Concurrency, Specification and Programming, Chemnitz, Germany, September 29–October 1, 2014, 2014, pp. 269–280, http://ceur-ws.org/Vol-1269/paper269.pdf.
[90] G.R. Weckman, C.V. Ganduri, D.A. Koonce, A neural network job-shop scheduler, Journal of Intelligent Manufacturing 19 (2) (Apr 2008) 191–201, https://doi.org/10.1007/s10845-008-0073-9.
[91] R.J. Williams, Simple statistical gradient-following algorithms for connectionist reinforcement learning, Machine Learning 8 (3) (May 1992) 229–256, https://doi.org/10.1007/BF00992696.
[92] M. Wojnarski, P. Gora, M.S. Szczuka, H.S. Nguyen, J. Swietlicka, D. Zeinalipour-Yazti, IEEE ICDM 2010 contest: Tomtom traffic prediction for intelligent GPS navigation, in: ICDMW 2010, the 10th IEEE International Conference on Data Mining Workshops, 2010, pp. 1372–1376.
[93] T. Wołoszynski, M. Kurzynski, A probabilistic model of classifier competence for dynamic ensemble selection, Pattern Recognition 44 (10–11) (2011) 2656–2668, http://www.sciencedirect.com/science/article/pii/S0031320311001245.

[94] D.H. Wolpert, W.G. Macready, No free lunch theorems for optimization, IEEE Transactions on Evolutionary Computation 1 (1) (1997) 67–82, https://doi.org/10.1109/4235.585893.

[95] J. Xu, R. Rahmatizadeh, L. Boloni, D. Turgut, Real-time prediction of taxi demand using recurrent neural networks, IEEE Transactions on Intelligent Transportation Systems 19 (8) (Aug 2018) 2572–2581.

[96] Z. Yang, J.-P. van Osta, B. van Veen, R. van Krevelen, R. van Klaveren, A. Stam, J. Kok, T. Bäck, M. Emmerich, Dynamic vehicle routing with time windows in theory and practice, Natural Computing 16 (1) (Mar 2017) 119–134, https://doi.org/10.1007/s11047-016-9550-9.

[97] N. Ye, Z. Wang, R. Malekian, Y. Zhang, R. Wang, A method of vehicle route prediction based on social network analysis, Journal of Sensors 2015 (2015) 210298, https://doi.org/10.1155/2015/210298.

[98] M. Zennaki, A. Ech-cherif, A new approach using machine learning and data fusion techniques for solving hard combinatorial optimization problems, in: 2008 3rd International Conference on Information and Communication Technologies: From Theory to Applications, April 2008, pp. 1–5.

[99] Z. Zhang, J. Geiger, J. Pohjalainen, A.E.-D. Mousa, W. Jin, B. Schuller, Deep learning for environmentally robust speech recognition: an overview of recent developments, ACM Transactions on Intelligent Systems and Technology 9 (5) (Apr. 2018) 49, https://doi.org/10.1145/3178115.

[100] Z. Zhao, P. Zheng, S. Xu, X. Wu, Object detection with deep learning: a review, CoRR, arXiv:1807.05511, 2018.

[101] B. Zhou, A. Lapedriza, J. Xiao, A. Torralba, A. Oliva, Learning deep features for scene recognition using places database, in: Proc. NIPS, 2014, pp. 487–495.

[102] Z.-H. Zhou, Ensemble Methods: Foundations and Algorithms, 1st edition, Chapman & Hall/CRC, 2012.

[103] Q. Zou, L. Ni, T. Zhang, Q. Wang, Deep learning based feature selection for remote sensing scene classification, IEEE Geoscience and Remote Sensing Letters 12 (11) (2015) 2321–2325.

Chapter 8

How to assess your Smart Delivery System?

Benchmarks for rich vehicle routing problems

Luis A.A. Meira, Paulo S. Martins, Mauro Menzori, and Guilherme A. Zeni
School of Technology (FT), University of Campinas (UNICAMP), Limeira, SP, Brazil

8.1 Introduction

Benchmarks are found in various fields of science, such as reservoir engineering [15,21], economics [30], and climatology [45], among other areas. They play a central role in computer science, for example, in image processing [18, 35,27], hardware performance [9], and optimization [41,34,7].

Within the context of optimization, Johnson [29] divided algorithm analysis into three approaches: the worst-case, the average-case, and the experimental analysis. Regarding experimental papers, he identifies four cases: *(i)* solving a real problem; *(ii)* providing evidence that one algorithm is superior to others; *(iii)* better understanding a problem; and *(iv)* studying the average case. He proposes the use of well-established benchmarks to provide evidence of the superiority of an algorithm (item *ii*). Such papers are called *horse-race papers*.

Johnson highlights that reproducibility and comparability are essential aspects of any experimental paper. The author mentions the difficulty in justifying experiments on problems with no direct application. Such problems have no real instances, and the researcher is forced to generate the data in a *vacuum*.

Barr et al. [5] indicate several purposes in an optimization research. The proposed algorithm may be fast, accurate, robust, simple, high-impact, generalizable, or innovative. They reinforce that results on standard benchmark problems and problems of practical interest should always be a part of an algorithmic experiment.

Our work deals with a variant of the Vehicle Routing Problem (VRP) based on a real mail delivery case in the city of Artur Nogueira. The post office receives thousands of mail items to be delivered on a daily basis. Such a mail is distributed to a set of 15–25 mail carriers for on-foot delivery. Each mail carrier is modeled as a vehicle, and each delivery point is a customer. The problem

Smart Delivery Systems. https://doi.org/10.1016/B978-0-12-815715-2.00007-5

is named here as Post Office Deliveries VRP (PostVRP). It is a multiobjective variant of the distance-constrained VRP.

Domain experts indicate that the PostVRP has three main objective functions to be minimized (while maintaining the feasibility of the solutions): *(i)* route length; *(ii)* unfairness, measured as the workload (i.e., route length) variance among the mail carriers; and *(iii)* number of mail carriers.

The PostVRP considers uncapacitated vehicles and constrained route length. Each mail carrier is allowed to carry a maximum load from 8 to 10 kg. A support truck restocks the mail carriers turning their capacities unlimited. Each mail carrier must follow a 6-to-8-hour working day, which implies a maximum capacity for the route length.

A limited route length is a constraint that models several real-world cases: A helicopter has a route length limited by the capacity of its fuel tank. Workers, in general, have a time window to operate the vehicle, which likewise limits the length of the route.

This work presents a Smart Delivery System (SDS) case study modeled in a Brazilian city located at $22°34'22''S47°10'22''W$. The proposed benchmark contains up to 30,000 customers. We make available the benchmark tool so that it is possible to create new arbitrarily large instances. The methodology can be applied to other cities and to other VRP variants.

The remainder of this chapter is organized as follows: the background and review of relevant work is provided in Section 8.2; in Section 8.3, we introduce the notation and definitions; Section 8.4 presents the model; and Section 8.5 addresses a real-world PostVRP benchmark case. Finally, we summarize and conclude in Section 8.6.

8.2 Literature review

One of the first references to the VRP dates back to 1959 [17], under the name Truck Dispatching Problem, a generalization of the Traveler Salesman Problem (TSP). The term VRP was first seen in the paper by Christofides [10]. Christophides defines VRP as a generic name, given to a class of problems that involves the visit of "customers" using vehicles.

Real-world aspects may impose variants to the problem. For example, the Capacitated-VRP (CVRP) considers a limit to the vehicle capacity [22], the VRP with Time Windows (VRPTW) accounts for the delivery time windows [32], and the Multi-Depot VRP (MDVRP) extends the number of depots [43].

Jozefowiez et al. [31] surveys the existing research related to multiobjective optimization in routing problems. Amaya et al. [1] introduces the capacitated arc routing problem with refill points. The vehicle servicing arcs must be refilled on the spot by using a second vehicle. Gussmagg-Pfliegl et al. [25] create heuristics for a real-world mail delivery problem. Four distinct vehicles are considered: car, moped, bicycle, and walking. Other VRP variants may be easily found in the literature.

Reinelt [41] created a benchmark for the TSP known as TSPLib. In his work he consolidated unsolved instances from 20 distinct papers. His repository, named TSPLIB95 [42], has instances of both symmetric and asymmetric Traveling Salesman Problems (TSP and aTSP) and three related problems: *(i)* CVRP; *(ii)* Sequential Ordered Problem (SOP); and *(iii)* Hamiltonian Cycle Problem (HCP).

The number of instances is 113, 19, 16, 41, 9 for TSP, aTSP, CVRP, SOP, and HCP, respectively. The number of vertices varies from 14 to 85,900 for the TSP, from 17 to 443 for the aTSP, from 7 to 262 for the CVRP, from 7 to 378 for the SOP, and from 1000 to 5000 for the HCP.

The optimum of all TSPLib instances was finally achieved in 2006, after fifteen years of notable progress in algorithm development. The optimum of the d15112 instance was found in 2001 [3]. This instance contains 15,112 German cities, and it required 22.6 years of processing split across 110 500 MHz processors [14]. The pla33810 instance was solved in March 2004 [3]. It represents a printed circuit board with 33,810 nodes (Fig. 8.1), and it was solved in 15.7 years of processing [19]. The last instance of the TSPLib, called pla85900, was solved in 2006 [3]. This instance contains 85,900 nodes representing a VLSI application.

Solomon [44] created a benchmark for the VRPTW in 1987. It is composed of 56 instances partitioned into six sets: R1, C1, RC1, R2, C2, and RC2. The data are randomly generated in problem sets R1 and R2, cluster-generated in problem sets C1 and C2, and generated as a mix of random and cluster structures in problem sets RC1 and RC2. The number of customers is 100 in all instances. The vehicle has a fixed capacity, and the customers have an integer demand. The number of vehicles is not fixed: it is derived from the fact that capacity is limited. Under this viewpoint, this can be considered a multiobjective problem. It aims to minimize the route and the number of vehicles.

The first optimum solution was published by Kohl et al. [33]. Chabrier [8] solved 17 open instances in the benchmark. Amini et al. [2] obtained solutions very close to the optimum, considering only the first 25 customers. In July 2015, 28 years after having launched the benchmark, Jawarneh and Abdullah [28] published a Bee Colony Optimization metaheuristic. Such algorithm reached 11 new results in extended Solomon's VRPTW instances. It is surprising that such small instances present a quite complex internal structure to be optimized. Fig. 8.2 shows a Solomon's instance composed of 100 customers and a given solution considering three vehicles.

Regardless of their complexity, the TSPLib and Solomon benchmarks have a number of customers between 100 and 262 for the VRP, which is currently a small value. Gehring and Homberger [23] extended the Solomon's instances, thus creating a benchmark for the VRPTW with the number of customers varying from 100 to 1000.

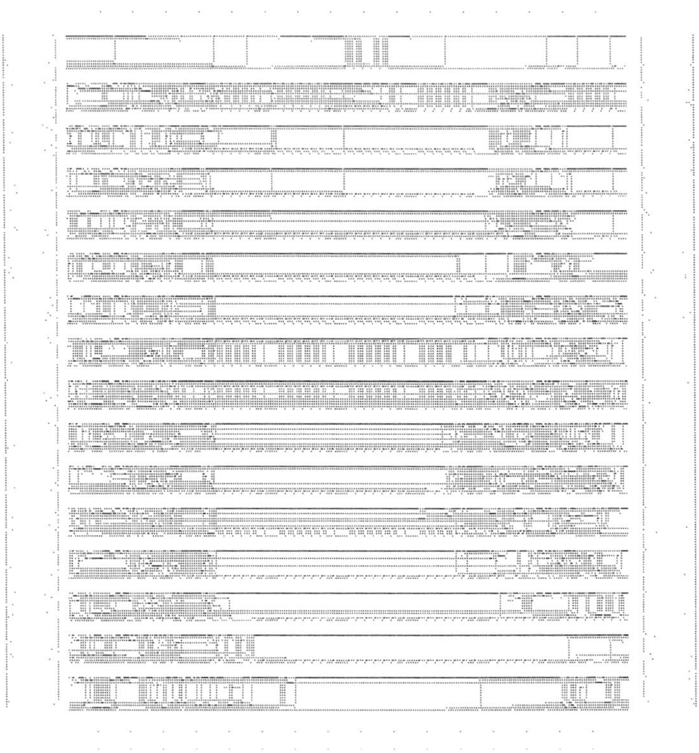

FIGURE 8.1 Instance pla33810.

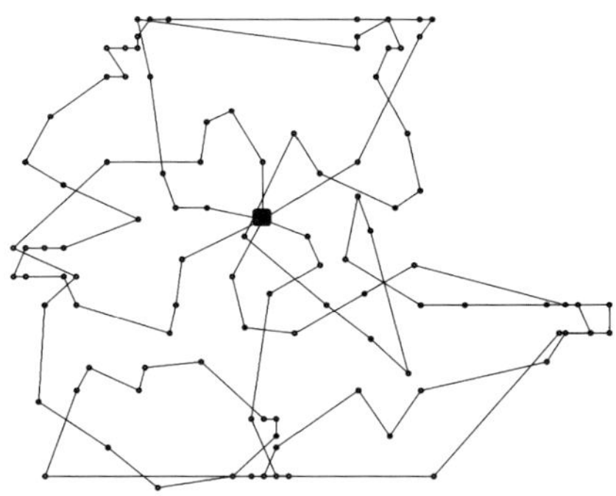

FIGURE 8.2 The Solomon's RC207 instance composed of 100 customers and one depot with a solution containing three routes [6]. The picture does not show the capacity and the time-window constraints.

Nguyen et al. [39] proposed a genetic algorithm to optimize the World TSP challenge. It is a TSP instance with 1,904,711 locations throughout the world. Each location is specified by its latitude and longitude.

For the CVRP, ABEFMP is a largely used set of instances, in which Augerat et al. [4] proposed the A, B, P classes in 1995, and Christofides and Eilon [12], Fisher [20], Christofides [11] proposed the E, F, and M classes in 1969, 1994, and 1979, respectively. In their benchmark the number of customers varies from 13 to 200, and the number of vehicles varies from 2 to 17.

Fukasawa et al. [22] and Contardo and Martinelli [13], among others, obtained the optimum in different ABEFMP instances. Pecin et al. [40] found the optimum solution for the last unsolved instance, named *M-n151-k12*, 35 years after its presentation by Christofides [11]. Despite that, most of those instances are very simple to solve nowadays.

Golden et al. [24] proposed new instances for the CVRP. It is a set of 20 instances, with the number of customers varying from 240 to 483. Such a benchmark remains entertaining, because most of its instances still do not have an optimum established [46]. Li et al. [37] created a set of instances with the number of customers between 560 and 1200. Currently, no optimum has been found for any of the instances [46].

Uchoa et al. [46] created the CVRPLib, where they consolidated the CVRP instances of [4,12,11,20,24,37]. In addition, Uchoa et al. [47] generated new instances with the number of customers between 100 and 1000. Their work indicates the lack of well-established challenging benchmarks for the VRP.

Uchoa et al. also highlight the fact that benchmarks are artificially created. Solomon and Uchoa et al. generated their own instances using random points. In the ABEFMP benchmark some random instances are generated, and other instances represent real problems. However, in all the instances the customers are points in the Euclidean space. The instances of Golden et al. [24] and Li et al. [37] are artificial as well.

Cwiek et al. [16] created a heuristic benchmark generator for VRPTW. Each delivery is $(x_i, y_i) \in \mathbb{R}^2$, and the distance between two deliveries is the Manhattan metric $c_{ij} = |x_i - x_j| + |y_i - y_j|$. Kytöjoki et al. [36] found solutions for artificial VRP instances with up to 20,000 customers and real-world instances with up to 12,000 customers. This is probably the largest VRP instance reported in the literature.

The SINTEF website contains best known results for standard benchmarks of several variants of the Vehicle Routing Problem (https://www.sintef.no/projectweb/top/).

8.3 Notation and definition

Consider a set of elements S where a *depot* is a special element $\pi \in S$. This work does not address multiple-depot variants to the VRP. The set of customers is defined by $C = S \setminus \{\pi\}$, and the number of customers is denoted by n, where $C = \{c_1, \ldots, c_n\}$. The number of vehicles in the fleet is represented by $k \in \mathbb{N}$. The value k is traditionally considered a constant, but it is possible to define variants to VRP where k is variable. Let $w : S \times S \to \mathbb{N}$ be the cost between any two elements in S. Let $\mathcal{S}(C,k) = (c_1, \ldots, c_n, \pi, \ldots, \pi)$. This sequence is created as follows: *(i)* all elements in C are inserted in $\mathcal{S}$; *(ii)* the depot vertex is inserted $k-1$ times.

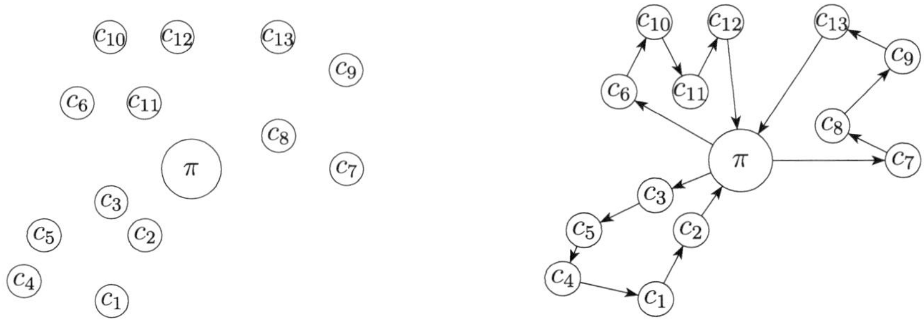

FIGURE 8.3 VRP vertices (left) and solution $\mathcal{S}' = (c_3, c_5, c_4, c_1, c_2, \pi, c_6, c_{10}, c_{11}, c_{12}, \pi, c_7, c_8, c_9, c_{13})$ (right).

Each permutation of $\mathcal{S}(C,k)$ represents a solution to the VRP. For example, consider the graph and vertices described in Fig. 8.3, and suppose that the number of vehicles is three (i.e., $k=3$). The $\mathcal{S}(C,3)$ sequence is given by $\mathcal{S}(C,3) = (c_1, \ldots, c_{13}, \pi, \pi)$. For example, the permutation $\mathcal{S}' = (c_3, c_5, c_4, c_1, c_2, \pi, c_6, c_{10}, c_{11}, c_{12}, \pi, c_7, c_8, c_9, c_{13})$ is the solution described in Fig. 8.3.

All routes begin and end at the depot. The $\mathcal{S}'$ solution represents a partition of the clients in three routes: $\mathcal{R}_1 = (c_3, c_5, c_4, c_1, c_2)$, $\mathcal{R}_2 = (c_6, c_{10}, c_{11}, c_{12})$, and $\mathcal{R}_3 = (c_7, c_8, c_9, c_{13})$. The vertex π is used to create a partition of the sequence in $k' \leq k$ routes. Let $Partition(\mathcal{S}) = (\mathcal{R}_1, \ldots, \mathcal{R}_{k'})$. By definition empty routes are not part of $Partition(\mathcal{S})$. Thus, $Partition(1, 2, \pi, \pi, 3, 4)$ is $\{(1,2), (3,4)\}$ and not $\{(1,2), (), (3,4)\}$, that is, $k' \leq k$.

The length of a route $\mathcal{R} = (r_1, \ldots, r_m)$ is given by

$$W(\mathcal{R}) = w(\pi, r_1) + w(r_m, \pi) + \sum_{i=1}^{m-1} w(r_i, r_{i+1}).$$

The length of a solution $\mathcal{S} = (s_1, \ldots, s_m)$ is the travel distance and is calculated as

$$W(\mathcal{S}) = \sum_{\mathcal{R} \in Partition(\mathcal{S})} W(\mathcal{R}).$$

The number of vehicles used in a given solution is equal to the number of nonempty routes $|Partition(\mathcal{S})|$. If the number of vehicles is k and nonempty routes are allowed, then we have the constraint $|Partition(\mathcal{S})| = k$. If the number of vehicles are at most k, or if empty routes are allowed, then we have $|Partition(\mathcal{S})| \leq k$. If the number of vehicles is not a part of the input, then the domain may be given by the permutation of the sequence $\mathcal{S}(C, n)$. In this case the number of vehicles is defined during the optimization step.

Given a feasible solution, it is necessary to calculate its costs. The most common objective function to be minimized is the length of the solution:

$$f_1(\mathcal{S}) = W(\mathcal{S}).$$

Another objective function consists in finding a feasible solution that minimizes the number of vehicles:

$$f_2(\mathcal{S}) = |Partition(\mathcal{S})|.$$

Suppose there are 25 available mail carriers and a feasible solution with 21 routes. In this case the post office may allocate the four available mail carriers to other internal tasks.

Finally, it is required that the solution meet the fairness criteria, that is, routes should be assigned in a way that balances out the workload (route length) among the mail carriers. The way we modeled fairness was through minimizing the variance of the route lengths:

$$f_3(\mathcal{S}) = \sqrt{\frac{\sum_{\mathcal{R} \in Partition(\mathcal{S})} \left(W(\mathcal{R}) - \overline{W(\mathcal{R})} \right)^2}{|Partition(\mathcal{S})| - 1}}.$$

VRP is a set of problems that consists of visiting customers using vehicles. Each variant has additional feasibility constraints, such as not allowing empty routes ($|Partition(\mathcal{S})| = k$). The PostVRP assumes that the length of the route is limited. Let $\mathcal{R}_{\max}$ be the maximum allowed route length, that is, $W(\mathcal{R}) \leq \mathcal{R}_{\max}$.

Definition 1 (PostVRP). Given a set of elements S, a weight function $w : S \times S \rightarrow \mathbb{N}$, a constant $k \in \mathbb{N}$ representing the maximum number of vehicles, a special vertex $\pi \in S$, and a maximum route length $\mathcal{R}_{\max} \in \mathbb{N}$. Let $C \leftarrow S \setminus \{\pi\}$. Consider the sequence $\mathcal{S}(C, k)$, and let Pe be the set of all feasible permutations of $\mathcal{S}(C, k)$ with respect to $\mathcal{R}_{\max}$. The PostVRP problem consists of minimizing $(f_1(\mathcal{S}), f_2(\mathcal{S}), f_3(\mathcal{S}))$ subject to $\mathcal{S} \in Pe$.

8.4 Model description

This section describes the PostVRP model. We start the process of creating the benchmark by mapping each street onto a street map graph. Each street St is modeled as a polygonal chain, which is defined as a set of planar coordinates. For example, suppose a University St. modeled as a polygonal chain $P = (c_1, \ldots, c_{n'})$, where $c \in \mathbb{R}^2$ for all $c \in P$. The complete map graph has a set $\mathcal{P} = (P_1, \ldots, P_{n''})$ of polygonal chains, one for each street.

We create a graph $G(V, E)$ based on $\mathcal{P}$. Each vertex $v \in V$ is associated with a Cartesian coordinate $(x_v, y_v) \in \mathbb{R}^2$, and each edge $e = (u, v)$ is a straight line segment between u and v. The edge weight is $w'(e) = \sqrt{(x_u - x_v)^2 + (y_u - y_v)^2}$. The vertices associated with corners are automatically built by a line segment intersection algorithm.

Fig. 8.4 (right) displays simple streets modeled as one polygonal chain and streets with islands, which are modeled by two parallel polygonal chains. Additional segments are added to allow shortcuts in footpaths.

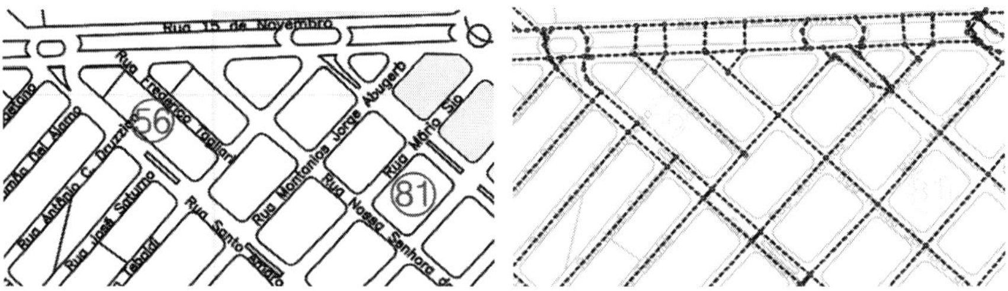

FIGURE 8.4 A section of the city map (left). Edges and vertices created over the city map (right).

Given an edge e, its street is denoted by $St(e)$. Each street St has an arbitrarily defined width $wth(St) \in \mathbb{R}^+$, which represents the cost of crossing the street. The value $wth(St)$ can be set to zero, thus resulting in no cost to cross the street.

A nonnormalized probability density $D(St)$ is assigned to each street. The probability of a street receiving a delivery workload per unit length is directly proportional to the density value D. Such a value is used to create a central street with a large workload compared to a distant one. The probabilities are outlined in the next subsection.

8.4.1 Generating delivery points

Consider a street map graph $G(V, E)$. The probability of one delivery being assigned to an edge e, denoted by $Prob(e)$, is

$$Prob(e) = \frac{D(St(e))w'(e)}{T}, \text{ where } T = \sum_{e' \in E} D(St(e'))w'(e').$$

$Prob(e)$ is directly proportional to the edge length $w'(e)$ and to the probability density $D(St(e))$, and it must be normalized to obtain $\sum_{e \in E} Prob(e) = 1$.

The location of a given delivery d, denoted by $loc(d)$, is composed of three attributes: an edge (u, v), a value $\alpha \in [0, 1]$, and a label $street\_side \in \{\oplus, \ominus\}$. The delivery is positioned at the affine combination of u and v in respect to α, that is, $(x_u, y_u)(\alpha) + (1 - \alpha)(x_v, y_v)$. The street of a delivery $d = (e, \alpha, street\_side)$, denoted by $St(d)$, is the street of the edge $St(e)$.

An integer n represents the number of deliveries, and an artificial delivery d_π is created for the depot. The value of α is randomly generated within the interval [0, 1]. The street side label is an equiprobable random choice in the set $\{\oplus, \ominus\}$. Algorithm 1 is then used to create the delivery set. Given that the number of deliveries is a part of the input, it is possible to set arbitrarily large instances.

```
Input: An integer n and a set of edges E with probabilities Prob(e),
       ∀e ∈ E
Output: A set of deliveries Del.
Del ← Ø
Partition all edge probabilities in the interval [0, 1]
for i=1 to n do
    Select a random value r ∈ [0, 1]
    if r is in the interval associated with Prob(e) then
        Select a random value α ∈ [0, 1]
        Select a random street side value s ∈ {⊕, ⊖}
        Del ← Del ∪ {(e, α, s)}
    end
end
return Del
```

Algorithm 1: Algorithm to create deliveries.

8.4.2 Defining the weight between a pair of deliveries

The street map graph $G(V, E)$ is used to compute the weight between deliveries. Given two deliveries d_a and d_b, the cost to cross the street is defined as

$$cross(d_a, d_b) = \begin{cases} wth(St(d_a)) \text{ if } (d_a, d_b)\ side \text{ labels are } \{(\oplus, \ominus), (\ominus, \oplus)\} \\ \quad \text{and } St(d_a) = St(d_b), \\ 0 \text{ otherwise.} \end{cases}$$

A constant $\beta \in \mathbb{R}^+$, which represents an additional fixed cost per delivery, must be defined. The weight between two deliveries $d_a = (e_a, \alpha_a, s_a)$ and $d_b = (e_b, \alpha_b, s_b)$ is given by $w(d_a, d_b)$. If $e_a = e_b$, then

$$w(d_a, d_b) = |\alpha_a - \alpha_b| w'(e_a) + cross(d_a, d_b) + \beta. \tag{8.1}$$

Let $G(V, E)$ be the original street map, $e_a = \{u_a, v_a\}$, $e_b = \{u_b, v_b\}$, and let $G^*(V^*, E^*)$ be defined as $V^* = V \cup \{d_a, d_b\}$, $E^* = E \cup \{(u_a, d_a), (v_a, d_a), (u_b, d_b), (v_b, d_b)\}$ (Fig. 8.5); thus

$$w(d_a, d_b) = minpath(d_a, d_b, G^*) + cross(d_a, d_b) + \beta. \tag{8.2}$$

The instance is composed of a matrix $w_{n \times n}$, an integer k, and an integer $\mathcal{R}_{\max}$. The first delivery represents the depot.

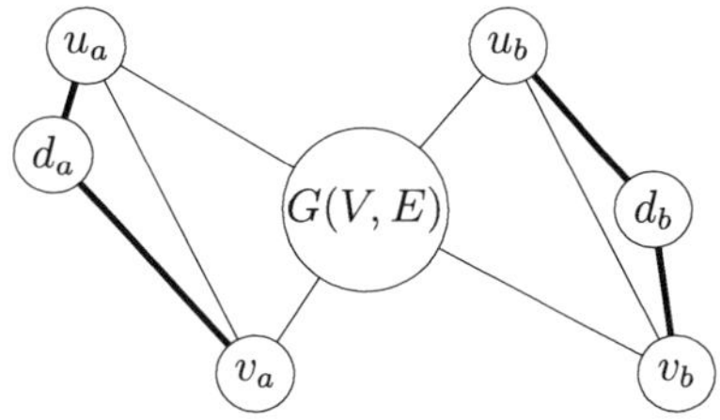

FIGURE 8.5 The cost between two deliveries d_a and d_b located at distinct edges e_a and e_b, respectively.

8.4.3 The benchmark tool

This subsection describes the tool that creates the benchmark. It has three configuration files, *background.png*, *model.txt,* and *instances.txt*. The background file contains an image used to improve visualization, and its resolution is used as the base for the model. The model file must contain the following information:

- Depot location: the coordinate position reference to the depot;
- Additional cost per delivery: cost to hand out the delivery;
- Decimal precision: number of digits after the fractional part;
- Pixel value: value used to convert a pixel into other units. This value may represent the speed of the vehicle;
- Attributes: attributes used to compute the street probability density and the cost to cross the street;
- Roadmap: the description of streets including the polygonal chain.

A researcher may perform experiments where pixel value, additional cost per delivery or decimal precision are variable. Consider a white background with 500 × 500 pixels and the model described in Fig. 8.6. The tool will process the model file and create a roadmap (Fig. 8.7). The depot is positioned at the closest edge. The probability density D of the 4th Av is 0.4 because it is a $[AVE, PERIPHERAL, RADIOACTIVE]$ with values $\{20, 0.2, 0.1\}$.

The last file is named *instance.txt* (Table 8.1). Each line corresponds to an instance in the benchmark. It must contain the instance ID, the directory and subdirectory, the maximum number of vehicles, and a comment line. Each line must also contain a pseudo-random generator seed and an MD5 signature.

The MD5 checksum value is used to ensure the instance identity. The tool will recreate the instances offline and verify the MD5 signature in the instance file. The tool will execute the files of Fig. 8.6 and Table 8.1 and create the instances shown in Fig. 8.8.

We can edit the instance file to create new instances for a given model. Once the new instances are created, it is necessary to manually update the MD5 signature. For instance, a new seed will create a new instance with another pseudo-random sequence.

```
#GENERAL
DEFINE DEPOT_X 250
DEFINE DEPOT_Y 250
DEFINE ADDITIONAL_COST_PER_DELIVERY 10
DEFINE DECIMAL_PRECISION 6
#ONE PIXEL IS PIXEL_VALUE UNITS (meters,sec,etc).
DEFINE PIXEL_VALUE 10.0
#PENALTIES
DEFINE PERIPHERAL .2
DEFINE RADIOACTIVE .1
#STR
DEFINE STR 10
DEFINE STR_CROSSCOST 20
#AVE
DEFINE AVE 20
DEFINE AVE_CROSSCOST 10

#ROADMAP
1th STR [STR][100,100]-[400,100]
2th STR [STR,PERIPHERAL][100,200]-[400,200]
3th STR [STR,RADIOACTIVE][100,300]-[400,300]
4th STR [STR,PERIPHERAL,RADIOACTIVE][100,400]-[400,400]
1th Av [AVE][100,100]-[100,400]
2th Av [AVE,PERIPHERAL][200,100]-[200,400]
3th Av [AVE,RADIOACTIVE][300,100]-[300,400]
4th Av [AVE,PERIPHERAL,RADIOACTIVE][400,100]-[350,250]-[400,400]
```

FIGURE 8.6 Example of a model file.

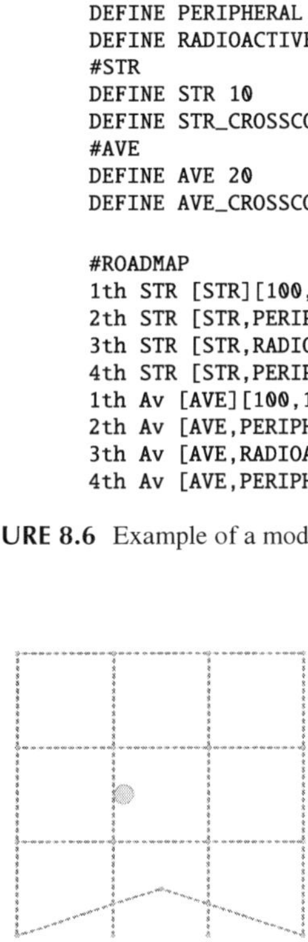

St	D(St)	wth(St) pixel
1th STR	10	20
2th STR	2	20
3th STR	1	20
4th STR	0.2	20
1th Av	20	10
2th Av	4	10
3th Av	2	10
4th Av	0.4	10

FIGURE 8.7 Map based on the model in Fig. 8.6.

TABLE 8.1 Instance file.

ID	Dir	Subdir	n	k	$\mathcal{R}_{max}$	Comment	Seed	MD5
0	ex	ex_0_0	0	0	2941.15	Max route [...]	100	4d...af
1	ex	ex_10_5	10	5	2941.15	The size [...]	101	e9...10
2	ex	ex_100_5	100	5	2941.15	Consider [...]	102	eb...a2
3	ex	ex_1000_5	1000	5	2941.15	Instance [...]	103	05...62
4	ex	ex_10000_5	10000	5	2941.15	Instance [...]	104	93...53

The project site [38] provides the source program that parses the instance file. The program executes a simple swap optimization and saves the route in

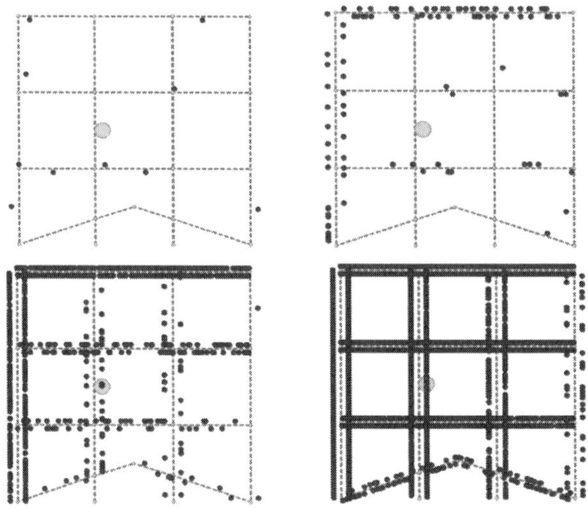

FIGURE 8.8 Resulting instances with 10, 100, 1000, and 10,000 deliveries.

both text and image files (Fig. 8.9). The purpose is to provide the researcher with a parse to the instance file and a visualization of the solution.

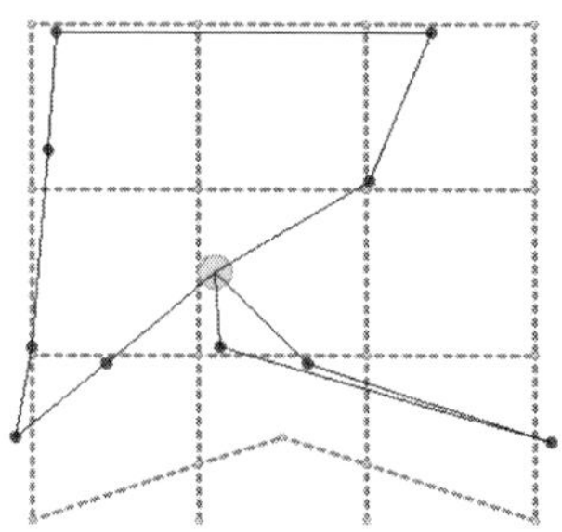

FIGURE 8.9 Solution to instance with 10 deliveries.

8.4.4 Modeling Manhattan (NY) streets

This section shows how the methodology can be adapted to other benchmark cases through a pilot sample that indicates its possible uses. We start off with an image of a section of Manhattan (Fig. 8.10).

The streets (34 in this case) are manually added as a polygonal chain in the model file. As the streets are generally straight, they are composed mostly by two pixel coordinates. We label the streets with four types with respective penalties and cost to cross the streets (Fig. 8.11).

Fig. 8.12 contains one instance with 1000 deliveries based on this model.

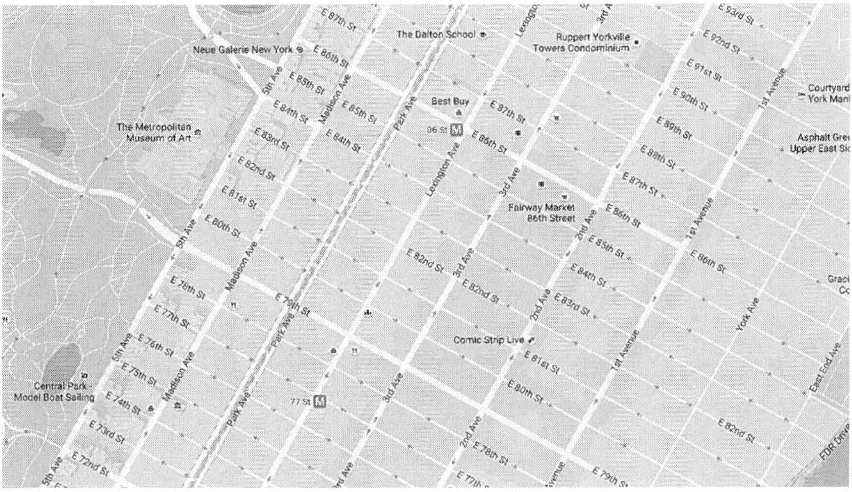

FIGURE 8.10 Section of the original map.

```
#GENERAL
DEFINE DEPOT_X 327
DEFINE DEPOT_Y 565
DEFINE ADDITIONAL_COST_PER_DELIVERY 3.2
DEFINE DECIMAL_PRECISION 3
#ONE PIXEL IS PIXEL_VALUE UNITS 1px=1.7m
DEFINE PIXEL_VALUE 1.7
DEFINE LARGE_AVENUE 100
DEFINE LARGE_AVENUE_CROSSCOST 0
DEFINE AVENUE 75
DEFINE AVENUE_CROSSCOST 10
DEFINE STREET 40
DEFINE STREET_CROSSCOST 5
DEFINE DRIVE 10
DEFINE DRIVE_CROSSCOST 5
#ROADMAP
Park Avenue [LARGE_AVENUE][654,246]
                              -[0,604]
Park Avenue B [LARGE_AVENUE][654,259]
                              -[0,620]
5th Avenue [AVENUE][654,54]-[0,416]
Madison Avenue [AVENUE][654,154]-[0,512]
Lexington Ave [AVENUE][654,352]-[0,712]
3rd Avenue [AVENUE][654,447]-[0,809]
2rd Avenue [AVENUE][654,583]-[0,942]
1rd Avenue [AVENUE][654,729]-[0,1090]
York Avenue [AVENUE][654,865]-[177,1130]
East End Ave [AVENUE][654,1010]-[432,1130]
E 71th street [STREET][645,60]-[654,78]
E 72th street [STREET][605,85]-[654,179]
E 73th street [STREET][567,108]-[654,274]
E 74th street [STREET][528,129]-[654,368]
E 75th street [STREET][489,152]-[654,462]
E 76th street [STREET][449,173]-[654,553]
E 77th street [STREET][410,195]-[654,646]
E 78th street [STREET][370,217]-[654,738]
E 79th street [STREET][328,239]-[654,835]
E 80th street [STREET][282,267]-[654,927]
E 81th street [STREET][247,285]-[654,1026]
E 82th street [STREET][208,306]-[629,1067]
                                  -[634,1075]
E 83th street [STREET][170,327]-[600,1110]
E 84th street [STREET][129,350]-[560,1130]
E 85th street [STREET][90,372]-[492,1099]
E 86th street [STREET][49,394]-[454,1130]
E 87th street [STREET][9,416]-[401,1130]
E 88th street [STREET][0,505]-[350,1130]
E 89th street [STREET][0,590]-[299,1130]
E 90th street [STREET][0,684]-[249,1130]
E 91th street [STREET][0,774]-[196,1130]
E 92th street [STREET][0,864]-[148,1130]
E 93th street [STREET][0,951]-[56,1058]
FDR Drive [DRIVE][654,1054]-[634,1075]
                  -[600,1110]-[573,1130]
```

FIGURE 8.11 Manhattan model file.

8.5 Real-world PostVRP benchmark (RWPostVRPB)

In this section the tool is used to model a real-world mail delivery in the city of Artur Nogueira, Brazil, namely RWPostVRPB. To make the instances as real-

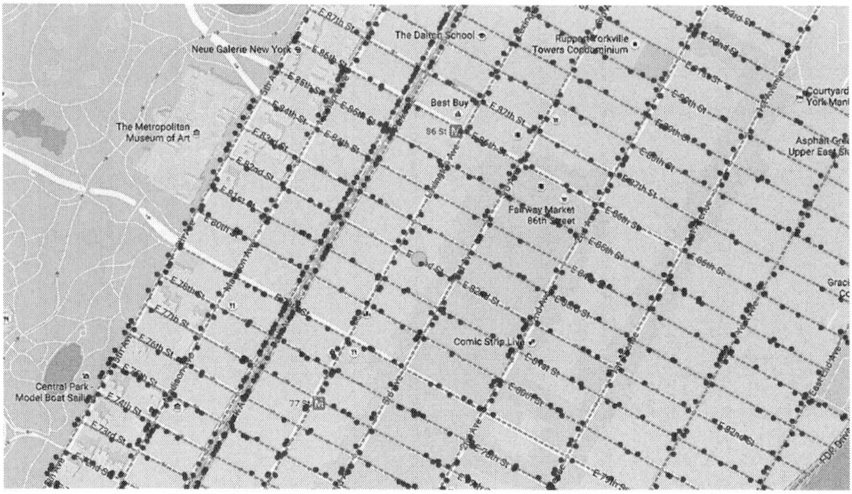

FIGURE 8.12 Instance image with 1000 deliveries.

istic as possible, the authors relied upon domain expertise from an actual post office in the aforementioned city.

We build the model from a PDF image instead of using existing street map graphs for the following reasons (Fig. 8.13): *(i)* the path on foot may differ from those available from graphs which prioritize delivery by vehicles; *(ii)* the number of streets in Artur Nogueira is sufficiently small to allow the manual creation of the graph (≈ 400 streets); and *(iii)* currently, public maps such as OpenStreetMap [26] are incomplete, that is, the city has a large number of streets not covered so far.

The tool automatically computes corners resulting in a graph with $|V| = 2111$ and $|E| = 3225$. Each vertex is associated with a pixel, and each edge is associated with a straight line between the two edge pixels. The cost of an edge (u, v) is directly proportional to the Euclidean distance between vertices u and v in $\mathbb{R}^2$.

Each street is classified using the Region (R), Type (T), and Zone (Z) attributes. Each attribute has a corresponding number of levels, and each attribute-level pair is associated with a multiplicative penalty (Pen) in $\mathbb{R}^+$. Table 8.2 contains the assignment of attribute, level, and penalty based on expert knowledge.

In the proposed model the streets located in the downtown area have a higher delivery rate per unit length than those located in the outskirts. Such behavior is captured by the Region attribute through four levels: *central, peripheral, distant,* and *isolated.* The Type attribute has also four levels, namely *avenue, street, way,* and *highway*, whereas the Zone attribute may be *commercial, mixed*, and *residential.* We used Google Maps as an auxiliary tool to classify streets. Each of the 422 streets received a value in $R \times T \times Z$ according to expert knowledge.

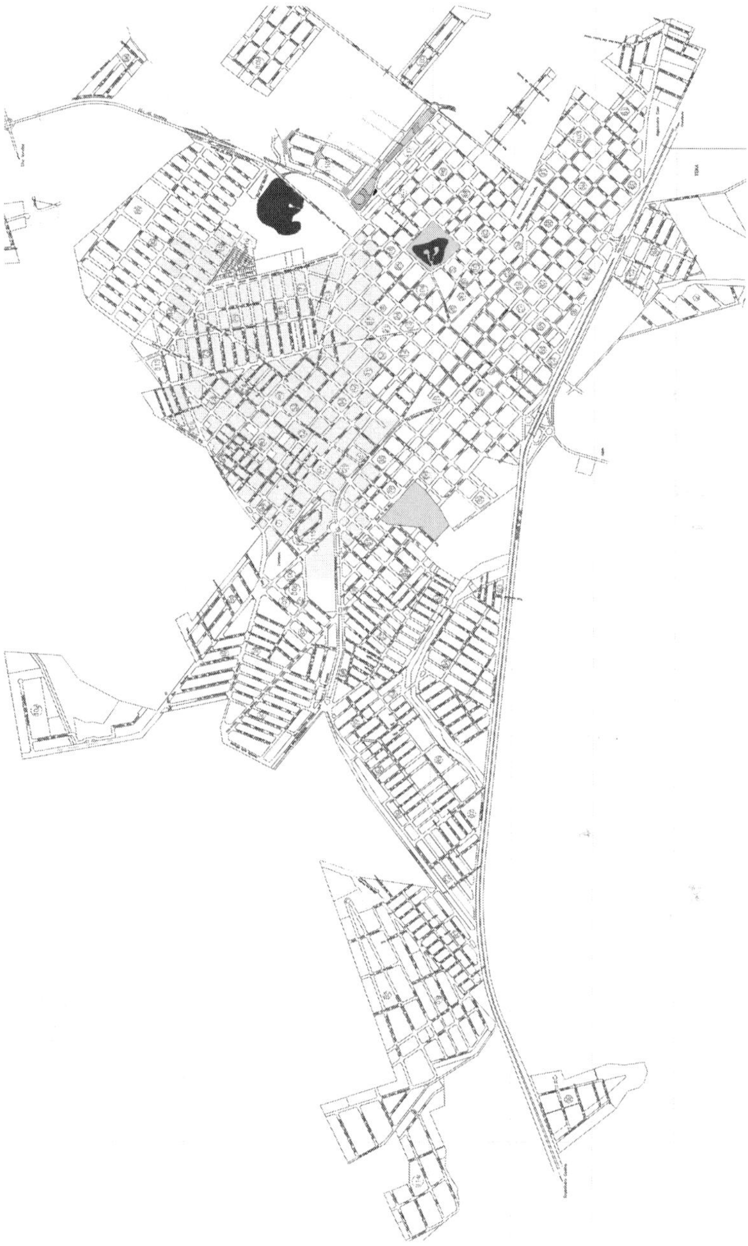

FIGURE 8.13 Original PDF map from Artur Nogueira.

The (nonnormalized) probability density $D : Streets \rightarrow \mathbb{R}^+$ is obtained from the multiplicative penalties. For example, the 15th Avenue is (*central*,

TABLE 8.2 Attribute, level and penalty values.

Attribute	Level $1_{(Pen)}$	Level $2_{(Pen)}$	Level $3_{(Pen)}$	Level $4_{(Pen)}$
Region (R)	$central_{(1.0)}$	$peripher_{(.75)}$	$distant_{(.4)}$	$isolated_{(.2)}$
Type (T)	$largeAvn_{(1.0)}$	$Avenue_{(.75)}$	$Street_{(.4)}$	$Drive_{(0)}$
Zone (Z)	$commercial_{(1.0)}$	$mixed_{(.75)}$	$residential_{(.4)}$	–

$largeAvn, mixed)$, and thus $D(15th) = 1 \times 1 \times 0.7 = 0.7$. On the other hand, *Jasmine* street is $(isolated, Street, residential)$, and thus $D(Jasmine) = 0.2 \times 0.2 \times 0.4 = 0.16$. In this example a random delivery to the 15th avenue is $\frac{0.7}{0.16}$ more probable than a delivery to $Jasmine$ way by unit length.

The RWPostVRPB contains 78 instances divided into four groups: *Toy*, *Normal*, *OnStrike*, and *Christmas* (Table 8.3). The *Toy* set contains 30 instances with a small number of deliveries, and it may be used to validate algorithms before their use with realistic and larger instances. The maximum number of vehicles is 30, 45, and 60 for *Normal*, *OnStrike*, and *Christmas*, respectively.

TABLE 8.3 RWPostVRPB instances description.

Set	# Instances	# Deliveries	Length (hrs)	# Vehicles (max)
Toy	30	3 to 5000	6	5 to 15
Normal	15	10,000 to 14,000	6	30
OnStrike	15	15,000 to 19,000	8	45
Christmas	18	20,000 to 30,000	8	60

The *Normal* set contains 15 instances from 10,000 to 14,000 deliveries. Artur Nogueira city has around 50,000 inhabitants and an average of 12,000 daily mail deliveries. The normal daily work of a mail carrier has eight hours a day. The mail carrier spends two hours preparing the deliveries inside the post office and six hours to complete the deliveries on foot. The post office has around 15 mail carriers to perform the deliveries. The number of vehicles is variable in PostVRP with a defined maximum value. If a feasible solution with 11 mail carriers is found, then the post office may assign four mail carriers $(15 - 11)$ to internal tasks.

The *OnStrike* and *Christmas* sets may be used to model contingencies. The *OnStrike* set is similar to *Normal*, but the number of deliveries is larger (from 15,000 to 19,000), and the maximum route length is eight hours. The *Christmas* set models special seasons with high delivery rates. The post office often hires extra mail carriers for the Christmas season. A feasible solution with 17 mail carriers represents two new hires $(15 + 2)$. For all sets, a minimum average route length is desired as well as a minimum variance between the lengths of the routes. Fig. 8.14 shows the distribution of 10,000 delivery points for a chosen area of Artur Nogueira.

FIGURE 8.14 Part of an instance generated with 10,000 delivery points.

The speed of the vehicles were set as 0.69 s/pixel, that is, `PIXEL_VALUE=0.69`. The fixed cost per delivery was defined as $5 \times 0.69\,\text{s} = 3.45\,\text{s}$, that is, `ADDITIONAL_COST_PER_DELIVERY=5`.

The full set of instances can be downloaded from the project website [38]. The website also includes a pilot sample modeling a section of Manhattan NY.

On July 10, 2017, each instance was available in the website as a matrix w, where $w(u, v)$ is the cost to make a delivery in v starting from u for all $u, v \in S$. For the largest instance, the matrix is $30{,}000 \times 30{,}000$ using 10.4 GB of memory. On January 17, 2018, we made available the instance represented as a roadmap graph $G(V, E, w)$. Each delivery is represented as a triple $(e, \alpha, street\_side)$, and the distance between two deliveries is computed by Eqs. (8.1) and (8.2). The first representation has a higher level of disorder than the second one. A third representation is the configuration files and the benchmark tool.

8.6 Final remarks and conclusion

In this chapter we have worked on the PostVRP, a multiobjective VRP variant with a route length constraint. The three objectives to be minimized were *(i)* the number of vehicles, *(ii)* the solution length, and *(iii)* the standard deviation of the length of the routes.

Informally, we list some desirable characteristics of a good benchmark: simple, generalizable, incremental complexity, well-defined objective function, it does not have private information that reduces search space and must have a model and a support tool.

A "simple benchmark" has a reasonable amount of information without excess. Thus we opted to model the map in the 2D plan instead of a 3D model. We could also have assumed that speed slows down with human fatigue, among other possibilities. Too much details can be negative.

The "generalization" concept refers to the application of the methodology to other situations. The proposed methodology is suitable for PostVRP, but it can be adapted to other variants of the VRP. The map is created as a set of polygonal chains, so it is relatively simple to model other cities. It is also possible, with an additional effort, to work with directed edges.

Incremental complexity is an advantage, since it is possible to analyze the limits of an algorithm. An exact algorithm can optimize relatively small instances compared to heuristic methods. A heuristic algorithm $O(n^3)$ must optimize smaller instances than another of complexity $O(n^2)$. An incremental complexity benchmark allows analyzing the boundaries of different algorithms.

For a well-defined objective function, we use rounding to work only with integer values. Even the variance can be converted to integer values. Minimizing σ is equivalent to minimizing $\sigma^2 n(n-1)$, with the second option having integer image. In addition, a java algorithm was provided that parses the input file and calculates the value of the objective functions.

To compare two algorithms, it is necessary that they receive the same information about the benchmark. If an algorithm has a hint that allows a warm start, then the results are not comparable. In this work we offer three representations of the instances. The first consists of a matrix $w_{n \times n}$, and the second, much more compact, has the roadmap graph and deliveries. An algorithm that uses the graph has a competitive advantage over an algorithm that works only with the w matrix. In this way we may say that the problems are different according to the available information.

In this research we proposed a benchmark where a model and a tool are both available. The tool provides a toy optimization algorithm with the calculation of the objective function and visualization mechanisms for the instances. This way of representing the instances can be generalized to other problems. The variables in the file *instaces.txt* contain parameters of the instances such as number of deliveries, maximum length of the route, and a seed.

By using the tool we created a benchmark that models a problem that comprised the mail delivery on foot in a Brazilian city. The instances were classified into four groups: *(i) Toy*, with up to 5000 deliveries, *(ii) Normal*, with up to 14,000 deliveries, *(iii) OnStrike*, with up to 19,000 deliveries, and *(iv) Christmas*, with up to 30,000 instances. The benchmark can be used both for comparison and validation of optimization algorithms for routing problems.

The application of the tool allows the generation of arbitrary and large instances in the proposed benchmark by changing the number of deliveries in the instance file. Likewise, new instances can be created by changing the seed of the instance. Additionally, researchers may also model new scenarios.

Acknowledgments

This research was partially supported by FAPESP (Grant Number 2015/11937-9).

References

[1] A. Amaya, A. Langevin, M. Trépanier, The capacitated arc routing problem with refill points, Operations Research Letters 35 (1) (2007) 45–53, https://doi.org/10.1016/j.orl.2005.12.009.

[2] S. Amini, H. Javanshir, R. Tavakkoli-Moghaddam, A PSO approach for solving VRPTW with real case study, International Journal of Research Reviews in Applied Sciences 4 (3) (2010) 118–126.

[3] D.L. Applegate, R.E. Bixby, V. Chvatal, W.J. Cook, The Traveling Salesman Problem: A Computational Study, Princeton University Press, 2011, pp. 506–507.

[4] P. Augerat, J.M. Belenguer, E. Benavent, A. Corberán, D. Naddef, G. Rinaldi, Computational Results With a Branch-and-Cut Code for the Capacitated Vehicle Routing Problem, Tech. Rep. RR 949-M, Université Joseph Fourier, Grenoble, 1998, http://citeseerx.ist.psu.edu/viewdoc/summary?doi=10.1.1.718.7812.

[5] R.S. Barr, B.L. Golden, J.P. Kelly, M.G. Resende, W.R. Stewart, Designing and reporting on computational experiments with heuristic methods, Journal of Heuristics 1 (1) (1995) 9–32, https://doi.org/10.1007/BF02430363.

[6] R. Bent, P. Van Hentenryck, A two-stage hybrid local search for the vehicle routing problem with time windows, Transportation Science 38 (4) (2004) 515–530, https://pubsonline.informs.org/doi/abs/10.1287/trsc.1030.0049.

[7] R.E. Burkard, S.E. Karisch, F. Rendl, QAPLIB – a quadratic assignment problem library, Journal of Global Optimization 10 (4) (1997) 391–403, https://doi.org/10.1023/A:1008293323270.

[8] A. Chabrier, Vehicle routing problem with elementary shortest path based column generation, Computers & Operations Research 33 (10) (2006) 2972–2990, https://doi.org/10.1016/j.cor.2005.02.029.

[9] S. Che, M. Boyer, J. Meng, D. Tarjan, J.W. Sheaffer, S.-H. Lee, K. Skadron, Rodinia: a benchmark suite for heterogeneous computing, in: IEEE International Symposium on Workload Characterization, 2009, IISWC 2009, IEEE, 2009, pp. 44–54.

[10] N. Christofides, The vehicle routing problem. Revue française d'automatique, d'informatique et de recherche opérationnelle, Recherche Opérationnelle 10 (1) (1976) 55–70, http://www.numdam.org/item?id=RO_1976__10_1_55_0.

[11] N. Christofides, The vehicle routing problem, in: N. Christofides, A. Mingozzi, P. Toth, C. Sandi (Eds.), Combinatorial Optimization, 1979.

[12] N. Christofides, S. Eilon, An algorithm for the vehicle-dispatching problem, Journal of the Operational Research Society (1969) 309–318, https://doi.org/10.1057/jors.1969.75.

[13] C. Contardo, R. Martinelli, A new exact algorithm for the multi-depot vehicle routing problem under capacity and route length constraints, Discrete Optimization 12 (2014) 129–146, https://doi.org/10.1016/j.disopt.2014.03.001.

[14] W. Cook, Traveling salesman problem site, (last check 01/30/2016), http://www.math.uwaterloo.ca/tsp/d15sol/, 2016.

[15] M. Correia, J. Hohendorff, A.T.F.S. Gaspar, D. Schiozer, Unisim-ii-d: benchmark case proposal based on a carbonate reservoir, in: SPE Latin American and Caribbean Petroleum Engineering Conference, Society of Petroleum Engineers, 2015.

[16] M. Cwiek, J. Nalepa, M. Dublanski, How to generate benchmarks for rich routing problems? in: Asian Conference on Intelligent Information and Database Systems, Springer, 2016, pp. 399–409.
[17] G.B. Dantzig, J.H. Ramser, The truck dispatching problem, Management Science 6 (1) (1959) 80–91, https://doi.org/10.1287/mnsc.6.1.80.
[18] J.H. du Buf, M. Kardan, M. Spann, Texture feature performance for image segmentation, Pattern Recognition 23 (3) (1990) 291–309, https://doi.org/10.1016/0031-3203(90)90017-F.
[19] D.G. Espinoza, On Linear Programming, Integer Programming and Cutting Planes, Ph.D. thesis, Georgia Institute of Technology, 2006, http://hdl.handle.net/1853/10482.
[20] M.L. Fisher, Optimal solution of vehicle routing problems using minimum k-trees, Operations Research 42 (4) (1994) 626–642, https://doi.org/10.1287/opre.42.4.626.
[21] R.M. Fonseca, C.R. Geel, O. Leeuwenburgh, Description of Olympus Reservoir Model for Optimization Challenge, Tech. rep., TNO, Delft University of Technology, ENI, Statoil and Petrobras, 2017.
[22] R. Fukasawa, H. Longo, J. Lysgaard, M.P. de Aragão, M. Reis, E. Uchoa, R.F. Werneck, Robust branch-and-cut-and-price for the capacitated vehicle routing problem, Mathematical Programming 106 (3) (2006) 491–511, https://doi.org/10.1007/s10107-005-0644-x.
[23] H. Gehring, J. Homberger, A parallel hybrid evolutionary metaheuristic for the vehicle routing problem with time windows, in: Proceedings of EUROGEN99, vol. 2, Springer, Berlin, 1999, pp. 57–64, http://citeseerx.ist.psu.edu/viewdoc/summary?doi=10.1.1.617.1364.
[24] B.L. Golden, E.A. Wasil, J.P. Kelly, I.-M. Chao, The impact of metaheuristics on solving the vehicle routing problem: algorithms, problem sets, and computational results, in: Fleet Management and Logistics, Springer, 1998, pp. 33–56.
[25] E. Gussmagg-Pfliegl, F. Tricoire, K.F. Doerner, R.F. Hartl, S. Irnich, Heuristics for a Real-World Mail Delivery Problem, Springer, Berlin, Heidelberg, 2011, pp. 481–490.
[26] M. Haklay, P. Weber, Openstreetmap: user-generated street maps, IEEE Pervasive Computing 7 (4) (2008) 12–18, https://doi.org/10.1109/MPRV.2008.80.
[27] G.B. Huang, M. Ramesh, T. Berg, E. Learned-Miller, Labeled Faces in the Wild: a Database for Studying Face Recognition in Unconstrained Environments, Tech. rep., Technical Report 07-49, University of Massachusetts, Amherst, 2007, https://hal.inria.fr/inria-00321923.
[28] S. Jawarneh, S. Abdullah, Sequential insertion heuristic with adaptive bee colony optimisation algorithm for vehicle routing problem with time windows, PLoS ONE 10 (7) (2015) 1–23, https://doi.org/10.1371/journal.pone.0130224.
[29] D.S. Johnson, A theoretician's guide to the experimental analysis of algorithms, in: Data Structures, Near Neighbor Searches, and Methodology: Fifth and Sixth DIMACS Implementation Challenges, vol. 59, 2002, pp. 215–250.
[30] P. Jorion, Value at Risk: The New Benchmark for Controlling Market Risk, Irwin Professional Pub., 1997.
[31] N. Jozefowiez, F. Semet, E.-G. Talbi, Multi-objective vehicle routing problems, European Journal of Operational Research 189 (2) (2008) 293–309, https://doi.org/10.1016/j.ejor.2007.05.055.
[32] B. Kallehauge, J. Larsen, O.B. Madsen, M.M. Solomon, Vehicle Routing Problem With Time Windows, Springer, 2005.
[33] N. Kohl, J. Desrosiers, O.B. Madsen, M.M. Solomon, F. Soumis, 2-path cuts for the vehicle routing problem with time windows, Transportation Science 33 (1) (1999) 101–116, https://doi.org/10.1287/trsc.33.1.101.
[34] R. Kolisch, A. Sprecher, PSPLIB – a project scheduling problem library: OR software – ORSEP operations research software exchange program, European Journal of Operational Research 96 (1) (1997) 205–216, https://doi.org/10.1016/S0377-2217(96)00170-1.
[35] A. Krizhevsky, I. Sutskever, G.E. Hinton, Imagenet classification with deep convolutional neural networks, in: Advances in Neural Information Processing Systems, 2012, pp. 1097–1105.
[36] J. Kytöjoki, T. Nuortio, O. Bräysy, M. Gendreau, An efficient variable neighborhood search heuristic for very large scale vehicle routing problems, Computers & Operations Research 34 (9) (2007) 2743–2757, https://doi.org/10.1016/j.cor.2005.10.010.

[37] F. Li, B. Golden, E. Wasil, Very large-scale vehicle routing: new test problems, algorithms, and results, Computers & Operations Research 32 (5) (2005) 1165–1179, https://doi.org/10.1016/j.cor.2003.10.002.

[38] L.A.A. Meira, G.A. Zeni, M. Menzori, P.S. Martins, PostVRP project site, (last checked July 2018), http://www.ft.unicamp.br/~meira/postvrp, 06 2017.

[39] H.D. Nguyen, I. Yoshihara, K. Yamamori, M. Yasunaga, Implementation of an effective hybrid GA for large-scale traveling salesman problems, IEEE Transactions on Systems, Man, and Cybernetics, Part B (Cybernetics) 37 (1) (2007) 92–99, https://doi.org/10.1109/TSMCB.2006.880136.

[40] D. Pecin, A. Pessoa, M. Poggi, E. Uchoa, Improved branch-cut-and-price for capacitated vehicle routing, in: Integer Programming and Combinatorial Optimization, Springer, 2014, pp. 393–403.

[41] G. Reinelt, TSPLIB – a traveling salesman problem library, ORSA Journal on Computing 3 (4) (1991) 376–384, https://doi.org/10.1287/ijoc.3.4.376.

[42] G. Reinelt, Tsplib95, Tech. rep., Interdisziplinäres Zentrum für Wissenschaftliches Rechnen (IWR), Heidelberg, 1995.

[43] J. Renaud, G. Laporte, F.F. Boctor, A tabu search heuristic for the multi-depot vehicle routing problem, Computers & Operations Research 23 (3) (1996) 229–235, https://doi.org/10.1016/0305-0548(95)O0026-P.

[44] M.M. Solomon, Algorithms for the vehicle routing and scheduling problems with time window constraints, Operations Research 35 (2) (1987) 254–265, https://doi.org/10.1287/opre.35.2.254.

[45] R.S. Tol, Estimates of the damage costs of climate change. Part 1: benchmark estimates, Environmental and Resource Economics 21 (1) (2002) 47–73, https://doi.org/10.1023/A:1014500930521.

[46] E. Uchoa, D. Pecin, A. Pessoa, M. Poggi, A. Subramanian, T. Vidal, CVRP library site (last checked July 2018), http://www.galgos.inf.puc-rio.br/vrp/, 2017.

[47] E. Uchoa, D. Pecin, A. Pessoa, M. Poggi, T. Vidal, A. Subramanian, New benchmark instances for the capacitated vehicle routing problem, European Journal of Operational Research 257 (3) (2017) 845–858, https://doi.org/10.1016/j.ejor.2016.08.012.

Chapter 9

Practical applications of smart delivery systems

Mine evacuation as an example of rich vehicle routing problem

Tomasz Jastrzab[a] **and Agata Buchcik**[b]

[a]*Institute of Informatics, Faculty of Automatic Control, Electronics and Computer Science, Silesian University of Technology, Gliwice, Poland,* [b]*Department of Mining Mechanization and Robotisation, Faculty of Mining, Safety Engineering and Industrial Automation, Silesian University of Technology, Gliwice, Poland*

9.1 Introduction

The number of active mines in Poland and their production capabilities are currently decreasing [50]. However, in the still active mines the working miners are at risk due to numerous threats. According to the study of Szlazak et al. [52] related to the accidents in Polish coal mines in the period 1990–2013, the main reasons for emergencies were:

- endogenous fires, which can be monitored and detected in advance as they grow slowly and are signaled by the increase in the levels of gases and fumes,
- egzogenous fires, which are more dynamic in nature and so harder to detect,
- methane ignitions, which are the most dynamic events and were the reasons for the only fatalities in the reported emergencies.

Although in case of methane ignitions, the chances of survival are small, for the remaining two cases, it is possible to help the miners in their evacuation. This can be partially achieved by following the rules and regulations regarding the suggested behavior in case of emergencies [44,45,13,34]. Furthermore, the design of new methods for finding safe exit paths in mines can help to make the number of accidents smaller and to increase the survival rates in case such accidents occur.

Let us note however that the main difficulty in establishing a safe evacuation route in the mine is the number of factors affecting the evacuation process. The factors can be divided into two groups, *objective factors* related to the physical characteristics of the escape route and *subjective factors* depending on the individual characteristics of the miners [52]. The objective factors include, among

Smart Delivery Systems. https://doi.org/10.1016/B978-0-12-815715-2.00012-9

other things:

- the dimensions (width, height, length) of the path and its slope,
- the direction of movement (upward or downward),
- obstacles existing on the path,
- the location and ease of following the signage system,
- the available equipment.

On the other hand, the subjective factors involve: the age, health, and experience of the given miner, the ability to remain calm in the chaotic situation of an emergency, the skills in using the equipment of individual protection, the knowledge of the mine ventilation system organization, the speed of communication between the miners and the rescue team, and the ability of the authorities to take fast and adequate actions [11,52].

Dziurzynski and Palka [11,12] distinguish also an additional group of factors, namely the factors related to the spread of fire. These factors affect the visibility, the amount of oxygen and sensory irritants in the air, and the temperature of the smoky air. According to the study of Kissel and Litton [26], the smoke and its effect on reduced visibility are the key elements hindering the evacuation process as they affect the miners much earlier than any problems with the lack of oxygen or excessive levels of sensory irritants.

With the aim to tackle the complex situation of underground mine evacuation, we consider the problem and its different aspects in the context of *rich* vehicle routing problems (RVRPs). Let us first review the information on the basic vehicle routing problem types to better understand what is meant by the term *rich* VRP. Let us recall that the notion of *capacitated* vehicle routing problem (CVRP) was introduced for the first time by Dantzig and Ramser [9] and since then has been actively developed in many forms and variants [55]. In its most basic formulation a CVRP is defined by the following elements [31]:

- the single depot, being the starting and the ending point of all routes,
- m homogenous vehicles (or, alternatively, at most m homogenous vehicles) constituting a fleet,
- n customers, each with a demand q_i, $i = 1, 2, \ldots, n$,
- the distance matrix C, where each element $c_{ij} \in C$ defines the cost of following the given route or the travel time,
- the vehicle capacity Q or the maximum travel duration (length) L.

The task is then to find exactly (or at most) m vehicle routes of minimum total cost such that the capacity and maximum length constraints are satisfied and all customers are visited only once. Additionally, each route has to begin and finish at the depot [9,31].

To summarize the most frequently discussed types of VRPs, we collected them in Table 9.1 [10,4]. Note that among some of the types presented in Table 9.1, there are also certain subtypes modifying slightly the general requirements specified by the given VRP. For instance, there are a Fleet Size and Mix VRP (FSMVRP), in which the number of vehicles is unlimited; Heterogenous

TABLE 9.1 Vehicle Routing Problem types.

Acronym	Problem type	Characteristics
AVRP	Asymmetric cost matrix VRP	the cost depends on the travel direction
DCVRP	Distance-Constrained VRP	limits the total path length
–	Dynamic VRP	planning and plan execution done simultaneously
GVRP	Green VRP	introduces environmental issues
HVRP	Heterogeneous fleet VRP	involves different types of vehicles
IVRP	Inventory VRP	goods delivered to avoid running out of stock
–	Location VRP	finds routes and depot locations
MDVRP	Multiple Depots VRP	multiple starting/ending points
OVRP	Open VRP	the route does not finish at the depot
PVRP	Periodic delivery VRP	planning horizon spans several days; not each customer has to be visited daily; frequency of deliveries differs
PDVRP	Pickup-and-delivery VRP	pickup and delivery demands at the customer; both demands cannot exceed the capacity
SDVRP	Split-delivery VRP	visits by multiple vehicles at the same customer
–	Stochastic VRP	introduces randomness/uncertainty
VRPB	VRP with Backhauls	all deliveries completed before the pickups
VRPTW	VRP with Time Windows	introduces service time intervals

Fleet VRP with Multiple use of vehicles (HVRPM), in which the vehicles can perform multiple trips; Simultaneous Pickup-and-delivery VRP (SPDVRP), in which the pickup and delivery at the customer happens at the same time; VRP with Multiple/Soft Time Windows, in which there are multiple/flexible service time intervals for the given customer [4].

All the VRP types mentioned in Table 9.1 have some clear definitions and requirements. Unfortunately, for the rich vehicle routing problems discussed in this paper the situation is not that clear. The first attempt to characterize rich VRPs was made by Toth and Vigo [55] as an extension of the existing formulations of the vehicle flow formulations. According to Pellegrini et al. [42], the rich vehicle routing problems try to model the reality more closely by including additional constraints or objectives. Following this line of thought, Crainic et al. [8] defined the rich VRPs as multiattribute optimization tasks. A quantitative approach toward defining the RVRPs was taken by Lahyani et al. [28]. They proposed to define a vehicle routing problem as rich if it combined either four or more strategic/tactical scenario characteristics or at least six routing/operational problem physical characteristics. The former characteristics pertain to the number of depots, the number of trips of a single vehicle, the operation types, etc., whereas the latter include the vehicle characteristics, the time constrains, the number of objective functions, and other things [28]. Throughout this chapter, we assume that the rich vehicle routing problem is the combination of several different VRP types having possibly certain additional characteristics. For more discussion on the various definitions of rich vehicle routing problems, see also [3,4].

The motivation for our research related to the evacuation of mines stems from the fact that although there are numerous methods dealing with safe exit paths searching, new methods and approaches are still much desired. Therefore, in this chapter, we aim to encourage other researchers and software developers working on rich VRPs to take interest in the presented problem and to devise some improved solutions. We believe that with the aid of new dedicated algorithms, it will be possible to increase the chances of miners' survival in case of underground emergencies.

The contributions of this chapter are as follows. Firstly, we propose a completely new rich vehicle routing problem. According to our best knowledge, this is the first time the problem of mine evacuation has been considered in the literature as an example of a rich VRP. Secondly, we lay the foundations for the definition of mine evacuation problem as the VRP by providing the sets of variables and constraints in the mixed-integer linear programming formulation of the problem. Finally, by analyzing the specific features of mine evacuation scenarios we indicate that to tackle such problems, both researchers and VRP software developers may need to change the way of thinking about VRPs. This is because in case of emergencies, it is not the financial cost that matters the most, but the peoples' lives that are at risk. Therefore, the speed of finding the solution, the fulfillment of all constraints (especially those related to the maximum travel duration or the dynamic time windows) and a sufficient accuracy of the solution are the most essential elements.

The remainder of this chapter is organized into four sections. In Section 9.2, we review the existing algorithms related to safe exit path determination and the selected approaches toward (rich) vehicle routing problems solving. In Section 9.3, we formally state the problem of mine evacuation in the context of vehicle routing problems. Next, in Section 9.4, we consider three example evacuation scenarios. Finally, in Section 9.5, we present the summary of the chapter and outline some future research perspectives.

9.2 Literature review

In this section, we first review the selected literature positions pertaining to the problems of evacuation. Note however that not all of the discussed references are directly related to the underground mines evacuation. Nevertheless, the proposed solutions are typically universal enough to be also applied in the mine escape context. In the second part of this section, we focus our attention on the chosen research results related to the rich vehicle routing problems.

9.2.1 Routing in emergencies

It has been stated by Grodzicka and Musiol [20] that the methods related to mine evacuation can be divided into the following three groups:

- statistical methods, which are based on the averaging of the evacuation times measured during multiple traversals of the paths by miners,
- simulation-based methods, typically using dedicated software packages,
- probabilistic methods, in which possible locations of the emergency are evaluated, and based on these estimations, a cost function is optimized.

In line with the first group of methods, Walus et al. [57] present the results of the statistical evaluation of a simulated evacuation scenario in Polish coal mine "Boleslaw Smialy". In their experiment a group of miners differing in age and posture traverses the evacuation path with and without goggles imitating smoky air. The measured escape times are then compared with the times obtained using an analytical method. In conclusion, the authors state that although it is possible to use analytical methods to estimate the evacuation times, a significant time overhead (of even 50%) should be taken into account to better reflect the reality. An important observation made in the paper is also that the fatigue level of miners should also be considered to make the estimations more accurate.

A simulation-based algorithm, implemented outside of a dedicated software package, is employed by Cisek and Kapalka [6]. The authors propose a building evacuation model based on the concepts inspired by the queuing theory. Their model deals with properties such as the number of people per unit surface, the dimensions of the doors, and the flow of people between locations. During the simulation, several different escape scenarios are evaluated simultaneously, and the one with the lowest cost is chosen. Cisek and Kapalka [6] propose also to use the results of the simulation as the source of information for a dynamic signage system directing the evacuees to available safe locations.

The simulation-based approach is also taken by Dziurzynski and Palka [11, 12]. The authors propose to use a software package VentGraph [22] to model the emergency in the mine and to aid the evacuation process. Dziurzynski and Palka [11,12] point out the variety and complexity of factors affecting the miners escape and propose to include in the estimations only the factors that can be assessed quantitatively. Furthermore, they suggest to include certain additional factors only if the estimated time is close to the time of operation of the equipment of individual protection. From the algorithmic point of view, the VentGraph software allows us to either determine all safe exit paths or to lead the miners to a particular location using the shortest path algorithms. The software uses graph-theoretic methods, both to guide the evacuees and to model the distribution of fire and fumes in the tunnels.

Another example of the simulation-based evacuation model is given by Adijski et al. [1]. The authors use two software tools, the fire dynamics simulator PyroSim [54] and the mine ventilation system simulator MineFire Pro+ [33], to model the spread of fire and to determine the affected parts of the mine. They also introduce the notion of an equivalent tunnel length, by including in their calculations the coefficients for the tunnel type, the tunnel slope, and the levels of gases and fumes. By applying these coefficients to the modeled situation Adijski et al. [1] compute the speed of miners movement and are thus able to better determine the optimal evacuation routes. Different groups of miners as well as different starting and ending locations are considered. Concluding, the authors underline that for the proposed methodology of safe path determination to be successful, it is necessary to constantly adapt the model to the changing layout of the mine.

The notion of the equivalent tunnel length appears also in the paper by Yan and Feng [61]. However, unlike Adijski et al. [1], who just focus on the tunnel type and slope as well as the effects of fire spread, Yan and Feng [61] include also additional coefficients, such as the wind speed, the tunnel particle concentrations, and the crowded degree. Additionally, by introducing the coefficient related to mine disaster they allow for more flexibility in modeling the different situations in the mine. To find the optimal escape route, apart from using the equivalent length, which affects the miners speed, Yan and Feng [61] propose to use an improved ant algorithm. The improvements are related to zoning of the tunnels (which speeds up the algorithm as unnecessary detours are avoided) and modified ant meeting and death rules. The downside of the proposed approach is that it only considers the single starting and ending points of the evacuation route.

Bioinspired algorithms for safe exit path searching are also proposed by Goodwin et al. [19,18]. The authors use Ant Colony Optimization (ACO) to find the shortest routes in buildings or on ships. The key point of their approach is that they model the hazard by a stochastic function involving, for example, the air temperature or heat radiation [18]. Additionally, they propose to stop the movement of the evacuees when the smoke levels reach the threshold above which the visibility is too low to safely proceed further. The aforementioned hazard function is applied to the static, dynamic, or capacitated environment. The environment is considered *static* when the hazard function does not change in time, it is *dynamic* when the probability of hazard occurrence changes regularly, and finally it is *capacitated* or control-flow based when capacity constraints are applied to the passages [19].

Apart from the aforementioned algorithms, there is also a number of methods originating from the graph theory. Among them, we may find the works of Jallali and Noroozi [24], who use the Floyd–Warshall algorithm supplemented with the π algorithm allowing us to find the actual routes to the safe exit points. On the other hand, Chen and Li [5], who also use the Floyd–Warshall algorithm, propose to include the crowd behavior characteristics in the model. The crowd behavior properties involve the maximization of the individual speed in the initial phase of the escape and its consecutive adjustment depending on the crowd density. Yet another solution is proposed by Jastrzab and Buchcik [25], who use the Dijkstra algorithm for finding the safe escape path. The authors consider also various speeds of miners, implicitly modeling the different factors affecting the evacuation process, discussed in [51,52,11].

To sum up the presented literature review, let us note that mine evacuation is a complex process with a multitude of factors affecting its course. Due to its complexity, most of the existing methods and algorithms assume certain simplifications of the real-life situation. Although such an approach allows us to find the solutions faster, their accuracy can be unsatisfactory. Therefore we conjecture that it is essential to continue the research on efficient methods dealing with as many factors influencing the evacuation process as possible. As a step toward satisfying this need, we propose to model the mine escape scenarios as rich vehicle routing problems discussed in the following sections.

9.2.2 Rich vehicle routing problems

To aid the design of new methods for solving rich VRPs, especially those related to the mine evacuation planning and execution, let us briefly review some of the existing methods dealing with VRPs. They can be divided into exact and approximate methods, the latter group including heuristics and metaheuristics [3].

Since vehicle routing problems are known to be NP-hard, exact algorithms can only find the optimal solutions for relatively small instances of even the basic VRPs. The following nonexhaustive list of exact algorithms for solving (rich) VRPs can be given [30,4]:

- branch-and-bound, branch-and-cut, branch-and-price, branch-and-cut-and-price algorithms, usually applicable to relatively small instances of basic CVRPs, modeled as (mixed-)integer linear programs,
- dynamic programming and column generation methods, which divide the larger problem into smaller subproblems,
- set partitioning and constraint programming methods, which are flexible enough to handle several constraints simultaneously within reasonable amount of time,
- A* and IDA* (A* with iterative deepening) algorithms, which are used in the context of shortest path searching problems.

In contrast to the exact methods, the heuristics and metaheuristics do not guarantee to find the optimal solution, but they allow us to find near-optimal solutions for much larger problem instances in a reasonable amount of time. Furthermore, the heuristic methods are usually tailored to solve some specific problems although there are also certain exceptions from this rule [7,43]. The metaheuristics in turn have some greater flexibility as to the range of problems they can solve [53]. Among the approximate methods for solving the basic and rich VRPs, we can find the following examples:

- nature inspired methods, such as the ant colony optimization [42], the bat algorithm [38], the genetic algorithms [40,41], the continuous and discrete versions of the firefly algorithm [39,37], and the continuous and discrete versions of the particle swarm optimization [48,17],

- greedy randomized adaptive search procedure [49], Clarke–Wright Savings method (CWS) [2], nearest, variable, and large neighborhood search methods [16,46], simulated annealing [23], tabu search [29,46],
- memetic algorithms, being the combination of the evolutionary algorithms performing the exploration of the search space, and local optimization techniques responsible for the exploitation of the search space [36,35].

For an excellent classification of other RVRP-related papers and solution methods, see also [3]. For an overview of exact and approximate methods used in different software packages dealing with VRPs, see [10]. Finally, to recall the general techniques used for solving the VRPs, such as the solution space size reduction or the temporary acceptance of the infeasible solutions, we refer to [58] and its references.

Let us now consider several examples of rich vehicle routing problems discussed in the literature. In general, they can be used to model the delivery of goods, the transportation of urban waste, the transportation of people (also with reduced mobility), the planning of electrical vehicle routes between different recharging stations, the ATM replenishment process, etc.

Masmoudi et al. [32] discuss a rich vehicle routing problem in the context of e-commerce logistics distribution system. They combine the heterogenous fleet approach with multicommodity demands from the customers and time windows determining the preferred delivery times. In their approach the fleet is composed of a limited number of vehicles of specialized types, which can carry only some selected types of goods. A single-depot and single-trip constraints are assumed. However, the given customer can be served by multiple vehicles, each dedicated to the transportation of the given type of commodities. The considered problem models the home delivery of products, often supported by super- and hypermarkets with online shopping possibilities. Masmoudi et al. [32] propose to use the mixed integer programming formulation of the problem. The problem is solved with the use of the CPLEX library [21].

Mixed-integer programming formulation of the problem is also applied by Souza Neto and Pureza [49]. The authors model the problem of delivery of drinks in Brazilian urban areas. They assume that the goods are delivered by multiple companies (from multiple depots), with vehicles performing multiple trips within the given time windows. Additionally, since the urban area is dense and multiple customers are located in close proximity, the authors propose to cluster them and increase the number of deliverymen to serve the demands. Furthermore, the regulations regarding the working hours of deliverymen or the risk factors resulting from visiting certain areas are taken into account. Regarding the algorithms for solving the given RVRP, Souza Neto and Pureza [49] use the Greedy Randomized Adaptive Search Procedure (GRASP) [14] combined with the heuristic and local search procedures and the branch-and-cut algorithm implemented in the GAMS/CPLEX [15] library. A hybrid approach combining GRASP and branch-and-cut algorithm is also considered.

The problem of both the vehicle routing and the driver (or deliverymen) scheduling is also discussed by Wen [58]. The author solves the VRP with Time Windows, in which the goods are delivered using a heterogenous fleet of vehicles over a multiperiod horizon and in a way conforming to the driving rules and drivers' working hours. The mixed-integer linear programming approach supplemented with a multilevel neighborhood search algorithm is applied to solve the given RVRP. Moreover, Wen [58] also considers two additional rich VRPs:

- the problem in which the goods to be delivered are first picked up at different locations and then consolidated at the depot, from which they are delivered further [60],
- the problem with uncertainty as to the types and amounts of future demands, modeled by a rolling planning horizon over which the demands are revealed incrementally [59].

The former problem is solved by means of tabu search heuristic, whereas for the latter, the mixed-integer programming formulation is applied.

A different rich vehicle routing problem is tackled by Osaba et al. [37]. Namely, the authors consider a newspaper distribution system with recycling policy. They model the problem as asymmetric clustered problem of simultaneous pickup and delivery with time windows, variable costs, and forbidden paths. The clustered nature of the problem arises from the company policy of disallowing multiple vehicles traveling to the same town or area. The recycling policy forces the problem to be modeled as SPDVRP, whereas the variable costs and forbidden path restrictions result from the real-life situations of "peak hours" in towns or one-way roads that cannot be used. The authors present the solution based on the bioinspired discrete firefly algorithm, in which the directions of fireflies' movement change in time and their light intensity is modified according to the absorption coefficient.

Lahyani [27] considered a problem of olive oil pickup in Tunisia. The problem is modeled as a multiproduct multiperiod multicompartment VRP with time windows and heterogenous fleet of vehicles. Similarly to the work of Osaba et al. [37], in which the variable costs and forbidden paths model the complexity of the real-life situation, Lahyani [27] introduces the constraints related to the required cleaning of compartments. The cleaning activity results from the regulations prohibiting the transportation of different types of olive oil in the same compartment without prior cleaning. The author proposes to solve the problem by means of mixed-integer linear programming supplemented with additional valid inequalities and an original branch-and-price algorithm.

A more general view of rich vehicle routing problems is taken in [56,47], where the authors do not focus on particular practical problems, but try to provide the frameworks for modeling various RVRPs. Vogel [56] introduces a solution framework for rich VRPs based on various metaheuristic algorithms. The author also prioritizes the features of a "good" VRP solving framework, indicating that the simplicity should be considered more important than the flexibility, which in turn should dominate the accuracy and the speed of the solver. On the other hand, Sim et al. [47] propose a model for a certain class of rich VRPs based on different scenario and problem physical characteristics (see also [28]). They also create a benchmark set of problems for the modeled RVRPs.

The rich vehicle routing problems are also discussed by Drexl [10] in the context of a serious gap between the mostly theoretical literature studies of VRPs and the practical, software-based solutions dealing with real-life VRPs. The author suggests that there are certain aspects in which the software developers lag behind the researchers, such as stochastic planning, time-dependent travel times, and mathematical programming approaches. On the other hand, the scientific community usually targets idealized benchmarks without regard for the "soft" or nondeterministic requests dealt with the software packages. Furthermore, the author points out that the main idealization of the VRPs is that the vehicles are independent of each other, which is not the case in reality (e.g., due to meeting points such as hubs and/or cross-docking locations).

To sum up the presented review of selected rich vehicle routing problems, we can state that there is a wide range of practical applications of rich VRPs. Most of them try to closely model some part of the reality, paying particular attention to some specific features of the given situation. These features can include, for example, legal regulations pertaining to the particular product being delivered [27] or additional requirements typical to the given region or city [49]. As for the algorithms for solving RVRPs, there is a certain number of general techniques that can be applied to different problems with little or no additional effort. Therefore it is rarely necessary to design heuristics dedicated to the given singular problem type.

9.3 Mine evacuation as a rich VRP

Let us now consider the features of mine evacuation process that correspond to different characteristics of rich vehicle routing problems. Let us note however that in the problem given, instead of routing the vehicles, we route the miners. Furthermore, the miners do not deliver or pick up any goods; they are just concerned with routing to the safe locations. The following features are then taken into account in the proposed problem formulation:

1. The fleet is *fixed* since there is a fixed number of miners working underground (and they constitute the fleet).
2. The fleet is also *heterogenous* since the miners differ in their age, fitness, experience, skills, etc. We assume here that the different subjective factors discussed by Szlazak et al. [52] can be modeled by means of different speeds of the miners [25], thus making the fleet heterogenous.
3. The problem is *asymmetric* since the movement downward is easier than the movement upward, especially in the case of steep or slippery slopes. Additionally, if one follows the suggestion of Yan and Feng [61] to also include

the wind speed and direction, the asymmetry of the problem may become even more evident (i.e., the weights of the cost matrix assigned to different directions of movement over the given path will differ significantly).

4. The problem is *open* as the path does not have to (or to be more precise, must not) end at the depot (starting location). The reason for not finishing the path at the depot is that the miners are trying to escape from the endangered zones and therefore should not return to their initial locations.
5. The problem is *dynamic* in the sense that the routing may change as the execution of the escape plan progresses. The reason why such a situation may occur is that the location of the fire and its spread cannot be determined in advance. So, it is possible that rerouting decisions will have to be taken later on during the escape.
6. We assume the *restricted route durations*, since the equipment of individual protection can be used for only limited amount of time (usually 50–60 minutes).
7. We assume *multiple depots* as the miners can work in different parts of the mine and start the evacuation from the locations in which they work.
8. We assume a *single trip* of each miner, as they just need to evacuate to some safe location. However, it is also possible to assume multiple trips when a rescue team needs to go and fetch trapped miners.
9. We assume the existence of *forbidden paths*. These could represent some collapsed tunnels or the tunnels in which some mining equipment is located, thus making the paths hardly traversable. Furthermore, some of the paths may dynamically become forbidden due to the spread of fire or the fumes and gases.
10. To handle the dynamic type of forbidden paths, we assume that each intersection has an associated *time window*. However, we are mainly concerned with the upper bound of the time window, that is, the latest time at which the intersection can be reached.
11. We assume that *multiple visits* in the given node are allowed because the miners working in the deeper areas of the mine may, for example, reach at some point the depot of the miners working in the areas closer to the ground. From there they may follow the optimal path of their coworkers (unless the dynamic forbidden paths prevent it). Furthermore, we assume that not all nodes have to be visited.

Given these assumptions, we can model the problem as a graph $G = (V, E)$, where V is the set of nodes representing tunnel intersection points, and E is the set of edges representing the tunnels. Since the cost matrix is asymmetric, we assume that whenever two different vertices v_i, v_j are connected, there are two connecting edges with associated costs.

We assume that there are three groups of nodes, denoted by V_{d}, V_{m}, and V_{s}, representing respectively the depots, the middle (or intermediate) nodes, and the safe nodes. These groups of nodes satisfy the following conditions:

$$V_d \cup V_m \cup V_s = V, \tag{9.1}$$

$$V_d \cap V_m = \emptyset,\ V_d \cap V_s = \emptyset,\ V_m \cap V_s = \emptyset, \tag{9.2}$$

meaning that all together they constitute the whole set of nodes and that they are mutually disjoint.

We assume that the costs correspond to the equivalent lengths of the tunnels. By *an equivalent* length of the tunnel we understand here the actual length multiplied by the coefficients representing the different features of the tunnel, such as its slope and type, the wind speed and direction, and others [61,1]. The cost of traversing the forbidden path is infinite to ensure that these paths are not included in the final solution.

In what follows, we assume the following notation:

- Sets:
  - V_d – a set of depot nodes, of size N_d;
  - V_m – a set of intermediate nodes being neither the depot nodes nor the safe nodes, of size N_m;
  - V_s – a set of safe nodes, of size N_s;
  - M – a set of miners to evacuate, of size K.
- Indices:
  - i, j, l – nodes (tunnel intersections). If node i belongs to the set of depot nodes, then $i = 0, 1, \ldots, N_d - 1$. If node i belongs to the set of intermediate nodes, then $i = N_d, N_d + 1, \ldots, N_d + N_m - 1$. If node i belongs to the set of safe nodes, then $i = N_d + N_m, N_d + N_m + 1, \ldots, N_d + N_m + N_s - 1$. Note that $N = N_d + N_m + N_s$ is the total number of nodes;
  - k – miners, $k = 0, 1, \ldots, K - 1$;
  - f – fire (or emergency in general).
- Parameters:
  - a_i, b_i – the earliest and latest times to reach node i, but we assume that $a_i = 0$ for all $i = 0, 1, \ldots, N - 1$;
  - l_{ij} – equivalent length of a path between nodes i and j (cost);
  - v_k – speed of miner k;
  - v_f – speed of fire spread;
  - K_i – number of miners initially at depot i, $0 \leq i \leq N_d - 1$.
  - T – maximum route duration, equivalent to the working time of the equipment of individual protection;
  - T_f – time to identify the fire location (nonnegative real number);
  - P – large positive number;
- Variables:
  - x_{ijk} – a binary variable denoting whether a path between nodes i and j was used by miner k;

- $x_{ij\text{f}}$ – a binary variable denoting whether the fire spread through a path between nodes i and j;
- t_{ijk} – time to traverse a path between nodes i and j by miner k;
- $t_{ij\text{f}}$ – time to traverse a path between nodes i and j by the fire;
- t_{ik} – time to reach node i by miner k;
- $t_{i\text{f}}$ – time to reach node i by the fire.

Note that we implicitly assume that $l_{ij} = 0$, $x_{ijk} = 0$, $x_{ij\text{f}} = 0$, $t_{ijk} = 0$, and $t_{ij\text{f}} = 0$ whenever $i = j$ for all $0 \le i, j \le N - 1$, $0 \le k \le K - 1$. Furthermore, $t_{ik} = 0$ for all K_i miners located at the ith depot node.

Given these elements, the problem is formulated as a mixed-integer linear programming problem with the objective function

$$\min \sum_{i=0}^{N-1} \sum_{j=0}^{N-1} \sum_{k=0}^{K-1} x_{ijk} t_{ijk}, \tag{9.3}$$

subject to the following constraints:

$$x_{ijk} \in \{0, 1\} \text{ for all } 0 \le i, j \le N - 1, 0 \le k \le K - 1; \tag{9.4}$$

$$x_{ij\text{f}} \in \{0, 1\} \text{ for all } 0 \le i, j \le N - 1; \tag{9.5}$$

$$t_{ik} \in \mathbb{R},\ t_{ijk} \in \mathbb{R} \text{ for all } 0 \le i, j \le N - 1, 0 \le k \le K - 1; \tag{9.6}$$

$$t_{ik} \ge 0, t_{ijk} \ge 0 \text{ for all } 0 \le i, j \le N - 1, 0 \le k \le K - 1; \tag{9.7}$$

$$t_{i\text{f}} \in \mathbb{R}, t_{ij\text{f}} \in \mathbb{R} \text{ for all } 0 \le i, j \le N - 1; \tag{9.8}$$

$$t_{i\text{f}} \ge 0, t_{ij\text{f}} \ge 0 \text{ for all } 0 \le i, j \le N - 1; \tag{9.9}$$

$$\sum_{i=0}^{N_\text{d}-1} K_i = K; \tag{9.10}$$

$$\sum_{j=0}^{N-1} \sum_{k=0}^{K-1} x_{ljk} - \sum_{i=0}^{N-1} \sum_{k=0}^{K-1} x_{ilk} = K_l \text{ for all } 0 \le l \le N_\text{d} - 1; \tag{9.11}$$

$$\sum_{i=0}^{N-1} x_{ilk} - \sum_{j=0}^{N-1} x_{ljk} = 0 \text{ for all } N_\text{d} \le l \le N_\text{d} + N_\text{m} - 1, 0 \le k \le K - 1; \tag{9.12}$$

$$\sum_{i=0}^{N-1} x_{ijk} \le 1 \text{ for all } N_\text{d} + N_\text{m} \le j \le N_\text{d} + N_\text{m} + N_\text{s} - 1, 0 \le k \le K - 1; \tag{9.13}$$

$$\sum_{j=0}^{N-1} \sum_{k=0}^{K-1} x_{ijk} = 0 \text{ for all } N_\text{d} + N_\text{m} \le i \le N_\text{d} + N_\text{m} + N_\text{s} - 1; \tag{9.14}$$

$$\sum_{i=0}^{N-1} \sum_{j=N_\text{d}+N_\text{m}}^{N-1} \sum_{k=0}^{K-1} x_{ijk} = K; \tag{9.15}$$

$$l_{ij}x_{ijk} < \infty \text{ for all } 0 \leq i, j \leq N-1, 0 \leq k \leq K-1; \tag{9.16}$$

$$\sum_{i=0}^{N-1}\sum_{j=0}^{N-1} x_{ijk}t_{ijk} \leq T \text{ for all } 0 \leq k \leq K-1; \tag{9.17}$$

$$a_i \leq t_{ik} \leq b_i \text{ for all } 0 \leq i \leq N-1, 0 \leq k \leq K-1; \tag{9.18}$$

$$b_i = \begin{cases} \infty \text{ for execution time } < T_\mathrm{f}, \\ t_{i\mathrm{f}} \text{ for execution time } \geq T_\mathrm{f}, \end{cases} \quad \text{for all } 0 \leq i \leq N-1; \tag{9.19}$$

$$t_{ik} + t_{ijk} - P(1 - x_{ijk}) \leq t_{jk} \text{ for all } 0 \leq i, j \leq N-1, 0 \leq k \leq K-1; \tag{9.20}$$

$$t_{i\mathrm{f}} + t_{ij\mathrm{f}} - P(1 - x_{ij\mathrm{f}}) \leq t_{j\mathrm{f}} \text{ for all } 0 \leq i, j \leq N-1; \tag{9.21}$$

$$t_{ijk} = l_{ij}/v_k \text{ for all } 0 \leq i, j \leq N-1, 0 \leq k \leq K-1; \tag{9.22}$$

$$t_{ij\mathrm{f}} = l_{ij}/v_\mathrm{f} \text{ for all } 0 \leq i, j \leq N-1. \tag{9.23}$$

The objective of the problem at hand, given by (9.3), is to minimize the total travel time of all evacuees. We assume here that the miners maintain constant speed on the whole route. So, the travel time can be calculated as a ratio of the distance traveled and the speed of miner. Constraints (9.4) to (9.9) define the domains of the problem variables, whereas (9.10) specifies that all miners are initially located at the depot nodes. Constraints (9.11) to (9.13) denote the conservation principles for the depot, intermediate and safe nodes, respectively. They state that the number of miners leaving each depot is the sum of the number of miners entering the depot and the number of miners initially located at the depot and that the miners do not stay at the intermediate nodes, and they may enter any safe node only once. Constraint (9.14) indicates that upon reaching the safe point, the miners do not move further, whereas constraint (9.15) specifies that all the miners collectively reach the safe nodes. Inequality (9.16) excludes forbidden paths, whereas inequality (9.17) restricts the maximum travel time of each miner. Constraints (9.18) and (9.19) are related to the dynamic time windows resulting from the spread of fire, fumes, and gases. Finally, constraints (9.20) to (9.23) ensure that all time variables are valid.

9.4 Evacuation scenario examples

Let us now consider a hypothetical mine model shown in Fig. 9.1. We assume that there are two depot nodes, denoted 0 and 1, and two safe nodes, marked 14 and 15. We also assume that there are five miners,[1] two of which are located at the depot 0, and the remaining three at the depot 1. Furthermore, we assume that the miners' speeds are as follows: $v_0 = 65$ m/min, $v_1 = 35$ m/min, $v_2 = 65$ m/min, $v_3 = 45$ m/min, and $v_4 = 55$ m/min [25]. We also set $T = 60$ min, and so the maximum route lengths for the five miners are respectively equal to 3900 m (for miners 0 and 2), 3300 m (for miner 4), 2700 m (for miner 3), and 2100 m (for miner 1).

1. The miners are marked in Fig. 9.1 with small circles within nodes 0 and 1.

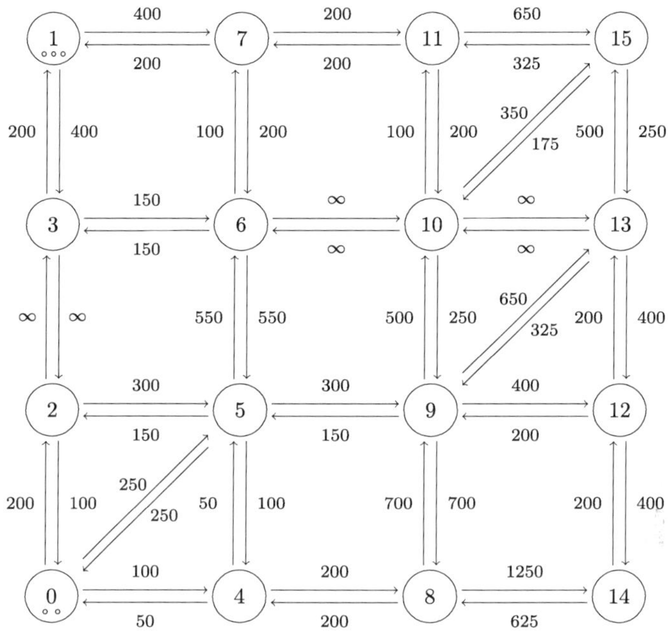

FIGURE 9.1 An example mine model. The nodes, numbered 0–15, denote the intersections. Nodes 0 and 1 are the initial depots, whereas nodes 14 and 15 are the destination (safe) depots. The arcs, labeled with equivalent lengths, denote the tunnels. The small circles in nodes 0 and 1 represent the miners.

The first scenario we consider is a scenario of pure routing without any emergencies. Therefore $T_f = \infty$, since no fire is present in the mine. As a consequence, the time windows for each node remain unchanged and are given by $(a_i, b_i) = (0, \infty)$ for all $0 \le i \le 15$.

It is then easy to observe that both miners located initially at depot 0 will move to the safe node number 14, following the path 0 – 4 – 5 – 9 – 12 – 14 and covering the total distance of 1250 m. On the other hand, the three miners located at the depot 1 should go to the safe node 15, visiting the nodes 1 – 7 – 11 – 10 – 15. The distance covered is equal to 1050 m. The objective function value is approximately equal to 113.5 minutes. Note also that the constraints of maximum route duration are satisfied for all miners.

In the second scenario, we assume that the fire appears in the tunnel between nodes 4 and 8 at the distance 100 m from each node. We also assume that the speed of fire $v_f = 50$ m/min and that the fire is detected immediately, which means that $T_f = 0$. Furthermore, only for simplicity of the example, we assume that the fire spreads toward nodes 4 and 8, and later also toward node 9. Upon reaching the aforementioned nodes, the fire ceases its spread.

Due to the spread of fire, some time windows no longer remain unchanged, and the situation in the mine is shown in Fig. 9.2 (the fire location is marked

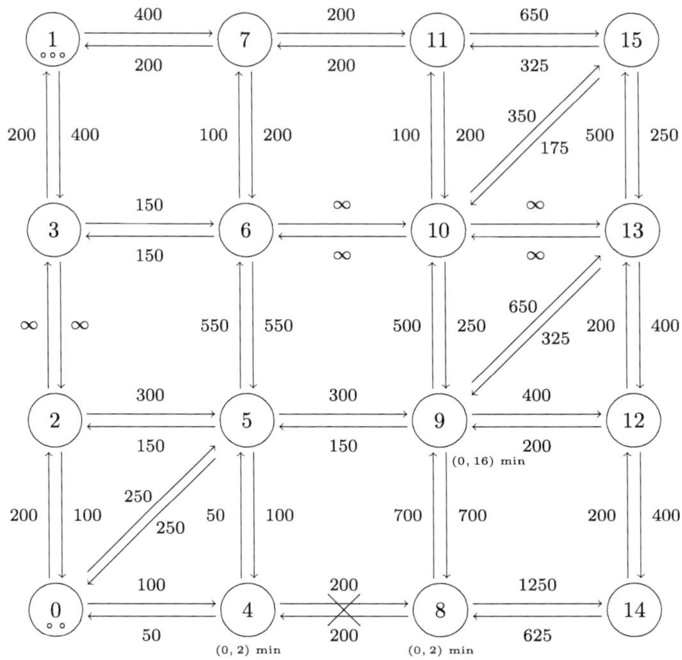

FIGURE 9.2 An example mine model with time windows. The nodes, numbered 0–15, denote the intersections. Nodes 0 and 1 are the initial depots, while nodes 14 and 15 are the destination (safe) depots. The numbers below nodes 4, 8 and 9 denote the safe time windows. The arcs, labeled with equivalent lengths, denote the tunnels. The small circles in nodes 0 and 1 represent the miners. The cross denotes the fire location.

with a cross). The time windows for nodes 4, 8, and 9, which are affected by the fire, are shown below the considered nodes.

The location of the fire and its spread do not affect the miners moving between depot 1 and safe node 15. However, due to the different speeds of miners located at the depot 0, miner 1 has to change the route taken in the first scenario to satisfy constraint (9.18) for node 4. It is so because he reaches node 4 after around 2.85 minutes, which is no longer allowed due to the applied time window. On the other hand, miner 0 reaches node 4 after 1.5 minute, which satisfies constraint (9.18). As a consequence, the route for miner 0 remains unchanged and includes nodes 0, 4, 5, 9, 12, 14. The route for miner 1 becomes 0 – 5 – 9 – 12 – 14 and is longer from the route determined in the first scenario by 100 m, which corresponds to around 3 minutes. The objective function value becomes then approximately equal to 116.4 minutes.

Finally, let us consider again the example shown in Fig. 9.2, but this time, let us assume that the fire location is determined after $T_f = 2$ min. Obviously, this does not change the time windows for nodes 4, 8, and 9, but it affects the behavior of miner 1. Since the miner does not know the location of the fire, he starts moving according to the path determined in the first scenario. However, after 2 minutes (70 m), he gets to know that the intersection at node 4 cannot be used due to the fire spread, and consequently, a rerouting decision has to be made. The miner returns to the depot and picks the route through node 5. However, the route through node 9 is also unavailable, since the loss of 4 minutes for rerouting makes it impossible to reach node 9 within the given time window. Hence the final route for miner 1 becomes 0 – 5 – 6 – 7 – 11 – 15. Let us note that the distance covered equals then 1790 m, which corresponds to 51 minutes, a result close to the maximum allowed duration of 60 minutes. The objective function value is consequently also much larger and is equal to almost 130 minutes. Finally, because of the delay in the determination of the fire location, the miner ends up in a different safe node than in the previous two scenarios.

To conclude, let us observe that using the concepts related to vehicle routing problems, we managed to successfully model the mine evacuation process. It should be also underlined that the crucial aspect of safe evacuation is to minimize the time to determine the actual location of the fire. As shown in the third example, even relatively small delay in finding the fire location may force the miners to take a much longer and possibly dangerous route through the mine.

9.5 Summary and future work

In the chapter, we have presented an overview of selected works in the field of practical applications of rich vehicle routing problems. We have also proposed to model the process of underground mine evacuation as a rich VRP. To this aim, we have reviewed the literature pertaining to the simulation and analysis of evacuation scenarios. We have also devised a mixed-integer linear programming formulation of the rich VRP version of the evacuation process. Finally, to better illustrate the proposed approach, we have presented three examples of mine evacuation scenarios.

As an outline for the future research, we propose to investigate:

1. The possible ways of modeling the spread of fire and gases in the tunnels. Fire dynamics simulators, such as the PyroSim simulator, or analytical methods could be named as just two examples of possible solutions. From the perspective of the presented problem, the research on fire spread models could help to determine better and more precise time windows.
2. The possibility of applying stochasticity to the determination of the time windows based on various probability distribution functions. Such an approach should help to establish more precise time windows in the period before the fire location is found.
3. The inclusion of interactions between miners, such as the crowded degree of a given tunnel or intersection, or the fact that the faster miner moving behind the slower one has to adjust his speed accordingly. This should make the problem even more dynamic, but at the same time closer to the reality.

References

[1] V. Adijski, D. Mirakovski, Z. Despodov, S. Mijalkovski, Simulation and optimization of evacuation routes in case of fire in underground mines, Journal of Sustainable Mining 14 (2015) 133–143.
[2] J. Caceres-Cruz, D. Riera, R. Buil, A. Juan, Applying a savings algorithm for solving a rich vehicle routing problem in a real urban context, in: Proceedings of the 5th International Conference on Applied Operational Research, in: LNMS, vol. 5, Tadbir Operational Research Group Ltd., 2013, pp. 84–92.
[3] J. Cacerez Cruz, Randomized Algorithms for Rich Vehicle Routing Problems: From a Specialized Approach to a Generic Methodology, Ph.D. thesis, Universitat Oberta de Catalunya, 2013.
[4] J. Cacerez Cruz, P. Arias, D. Guimarans, D. Riera, A. Juan, Rich vehicle routing problem: survey, ACM Computing Surveys 47 (2) (2014) 1–28.
[5] X. Chen, Y. Li, Geometry-based virtual simulation for fire escape in emergency environment, in: L. Huang, K. Lai, J. Cao, M. Li (Eds.), Fourth International Conference on Networking and Distributed Computing, IEEE, 2013, pp. 56–59.
[6] M. Cisek, M. Kapalka, Evacuation route assessment model for optimization of evacuation in buildings with active dynamic signage system, in: W. Daamen, D. Duives, S. Hoogendoorn (Eds.), The Conference on Pedestrian and Evacuation Dynamics 2014, PED2014, in: Transportation Research Procedia, vol. 2, Elsevier, 2014, pp. 541–549.
[7] J. Cordeau, G. Laporte, A. Mercier, A unified tabu search heuristic for vehicle routing problems with time windows, Journal of the Operational Research Society 52 (8) (2001) 928–936.
[8] T. Crainic, G. Crisan, M. Gendreau, N. Lahrichi, W. Rei, A concurrent evolutionary approach for rich combinatorial optimization, in: F. Rothlauf (Ed.), Proceedings of the 11th Annual Conference Companion on Genetic and Evolutionary Computation Conference: Late Breaking Papers, ACM, New York, NY, 2009, pp. 2017–2022.
[9] G. Dantzig, J. Ramser, The truck dispatching problem, Management Science 6 (1) (1959) 80–91.
[10] M. Drexl, Rich Vehicle Routing in Theory and Practice, Tech. rep., Johannes Gutenberg University, Mainz, Germany, 2011.
[11] W. Dziurzynski, T. Palka, Optimization of Escape Routes in Case of Fire With Respect to Crew Withdrawal, Tech. rep., Instytut Mechaniki Gorotworu PAN, 2013, in Polish.
[12] W. Dziurzynski, T. Palka, Determining the fire exit routes in underground mines – new potentials of the VentGraph system, Prace Instytutu Mechaniki Gorotworu PAN 16 (1–2) (2014) 3–16, in Polish.
[13] Escape and Rescue Operations in Mines Working Group, Guidance and Information on the role and design of safe havens in arrangements for escape from mines, Health and Safety Executive, 2007.
[14] T. Feo, M. Resende, Greedy randomized adaptive search procedures, Journal of Global Optimization 6 (2) (1995) 109–134.
[15] GAMS Development Corporation, CPLEX 12 Documentation, 2018.
[16] A. Goel, V. Gruhn, Large Neighborhood Search for rich VRP with multiple pickup and delivery locations, in: P. Hansen, N. Mladenovic, J. Moreno Perez (Eds.), Proceedings of the 18th Mini Euro Conference on VNS, 2005, pp. 1–11.
[17] Y. Gong, J. Zhang, O. Liu, R. Huang, H. Chung, Y. Shi, Optimizing the vehicle routing problem with time windows: a discrete particle swarm optimization approach, IEEE Transactions on Systems Man and Cybernetics Part C (Applications and Reviews) 42 (2) (2012) 254–267.
[18] M. Goodwin, O. Granmo, J. Razianti, Escape planning in realistic fire scenarios with ant colony optimisation, Applied Intelligence 42 (1) (2014) 24–35.
[19] M. Goodwin, O. Granmo, J. Razianti, P. Sarshar, S. Glimsdal, Ant colony optimisation for planning safe escape routes, in: M. Ali, T. Bosse, K. Hindriks, M. Hoogendoorn, C. Jonker, J. Treur (Eds.), Proceedings for 26th International Conference on Industrial, Engineering and Other Applications of Applied Intelligent Systems, IEA/AIE 2013, in: LNAI, vol. 7906, Springer, Berlin-Heidelberg, 2013, pp. 53–62.

[20] A. Grodzicka, D. Musiol, Evacuation routes miners – selected results of the survey, Gornictwo i Geologia 8 (1) (2013) 41–52, in Polish.

[21] IBM ILOG, CPLEX Callable Library (C API) Reference Manual, 2018.

[22] Instytut Mechaniki Gorotworu PAN, System of computer programs of ventilation engineer VentGraph Reference Manual, 2009.

[23] R. Jachimowski, M. Klodawski, Simulated annealing algorithm for the multi-level vehicle routing problem, Transport 97 (2013) 195–204.

[24] S. Jallali, M. Noroozi, Determination of the optimal escape routes of underground mine networks in emergency cases, Safety Science 47 (8) (2009) 1077–1082.

[25] T. Jastrzab, A. Buchcik, A parallel algorithm for finding the shortest exit paths in mines, in: Proceedings of the International Conference of Computational Methods in Sciences and Engineering, ICCMSE 2017, in: AIP Conference Proceedings, vol. 1906, American Institute of Physics, 2017, pp. 1–4, article no. 200013.

[26] F. Kissel, C. Litton, How smoke hinders escape from coal mine fires, Mining Engineering 44 (1) (1992) 79–83.

[27] R. Lahyani, Unified Matheuristic for Solving Rich Vehicle Routing Problems, Ph.D. thesis, École Centrale de Lille, 2015.

[28] R. Lahyani, M. Khemakhem, F. Semet, Rich vehicle routing problems: from a taxonomy to a definition, European Journal of Operational Research 241 (2015) 1–14.

[29] D. Lai, O. Demirag, J. Leung, A tabu search heuristic for the heterogeneous vehicle routing problem on a multigraph, Transportation Research Part E 86 (2016) 32–52.

[30] G. Laporte, The vehicle routing problem: an overview of exact and approximate algorithms, European Journal of Operational Research 59 (1992) 345–358.

[31] G. Laporte, Fifty years of vehicle routing, Transportation Science 43 (4) (2009) 408–416.

[32] M. Masmoudi, M. Benaissa, H. Chabchoub, Mathematical modeling for a rich vehicle routing problem in e-commerce logistics distribution, in: M. Abed, M. Benaissa (Eds.), Proceedings of the 2013 International Conference on Advanced Logistics and Transport, IEEE, 2013, pp. 290–295.

[33] Mine Ventilation Services, Inc., MineFire Pro+ User Manual, 2013.

[34] Mines Rescue Working Group and Mine Safety Operations Division of Industry & Investment NSW, Guidelines for underground emergency escape systems and the provision of self rescuers, Guidelines for determining withdrawal conditions from underground coal mines, Guidelines for in-seam response using CABA for events when life is at risk, 2010.

[35] J. Nalepa, M. Blocho, Adaptive cooperation in parallel memetic algorithms for rich vehicle routing problems, International Journal of Grid and Utility Computing 9 (2) (2018) 179–192.

[36] J. Nalepa, Z. Czech, A parallel memetic algorithm to solve the vehicle routing problem with time windows, arXiv:1402.6942v1, 2014. (Accessed 30 June 2018).

[37] E. Osaba, X. Yang, F. Diaz, E. Onieva, A. Masegosa, A. Perallos, A discrete firefly algorithm to solve a rich vehicle routing problem modelling a newspaper distribution system with recycling policy, Soft Computing 21 (18) (2016) 5295–5308.

[38] E. Osaba, X. Yang, I. Fister Jr., J. Del Ser, P. Lopez-Garcia, A. Vazquez-Pardavila, A discrete and improved bat algorithm for solving a medical goods distribution problem with pharmacological waste collection, Swarm and Evolutionary Computation 44 (2019) 273–286, https://doi.org/10.1016/j.swevo.2018.04.001.

[39] F. Pan, C. Ye, K. Wang, J. Cao, Research on the vehicle routing problem with time windows using firefly algorithm, Journal of Computers 8 (9) (2013) 2256–2261.

[40] I. Panessa, M. Baba, N. Iksan, A hybrid genetic algorithm for the rich vehicle routing problem, in: S. Yenduri (Ed.), Advances in Computer Science, World Scientific and Engineering Academy and Society, 2012, pp. 450–456.

[41] Y. Park, J. Yoo, H. Park, A genetic algorithm for the vendor-managed inventory routing problem with lost sales, Expert Systems with Applications 53 (2016) 149–159.

[42] P. Pellegrini, D. Favaretto, E. Moretti, Multiple ant colony optimization for a rich vehicle routing problem: a case study, in: B. Apolloni, R. Howlett, L. Jain (Eds.), Knowledge-Based Intelligent Information and Engineering Systems, in: LNCS, vol. 4693, Springer, Berlin-Heidelberg, 2007, pp. 627–634.

[43] D. Pisinger, S. Ropke, A general heuristic for vehicle routing problems, Computers & Operations Research 34 (2007) 2403–2435.

[44] R.P. Sejm, Regulation of the Minister of Economy of 12 June 2002, on mine rescue, JoL no. 94, pos. 838, 2002.

[45] R.P. Sejm, Regulation of the Minister of Economy of 28 June 2002, on occupational health and safety, traffic management and specialized fire protection in underground mines, JoL no. 139, pos. 1169, no. 124, pos. 863 of 2006, and no. 126, pos. 855 of 2010.

[46] J. Sicilia, C. Quemada, B. Royo, D. Escuin, An optimization algorithm for solving the rich vehicle routing problem based on variable neighborhood search and tabu search metaheuristics, Journal of Computational and Applied Mathematics 291 (2016) 468–477.

[47] K. Sim, E. Hart, N. Urquhart, T. Pidgen, A new rich vehicle routing problem model and benchmark resource, in: E. Minisci, M. Vasile, J. Periaux, N. Gauger, K. Giannakoglou, D. Quagliarella (Eds.), Advances in Evolutionary and Deterministic Methods for Design, Optimization and Control in Engineering and Sciences, in: Computational Methods in Applied Sciences, vol. 48, Springer, Cham, 2018, pp. 503–518.

[48] P. Sombuntham, V. Kachitvichayanukul, A particle swarm optimization algorithm for multi-depot vehicle routing problem with pickup and delivery requests, in: S. Ao, O. Castillo, C. Douglas, D. Feng, J. Lee (Eds.), Proceedings of the International MultiConference of Engineers and Computer Scientists, IMECS 2010, Newswood Limited, 2010, pp. 1998–2003.

[49] J. Souza Neto, V. Pureza, Modeling and solving a rich vehicle routing problem for the delivery of goods in urban areas, Pesquisa Operacional 36 (3) (2016) 421–446.

[50] W. Suwala, Coal sector in Poland, light in the tunnel or dimming candle?, Mineral Economics 31 (2018) 263–268.

[51] J. Szlazak, A. Grodzicka, D. Musiol, Evaluation of methods for determining escape routes in case of a fire hazard in the mine, Wiadomości Górnicze 65 (7–8) (2014) 378–385, in Polish.

[52] J. Szlazak, D. Musiol, H. Badura, Analysis of mining crew evacuation on the escape routes in hard coal mines, Przeglad Gorniczy 71 (2) (2015) 72–78, in Polish.

[53] E. Talbi, Metaheuristics: From Design to Implementation, John Wiley & Sons, Inc., 2009.

[54] Thunderhead Engineering Consultants, Inc., PyroSim User Manual, 2012.

[55] P. Toth, D. Vigo, The Vehicle Routing Problem, SIAM Monographs on Discrete Mathematics and Application, SIAM, Philadelphia, PA, 2002.

[56] U. Vogel, A Flexible Metaheuristic Framework for Solving Rich Vehicle Routing Problems, Ph.D. thesis, Universität zu Köln, 2011.

[57] K. Walus, K. Slota, Z. Slota, The investigation with range of measurement of real times evacuation crew from menaced region in conditions of "Boleslaw Smialy" coal mine, Górnictwo i Geologia 8 (4) (2013) 183–191, in Polish.

[58] M. Wen, Rich Vehicle Routing Problems and Applications, Ph.D. thesis, DTU Management Engineering, 2010.

[59] M. Wen, J. Cordeau, G. Laporte, J. Larsen, The dynamic multi-period vehicle routing problem, Computers & Operations Research 37 (9) (2010) 1615–1623.

[60] M. Wen, J. Larsen, J. Clausen, J. Cordeau, G. Laporte, Vehicle routing with cross-docking, Journal of the Operational Research Society 60 (12) (2009) 1708–1718.

[61] G. Yan, D. Feng, Escape-route planning of underground coal mine based on improved ant algorithm, Mathematical Problems in Engineering 2013 (2013) 687969.

Index

D

L

M

N